T0130732

Statistics for Ecologists Using R and Excel

Statistics for Ecologists Using R and Excel

Data collection, exploration, analysis and presentation

Mark Gardener

DATA IN THE WILD SERIES

Pelagic Publishing | www.pelagicpublishing.com

Published by Pelagic Publishing
www.pelagicpublishing.com
PO Box 725, Exeter, EX1 9QU, UK

Statistics for Ecologists Using R and Excel: Data collection, exploration, analysis and presentation

ISBN 978-1-78427-139-8 (Pbk)
ISBN 978-1-78427-140-4 (Hbk)
ISBN 978-1-78427-141-1 (ePub)
ISBN 978-1-78427-142-8 (mobi)
ISBN 978-1-78427-143-5 (ePDF)

A catalogue record for this book is available from the British Library.

Cover image: *Arizona Sunset* © istockphoto.com/tonda

Printed in India by Imprint Press

About the author

Mark is an ecologist, lecturer and writer and has worked as a teacher and supervisor around the world. He runs courses in ecology, data analysis and R: the statistical programming language, for a variety of organizations.

Acknowledgements

Thanks to Nigel Massen at Pelagic Publishing for encouragement and support in the production of this new edition.

With a book of this nature data examples are always useful. Some of the data illustrated here were collected by students, and I gratefully acknowledge their efforts and send thanks for allowing me to use these data as examples.

Software used

Various versions of Microsoft's Excel® spreadsheet were used in the preparation of this manuscript. Most of the examples presented show version 2013 for Microsoft Windows® although other versions may also be illustrated (including Excel 2010).

Several versions of the R program were used and illustrated, including 2.15.1, for Windows and 3.0.2 for Macintosh: R Foundation for Statistical Computing, Vienna, Austria. ISBN 3-900051-07-0, URL http://www.R-project.org/.

Who this book is for

Students of ecology and environmental science will find this book aimed at them although many other scientists will find the text useful as the principles and data analysis are the same in many disciplines. No prior knowledge is assumed and the reader can develop their skills up to degree level.

What you will learn in this book

This is a book about the scientific process and how you apply it to data in ecology. You will learn how to plan for data collection, how to assemble data, how to analyze data and finally how to present the results. The book uses Microsoft Excel and the powerful Open Source R program to carry out data handling as well as to produce graphs. Specific topics include:

- How to plan ecological projects.
- How to record and assemble your data.
- How to use Excel for data analysis and graphs.
- How to use R for data analysis and graphs.

- How to carry out a wide range of statistical analyses, including analysis of variance and regression.
- How to create professional looking graphs.
- How to present your results.

What's new in the second edition?

The changes from the first edition can be summarized as follows:

- Completely revised chapter on graphics. The chapter is now a one-stop resource for all graphics-related topics.
  - New: graph types and their uses.
  - New: Excel Chart Tools.
  - New: R graphics commands.
  - New: producing different chart types in Excel and in R.
- More support material online, including example data, exercises and additional notes and explanations.
- New: chapter on basic community statistics, biodiversity and similarity.
- New: chapter summaries.
- New: end of chapter exercises.

How this book is arranged

The book is broadly laid out in four sections, roughly corresponding to the topics: planning, recording, analysing and reporting. The sections are rather unequal in length, with the focus on the analysis chapters and the production of graphics.

Throughout the book you will see example exercises that are intended for you to try out. In fact they are expressly aimed at helping you on a practical level; reading how to do something is fine but you need to do it for yourself to learn it properly. The Have a Go exercises are hard to miss.

Have a Go: Learn something by doing it

The Have a Go exercises are intended to give you practical experience at various tasks, such as preparing graphs or analysing data. Many exercises will refer to supplementary data, which you can get from the companion website.

You will also see tips and notes, which will stand out from the main text. These are 'useful' items of detail pertaining to the text but which I felt were important to highlight.

Tips and Notes: Useful additional information

At certain points in the text you'll see tips and notes highlighted like this. These items contain things that I felt were important to highlight and mention especially.

At the end of each chapter there is a summary table to help give you an overview of the material in that chapter. There are also some self-assessment exercises for you to try out. The answers are in the Appendix.

Support files

Most of the examples illustrated in the text are available on the support website. You can download data files and explore additional examples and notes. In most cases you'll see the following text.

Go to the website for support material.

The support website can be accessed at:

http://www.gardenersown.co.uk/Publications/

Contents

Preface to Edition 1

This is not just a statistics textbook! Although there are plenty of statistical analyses here, this book is about the processes involved in looking at data. These processes involve planning what you want to do, writing down what you found and writing up what your analyses showed. The statistics part is also in there of course but this is not a course in statistics. By the end I hope that you will have learnt some statistics but in a practical way, i.e. what statistics can do for you. In order to learn about the methods of analysis, you'll use two main tools: a Microsoft Excel spreadsheet (although Open Office will work just as well) and a computer program called R. The spreadsheet will allow you to collect your data in a sensible layout and also do some basic analyses (as well as a few less basic ones). The R program will do much of the detailed statistical work (although you will also use Excel quite a bit). Both programs will be used to produce graphs. This book is not a course in computer programming; you'll learn just enough about the programs to get the job done.

It is important to recognise that there is a process involved. This is the scientific process and may be summarized by four main headings:

- Planning.
- Data recording.
- Data exploration.
- Reporting results.

The book is arranged into these four broad categories. The sections are rather uneven in size and tend to focus on the analysis. The section on reporting also covers presentation of analyses (e.g. graphs).

Although the emphasis is on ecological work and many of the data examples are of that sort, I hope that other scientists and students of other disciplines will see relevance to what they do.

Mark Gardener
2011

Preface to Edition 2

The first edition of *Statistics For Ecologists* was the first book-length work I had written since my PhD. The process was illuminating, and overall I am happy with what I achieved. However, I also recognize that there are many shortcomings and set out to produce a new edition that was a better textbook.

Although this is primarily a book about statistics it is important to realize that the whole scientific process is involved. If you plan correctly you will record and arrange your data most appropriately. This will allow you to carry out the appropriate data exploration more easily. I revised the chapter on graphics most heavily and essentially gathered the majority of the information about graphs into an earlier chapter than before. Visualizing your data is really important, so I thought that bringing the material about this topic into play sooner would be helpful.

I added chapter summaries to make the book more useful as a quick revision tool. I copied the tables of critical values to the Appendix so that you can find them more easily. There are also some self-assessment questions (and answers). I also added a lot more material to the support website. There were many topics that I would have liked to expand but I thought that they might make the book too unwieldy. There is a new chapter about community ecology. This is a large topic but I had a few requests to incorporate some of the more basic analyses, such as diversity and similarity.

There are many other tweaks and revisions. I hope that you will find the book useful as a learning tool and also as a resource to return to time and again. Try to remember that the data exploration part (the statistics) should be the exciting bit of your project, not just the dull number-crunching bit!

Mark Gardener
2016

1. Planning

The planning process is important, as it can save you a lot of time and effort later on.

What you will learn in this chapter

» Steps in the scientific method
» How to plan your projects
» The different types of experiment/project
» How to recognize different types of data
» How to phrase a hypothesis and a null hypothesis
» When to use different sampling strategies
» How to install R, the statistical programming environment
» How to install the *Analysis ToolPak* for Excel

1.1 The scientific method

Science is a way of looking at the natural world. In short, the process goes along the following lines:

- You have an idea about something.
- You come up with a hypothesis.
- You work out a way of testing this hypothesis/idea.
- You collect appropriate data in order to apply a test.
- You test the hypothesis and decide if the original idea is supported or rejected.
- If the hypothesis is rejected, then the original idea is modified to take the new findings into account.
- The process then repeats.

In this way, ideas are continually refined and your knowledge of the natural world is expanded. You can split the scientific process into four parts (more or less): planning, recording, analysing and reporting (summarized in Table 1.1).

Table 1.1 Stages in the scientific method.

Planning	Recording	Analysing	Reporting
This is the stage where you work out what you are going to do. Formulate your idea(s), undertake background research, decide what your hypothesis will be and determine a method of collecting the appropriate data and a means by which the hypothesis may be tested.	The means of data collection is determined at the planning stage although you may undertake a small pilot study to see if it works out. After the pilot stage you may return to the planning stage and refine the methodology. Data is finally collected and arranged in a manner that allows you to begin the analysis.	The method of analysis should have been determined at the planning stage. Analytical methods (often involving statistics) are used to test the null hypothesis. If the null hypothesis is rejected then this supports the original idea/hypothesis.	Disseminating your work is vitally important. Your results need to be delivered in an appropriate manner so they can be understood by your peers (and often by the public). Part of the reporting process is to determine what the future direction needs to be.

1.1.1 Planning stage

This is the time to get the ideas. These may be based on previous research (by you or others), by observation or stem from previous data you have obtained. On the other hand, you might have been given a project by your professor, supervisor or teacher. If you are going to collect new data, then you will determine what data, how much data, when it will be collected, how it will be collected and how it will be analysed, all at this planning stage. Looking at previous research is a useful start as it can tell you how other researchers went about things. If you already have old data from some historic source then you still need to plan what you are going to do with it. You may have to delve into the data to some extent to see what you have – do you have the appropriate data to answer the questions you want answered? It may be that you have to modify your ideas/questions in light of what you have. A *hypothesis* is a fancy term for a research question. A hypothesis is framed in a certain scientific way so that it can be tested (see more about hypotheses in Section 1.4).

1.1.2 Recording stage

Finally, you get to collect data. The planning step will have determined (possibly with the help of a pilot study) how the data will be collected and what you are going to do with it. The recording stage nevertheless is important because you need to ensure that at the end you have an accurate record of what was done and what data were collected. Furthermore, the data need to be arranged in an appropriate manner that facilitates the analysis. It is often the case, especially with old data, that the researcher has to spend a lot of time rearranging numbers/data into a new configuration before anything can be done.

Getting the data layout correct right at the start is therefore important (see more about data layout in Chapter 2).

1.1.3 Analysis stage

The means of undertaking your analysis should have been worked out at the planning stage. The analysis stage is where you apply the statistics and data handling methods that make sense of the numbers collected. Helping to understand data is vastly aided by the use of graphs. As part of the analysis, you will determine if your original hypothesis is supported or not (see more about kinds of analysis in Chapter 5).

1.1.4 Reporting stage

Of course there is some personal satisfaction in doing this work, but the bottom line is that you need to tell others what you did and what you found out. The means of reporting are varied and may be informal, as in a simple meeting between colleagues. Often the report is more formal, like a written report or paper or a presentation at a meeting. It is important that your findings are presented in such a way that your target audience understands what you did, what you found and what it means. In the context of conservation, for example, your research may determine that the current management is working well and so nothing much needs to be done apart from monitoring. On the other hand, you may determine that the situation is not good and that intervention is needed. Making the results of your work understandable is a key skill and the use of graphs to illustrate your results is usually the best way to achieve this. Your audience is much more likely to dwell on a graph than a page of figures and text. You'll see examples of how to report results throughout the text, with a summary in Chapter 13.

1.2 Types of experiment/project

As part of the planning process, you need to be aware of what you are trying to achieve. In general, there are three main types of research:

- *Differences*: you look to show that *a* is different to *b* and perhaps that *c* is different again. These kinds of situations are represented graphically using bar charts and box–whisker plots.
- *Correlations*: you are looking to find links between things. This might be that species *a* has increased in range over time or that the abundance of species *a* (or environmental factor *a*) affects the abundance of species *b*. These kinds of situations are represented graphically using scatter plots.
- *Associations*: similar to the above except that the type of data is a bit different, e.g. species *a* is always found growing in the same place as species *b*. These kinds of situations are represented graphically using pie charts and bar charts.

Studies that concern whole communities of organisms usually require quite different approaches. The kinds of approach required for the study of *community ecology* are dealt with in detail in the companion volume to this work (*Community Ecology, Analytical Methods Using R and Excel*, Gardener 2014).

In this volume you'll see some basic approaches to community ecology, principally diversity and sample similarity (see Chapter 12). The other statistical approaches dealt with in this volume underpin many community studies.

Once you know what you are aiming at, you can decide what sort of data to collect; this affects the analytical approach, as you shall see later. You'll return to the topic of project types in Chapter 5.

1.3 Getting data – using a spreadsheet

A spreadsheet is an invaluable tool in science and data analysis. Learning to use one is a good skill to acquire. With a spreadsheet you are able to manipulate data and summarize it in different ways quite easily. You can also prepare data for further analysis in other computer programs in a spreadsheet. It is important that you formalize the data into a standard format, as you'll see later (in Chapter 2). This will make the analysis run smoothly and allow others to follow what you have done. It will also allow you to see what you did later on (it is easy to forget the details).

Your spreadsheet is useful as part of the planning process. You may need to look at old data; these might not be arranged in an appropriate fashion, so using the spreadsheet will allow you to organize your data. The spreadsheet will allow you to perform some simple manipulations and run some straightforward analyses, looking at means, for example, as well as producing simple summary graphs. This will help you to understand what data you have and what they might show. You'll look at a variety of ways of manipulating data later (see Section 3.2).

If you do not have past data and are starting from scratch, then your initial site visits and pilot studies will need to be dealt with. The spreadsheet should be the first thing you look to, as this will help you arrange your data into a format that facilitates further study. Once you have some initial data (be it old records or pilot data) you can continue with the planning process.

1.4 Hypothesis testing

A *hypothesis* is your idea of what you are trying to determine. Ideally it should relate to a single thing, so "Japanese knotweed and Himalayan balsam have increased their range in the UK over the past 10 years" makes a good overall aim, but is actually two hypotheses. You should split up your ideas into parts, each of which can be tested separately:

"Japanese knotweed has increased its range in the UK over the past 10 years."

"Himalayan balsam has increased its range in the UK over the past 10 years."

You can think of hypothesis testing as being like a court of law. In law, you are presumed innocent until proven guilty; you don't have to prove your innocence.

In statistics, the equivalent is the *null hypothesis*. This is often written as H0 (or H_0) and you aim to reject your null hypothesis and therefore, by implication, accept the alternative (usually written as H1 or H_1).

The H0 is not simply the opposite of what you thought (called the *alternative hypothesis*, H1) but is written as such to imply that no difference exists, no pattern (I like to think of it as the *dull* hypothesis). For your ideas above you would get:

> "There has been no change in the range of Japanese knotweed in the UK over the past 10 years."

> "There has been no change in the range of Himalayan balsam in the UK over the past 10 years."

So, you do not say that the range of these species is shrinking, but that there is no change. Getting your hypotheses correct (and also the null hypotheses) is an important step in the planning process as it allows you to decide what data you will need to collect in order to reject the H0. You'll examine hypotheses in more detail later (Section 5.2).

1.4.1 Hypothesis and analytical methods

Allied to your hypothesis is the analytical method you will use to help test and support (or otherwise) your hypothesis. Even at this early stage you should have some idea of the statistical test you are going to apply. Certain statistical tests are suitable for certain kinds of data and you can therefore make some early decisions. You may alter your approach, change the method of analysis and even modify your hypothesis as you move through the planning stages: this all part of the scientific process. You'll look at ways to choose which statistical test is right for your situation in Section 5.3, where you will see a decision flow-chart (Figure 5.1) and a key (Table 5.1) to help you. Before you get to that stage, though, you will need to think a little more about the kind of data you may collect.

1.5 Data types

Once you have sorted out more or less what your hypotheses are, the next step in the planning process is to determine what sort of data you can get. You may already have data from previous biological records or some other source. Knowing what sort of data you have will determine the sorts of analyses you are able to perform.

In general, you can have three main types of data:

- *Interval*: these can be thought of as "real" numbers. You know the sizes of them and can do "proper" mathematics. Examples would be counts of invertebrates, percentage cover, leaf lengths, egg weights, or clutch size.
- *Ordinal*: these are values that can be placed in order of size but that is pretty much all you can do. Examples would be abundance scales like DAFOR or *Domin* (named after a Czech botanist). You know that A is bigger than O but you cannot say that one is twice as big as the other (or be exact about the difference).
- *Categorical* (sometimes called *nominal* data): this is the tricky one because it can be confused with ordinal data. With categorical data you can only say that things are different. Examples would be flower colour, habitat type, or sex.

With *interval* data, for example, you might count something, keep counting and build up a sample. When you are finished, you can take your list and calculate an average, look to see how much larger the biggest value is from the smallest and so on. Put another way, you have a scale of measurement. This scale might be millimetres or grams or anything else. Whenever you measure something using this scale you can see how it fits into the scheme of things because the interval of your scale is fixed (10 mm is bigger than 5 mm, 4 g is less than 12 g). Compare this to the *ordinal* scales described below.

With *ordinal* data you might look at the abundance of a species in quadrats. It may be difficult or time consuming to be exact so you decide to use an abundance scale. The Domin scale shown in Table 1.2, for example, converts percentage cover into a numerical value from 0 to 10.

Table 1.2 The Domin scale; an example of an ordinal abundance scale.

Domin	% cover
10	91–100
9	75–90
8	51–74
7	34–50
6	26–33
5	11–25
4	4–10
3	<4% many (>10 individuals)
2	<4 some (4–10 individuals)
1	<4 few (1–3 individuals)

The Domin scale is generally used for looking at plant abundance and is used in many kinds of study. You can see by looking at Table 1.2 that the different classifications cover different ranges of abundance. For example, a Domin of 8 represents a range of values from about half to three-quarters coverage (51–74%). A value of 6 represents a range from about a quarter to a third coverage (26–33%). The first three divisions of the Domin scale all represent less than 4% coverage but relate to the number of individuals found. The Domin scale is useful because it allows you to collect data efficiently and still permits useful analysis. You know that 10 is a greater percentage coverage than 8 and that 8 is bigger than 6; it is just that the intervals between the divisions are unequal.

Table 1.3 An example of a generalized DAFOR scale for vegetation, an example of an ordinal abundance scale.

Abundance	Scale	Description
Dominant	D	The dominant vegetation/species highly visible, usually more than 70% cover
Abundant	A	Many individuals or patches visible, usually 30–50% cover
Frequent	F	Several individuals or few patches, cover usually 10–20%
Occasional	O	A small patch or a few individuals, cover usually around 5–8%
Rare	R	Single very small patch or individual, cover usually around 1–3%

There are many other abundance scales, and various researchers have at times worked out useful ways to simplify the abundance of organisms. The DAFOR scale is a general phrase to describe abundance scales that convert abundance into a letter code. There are many examples. Table 1.3 shows a generalized scale for vegetation analysis.

There are other letters that might be used to extend your scale. For example C for "common" might be inserted between A and F (ACFOR is a commonly used ordinal scale). You might add E and/or S for "extremely abundant" and "super abundant". You might also add N for "not found". The DAFOR type of scale can be used for any organism, not just for vegetation.

When you are finished, you can convert your DAFOR scale into numbers (ranks) and get an average, which can be converted to a DAFOR letter, but you cannot tell how much larger the biggest is from the smallest – the interval between the values is inexact.

Many of the abundance scales used are derived from the work of Josias Braun-Blanquet, an eminent Swiss botanist. Table 1.4 shows a basic example of a Braun-Blanquet scale for vegetation cover.

Table 1.4 The basic Braun-Blanquet scale, an ordinal abundance scale. There are many variations on this scale.

Scale	% cover
5	>75
4	51–75
3	26–50
2	5–25
1	1–5
+	<1

With *categorical* data it is useful to think of an example. You might go out and look to see what types of insect are visiting different colours of flower. Every time you spot an insect, you record its type (bee, fly, beetle) and the flower colour. At the end you can make a table with numbers of how many of each type visited each colour. You have numbers but each value is uniquely a combination of two categories.

Table 1.5 shows an example of categorical data laid out in what is called a contingency

Table 1.5 An example of categorical data. This type of table is also called a contingency table. The rows and columns are each sets of categories. Each cell of the table represents a unique combination of categories.

	Bee	Butterfly	Beetle	Fly	Moth
Red	87	39	58	46	12
White	55	56	34	64	120
Blue	38	112	11	78	14
Yellow	59	65	23	56	45

table. The rows are one category (colour) and the columns another category (type of insect).

1.6 Sampling effort

Sampling effort refers to the way you collect data and how much to collect. For example, you have decided that you need to determine the abundance of some plant species in meadows across lowland Britain. How many quadrats will you use? How large will the quadrats need to be? Do you need quadrats at all?

Sample is the term used to describe the set of data that you have. Because you generally cannot measure "everything", you will usually have a subset of stuff that you've measured (or weighed or counted). Think about a field of buttercups as an example. You wish to know how many there are in the field, which is a hectare in size (i.e. 100 m × 100 m). You aren't really going to count them all (that would take too long) so you make up a square that has sides of 1 metre and count how many buttercups there are in that. Now you can estimate how many buttercups there are in the whole field. Your sample is 1/10,000th of the area, which is pretty small. The estimate is not likely to be very good (although by random chance it could be). It seems reasonable to count buttercups in a few more 1 m^2 areas. In this way your estimate is likely to get more "on target". Think of it this way: if you carried on and on and on, eventually you would have counted buttercups in every 1 m^2 of the field. Your estimate would now be spot on because you would have counted everything. So as you collect more and more data, your estimate of the true number of buttercups will likely become more and more like the true number.

The problem is, how many 1 m^2 areas will you have to count in order to get a good estimate of the true number? You will return to this issue a little later. Another problem – where do you put your 1 m^2 areas? Will it make a difference? Is a 1 m^2 quadrat the right size? You will look at these themes now.

1.6.1 Quadrat size

If you are doing a British NVC survey, then the size and number of quadrats is predetermined; the NVC methodology is standardized. Similarly, if you are making bird species lists for different sites, the methodology already exists for you to follow. Don't reinvent the wheel!

Whenever you collect data, you cannot measure everything, so you take a sample, essentially a representative subset of the whole. What you are aiming for is to make your sample as representative as possible. If, for example, you were counting the frequency of spider orchids across a site, you would aim to make your quadrat a reasonable size and in line with the size and distribution of the organism – you would not have the same size quadrat to look at oak trees as you would to look at lichens.

1.6.2 Species area rule

If you are looking at communities, then the wider the area you cover the more species you will find. Imagine you start off with a tiny quadrat: you might just find a few species. Make the quadrat double the size and you will find more. Keep doubling the quadrat and you will keep finding more species. If, however, you draw a graph of the cumulative number of species, you will see it start to level off and eventually you won't find any more species. Even well before this, the number of new species will be so small that it is not worth the extra effort of the larger quadrat. This idea is called the *species area rule*. You'll see more details about community studies in Chapter 12.

You can extend the same idea to kick-sampling, which is a method for collecting freshwater invertebrates. You use a standard net for freshwater invertebrate sampling but can vary the time you spend sampling. This is akin to using a bigger quadrat. The longer you sample, the more you get. You can easily see that it is not worth spending 20 minutes to get the 101st species when during 3 minutes you net the first 100.

1.6.3 How many replicates?

When you go out to collect your data, how much work do you have to do? If you are counting the abundance of a plant in a field and are unlikely to count every plant, you take a sample. The idea of sampling is to be representative of the whole without having the bother of counting everything. Indeed, attempting to count everything is often difficult, time consuming and expensive.

As you shall see later when you look at statistical tests (starting with Chapter 5), there are certain minimum amounts of data that need to be collected. Now, you should not aim to collect just the minimum that will allow a result to be calculated, but aim to be representative. If you are sampling a field, you might try to sample 5–10% of the area; however, even that might be a huge undertaking. You should estimate how long it is likely to take you to collect various amounts of data. A short pilot study or personal experience can help with this.

Whenever you sample something from a larger population you are aiming to gain insights into that larger population from your smaller sample. You are going to work out an average of some sort; this might be average abundance, size, weight or something else. You'll see different averages later on (Section 4.1.1). You can use something called a running mean to help you determine if you are reaching a good representation of the sample (Section 4.7). In brief, what you do is take each successive number from a quadrat or net and work out the average. Each time you get a new value you can work out a new average. You can then plot these values on a simple graph. When you have only a few values, the running mean is likely to "wobble" quite a bit. After you collect more data, however, the average is likely to settle down. Once your running mean reaches this point, you can see that you've probably collected enough data. You will see running means in more detail in Section 4.7.

1.6.4 Sampling method

You need to think how you are going to select the things you want to measure. In other words you need a sampling strategy.

Remember the field of buttercups? You can see that it is good to have a lot of data items (a large sample) in terms of getting close to the true mean, but exactly where do you put your sample squares (called quadrats: they do not really need to be square but it is convenient) in order to count the buttercups? Does it even matter? It matters of course because you need your sample to be representative of the larger population. You want to eliminate bias as far as possible. If you placed your quadrat in the buttercup field you might be tempted to look for patches of buttercups. On the other hand, you may wish to minimize your counting effort and look for areas of few buttercups! Both would introduce bias into your methods.

What you need is a sampling strategy that eliminates bias; there are several:

- Random.
- Systematic.
- Mixed.
- Haphazard.

Each method is suitable for certain situations, as you'll see now.

Random sampling

In a random sampling method, you use predetermined locations to carry out your sampling. If you were looking at plants in a field, for example, you could measure the field and use random numbers to generate a series of x, y co-ordinates. You then place your quadrats at these co-ordinates. This works nicely if your field is square. If your field is not square you can measure a large rectangle that covers the majority of the field and ignore co-ordinates that fall outside the rectangle. For other situations you can work out a method that provides co-ordinates to place your quadrats. Basically, the locations are predetermined before you start, which is more efficient and saves a lot of wandering about.

In theory, every point within your area should have an equal chance of being selected and your method of creating random positions should reflect this. What happens if you get the same location twice (or more)? There are two options:

- *Random sampling without replacement*. If you get duplicate locations you skip the duplicate and create another random co-ordinate instead.
- *Random sampling with replacement*. If you get duplicate locations you use them again.

In *random sampling without replacement* you never use the same point twice, even if your random number generator comes up with a duplicate.

In *random sampling with replacement* you use whatever locations arise, even if duplicated. In practice, this means that you use the same data and record it both times. It is important that you do not ignore duplicate co-ordinates. If you have ten co-ordinates, which include duplication, then you will still need to get ten values when you have finished. Obviously you do not need to place the quadrat a second time and count the buttercups again, you simply copy the data.

Random sampling is good for situations where there is no detectable pattern. In other cases a pattern may exist. For example, if you were sampling in medieval fields you might have a ridge and furrow system. The old methods of ploughing the field create high and low points at regular intervals. These ridges and furrows may affect the growth of the plants (you assume the ridges are drier and the furrows wetter for instance). If you sampled randomly, you may well get a lot more data from ridges than from furrows. Consequently, you are introducing unwanted bias.

In other cases you may be deliberately looking at a situation where there is an environmental gradient of some sort. For example, this might be a slope where you suspect that the top is drier than the bottom. If you sample randomly then you may once again get bias data because you sampled predominantly in the wetter end of the field (or the drier end). You need to alter your sampling strategy to take into account the situation.

Systematic sampling

In some cases you are deliberately targeting an area where an environmental gradient exists. What you want is to get data from right across this gradient so that you get samples from all parts. Random sampling would not be a good idea (by chance all your observations could be from one end) so you use a set system to ensure that you cover the entire gradient.

Systematic sampling often involves *transects*. A transect is simply the term used to describe a slice across something. For example, you might wish to look at the abundance of seaweed across a beach. The further up the beach, the drier it gets because of the tide so what you do is to create a transect that goes from the top of the beach (high water) to the bottom of the beach (low water). In this way you cover the full range of the gradient from very dry (only covered by water at high tides) to very wet (in the sea).

There are several kinds of transect:

- *Line*: this is exactly what it sounds like. You run a line along your sampling location and record everything along it.
- *Belt*: this kind of transect has definite width! This may be a quadrat or possibly a line of sight (used in butterfly or bird surveys). The transect is sampled continuously along its entire length.
- *Interrupted belt*: this kind of transect is most commonly used when you have quadrats (or their equivalent). Rather than sample continuously you sample at intervals. Often the intervals are fixed but this is not always necessary.

You take your samples along the transect, either continuously (line, belt) or at intervals (interrupted). You do not necessarily have to measure at regular fixed intervals along the transect (although it is common to do so).

One transect might not be enough because you may miss a wider pattern (Figure 1.1). You ought to place several transects and combine the data from them all. In this way you are covering a wider part of the habitat and being more representative of the whole, which is the point.

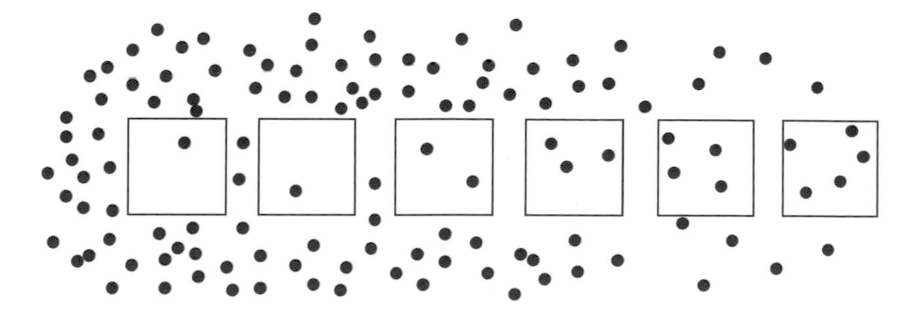

Figure 1.1 One transect may not be enough to see the true pattern. In this case several transects would give a truer representation.

You also need to determine how long the transect should be. You might, for example, be looking at a change in abundance of a plant species along a transect, which may relate to an environmental factor. You need to make sure that you make the transect long enough to cover the change in abundance but not so long that you over-run too far.

Mixed sampling

There are occasions where you may wish to use a combination of systematic and random sampling. In essence, what you do is set up several transects and sample at random intervals along them. Think for example of a field where you wish to determine the height of some plant species. You could set up random co-ordinates but once you get to each co-ordinate how do you select the plant to measure? One option would be to measure the height of the plant nearest the top left corner. Each quadrat is placed randomly but you have a system to pick which plant to measure. You've eliminated bias because you determined this strategy before you started. Another option would be to place transects (a simple piece of string would do; the transect would then be a *line transect*) at intervals across the field. You then measure plants that touch the string (transect) or the nearest to the string, at some random distance along. There are many options of course and you must decide what seems best at the time. The point is that you are trying to eliminate bias and get the most representative sample you can.

Another example might be in sampling for freshwater invertebrates in a stream. You decide that you wish to look for differences between fast-running riffles and slow-moving pools. You need some systematic approach to get a balance between riffles and pools. On the other hand, you do not want to pick the "most likely" locations; you need an element of randomness. You might identify each pool and riffle and assign a number to each one, which you then select at random for sampling. Again the idea is to eliminate any element of bias.

Haphazard sampling

There are times when it is not easy to create an area for co-ordinate sampling, for example, you may be examining leaves on various trees or shrubs that have a "dark" and a "light" side. It is quite difficult to come up with a quadrat that balances in the foliage and you might attempt to grab leaves at random. Of course you can never be really random. In this case you say that the leaves were collected *haphazardly*. To further eliminate bias, you might grab branches haphazardly and then select the leaf nearest the end.

Whenever you get a situation where you stray from either a set system or truly random, you describe your collection method as haphazard.

1.6.5 Precision and accuracy

Whenever you measure something you use some appropriate device. For example, if you were looking at the size of water beetles in a pond you would use some kind of ruler. When you record your measurement, you are saying something about "how good" your recording device is. You might record beetle sizes as 2 cm, 2.3 cm or perhaps 2.36 cm. In the first instance you are implying that your ruler only measures to the nearest centimetre. In the second case you are saying that you can measure to the nearest millimetre. In the third case you are saying that your ruler can measure to 1/10th of a millimetre. If you were to write the first measurement as 2.0 cm then you'd be saying that your beetle was between 1.9 and 2.1 cm.

What you are doing by recording your results in this way is setting the level of *precision*. If you used a different ruler you might get a slightly different result; for example, you could measure a beetle with two rulers and get 2.36 and 2.38 cm. The level of precision is the same in both cases (0.01 cm) but they cannot both be correct (the problem may lie with the ruler or the operator). Imagine that the real size of the beetle was 2.35 cm. The first ruler is more accurate than the second ruler.

So precision is how fine the divisions on your scale of measurement are. Accuracy is how close to the real result your measurement actually is. Ideally you should select a level of precision that matches the equipment you have and the scale of the thing you are measuring. It seems a little pointless to measure the weight of an elephant to the nearest gram for example.

1.7 Tools of the trade

Learning to use your spreadsheet is time well spent. It is important that you can manipulate data and produce summaries, including graphs. You'll see a variety of aspects of data manipulation as well as the production of graphs later. Many statistical tests can be performed using a spreadsheet but there comes a point when it is better to use a dedicated computer program for the job. There are many on the market, some are cheap (or even free) and others are expensive. Some programs will interface with your spreadsheet and others are totally separate. Some programs are specific to certain types of analysis and others are more general.

Here you will focus on two programs. The spreadsheet you will see used is Microsoft Excel. This is common and widely available. There are alternatives and indeed the Open Office spreadsheet uses the same set of formulae and can be regarded as equivalent. The analytical program you'll see is called R; this is described first.

1.8 The R program

The program called R is a powerful environment for statistical computing and graphics. It is available free at the Comprehensive R Archive Network (CRAN) on the Internet. It is open source and available for all major operating systems.

R was developed from a commercial programming language called S. The original authors were called Robert and Ross so they called their program R as a sort of joke.

Because R is a programming language it can seem a bit daunting; you have to type in commands to get it to work; however, it does have a Graphical User Interface (GUI) to make things easier and it is not so different from typing formulae into Excel. You can also copy and paste text from other applications (e.g. word processors). So if you have a library of these commands, it is easy to pop in the ones you need for the task at hand.

R will cope with a huge variety of analyses and someone will have written a routine to perform nearly any type of calculation. R comes with a powerful set of routines built in at the start but there are some useful extra "packages" available on the CRAN website. These include routines for more specialized analyses covering many aspects of scientific research as well as other fields (e.g. economics).

There are many advantages in using R:

- It is free, always a consideration.
- It is open source; this means that many bugs are ironed out.
- It is extremely powerful and will handle very complex analyses as easily as simple ones.
- It will handle a wide variety of analyses. This is one of the most important features: you only need to know how to use R and you can do more or less any type of analysis; there is no need to learn several different (and expensive) programs.
- It uses simple text commands. At first this seems hard but it is actually quite easy. The upshot is that you can build up a library of commands and copy/paste them when you need them.
- Documentation. There is a wealth of help for R. The CRAN site itself hosts a lot of material but there are also other websites that provide examples and documentation. Simply adding CRAN to a web search command will bring up plenty of options.

1.8.1 Getting R

Getting R is easy via the Internet. The R Project website is a vast enterprise and has local mirror sites in many countries. The first step is to visit the main R Project webpage (http://www.r-project.org) where you can select the most local site to you (this speeds up the download process a bit).

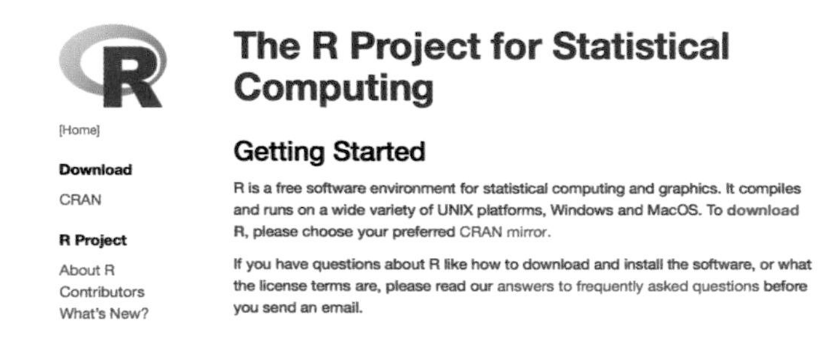

Figure 1.2 Getting R from the R Project website. Click the download link and select the nearest mirror site.

Once you have clicked the download link (Figure 1.2), you have the chance to select a mirror site. These mirrors sites are hosted in servers across the world and using a local one will generally result in a speedier download.

Once you have selected a mirror site, you can click the link that relates to your operating system (Figure 1.3). If you use a Mac then you will go to a page where you can select the best option for you (there are versions for various flavours of OSX). If you use Windows then you will go to a Windows-specific page. If you are a Linux user then read the documentation; you generally install R through the terminal rather than via the web page.

Now the final step is to click the link and download the installer file, which will download in the usual manner according to the setup of your computer.

The Comprehensive R Archive Network

Download and Install R

Precompiled binary distributions of the base system and contributed packages, **Windows and Mac** users most likely want one of these versions of R:

- Download R for Linux
- Download R for (Mac) OS X
- Download R for Windows

R is part of many Linux distributions, you should check with your Linux package management system in addition to the link above.

Figure 1.3 Getting R from the R Project website. You can select the version that is specific to your operating system.

1.8.2 Installing R

Once you have downloaded the install file, you need to run it to get R onto your computer. If you use a Mac you need to double-click the disk image file to mount the virtual disk. Then double-click the package file to install. If you use Windows, then you need to find the EXE file and run it. The installer will copy the relevant files and you will soon be ready to run R.

Now R is ready to work for you. Launch R using the regular methods specific to your operating system. If you added a desktop icon or a quick launch button then you can use these or run R from the Applications or Windows *Start* button.

1.9 Excel

There are many versions of Excel and your computer may already have had a version installed when you purchased it. The basic functions that Excel uses have not changed for quite some while so even if your version is older than is described here, you should be able to carry out the same manipulations. You will mainly see Excel 2013 for Windows illustrated here. If you have purchased a copy of Excel (possibly as part of the Office

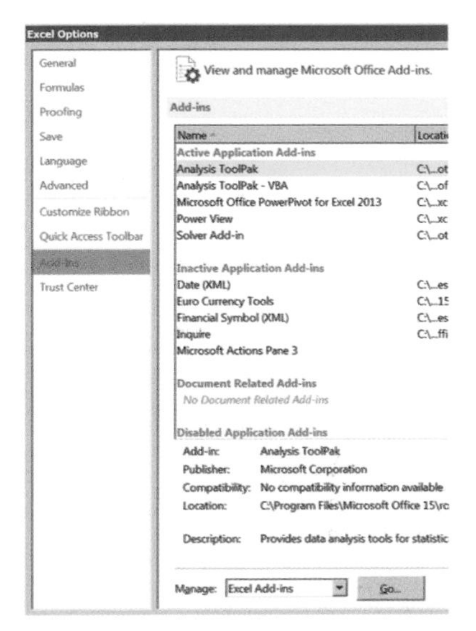

Figure 1.4 Selecting Excel add-ins from the Options menu.

suite) then you can install this following the instructions that came with your software. Generally, the defaults that come with the installation are fine although it can be useful to add extra options, especially the *Analysis ToolPak*, which you will see described next.

1.9.1 Installing the *Analysis ToolPak*

The *Analysis ToolPak* is an add-in for Excel that allows various statistical analyses to be carried out without the need to use complicated formulae. The add-in is not installed as standard and you will need to set up the tool before you can use it. The add-ins are generally ready for installation once Excel is installed and you usually do not require the original disk.

In order to install the *Analysis ToolPak* (or any other add-in) you need to click the *File* button (at the top left of the screen) and select *Options*, then choose *Add-Ins* from the sidebar menu.

In Figure 1.4 you can see that there are several add-ins already active and some not yet ready. To activate (i.e. install) the add-in, you click the *Go* button at the bottom of the screen. You then select which add-ins you wish to activate (Figure 1.5).

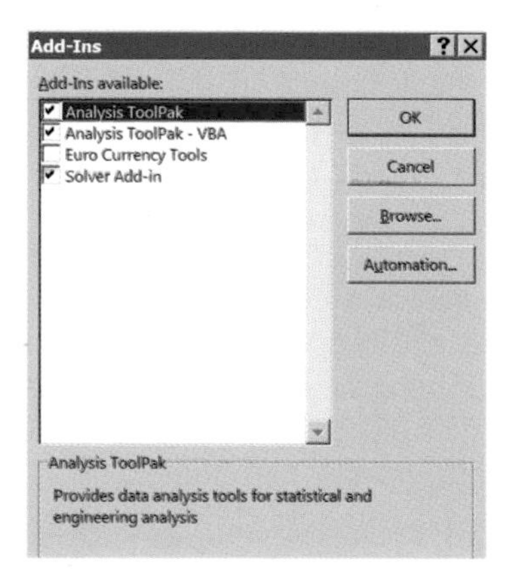

Figure 1.5 Selecting the add-ins for Excel.

Once you have selected the add-ins to activate, you click the *OK* button to proceed. The add-ins are usually available to use immediately after this process.

To use the *Analysis ToolPak* you use the *Data* button on the ribbon and select the *Data Analysis* button (Figure 1.6). Once you have selected this, you are presented with various analysis tools (Figure 1.7).

Figure 1.6 The *Analysis ToolPak* is available from the *Data Analysis* button on the Excel *Data* ribbon.

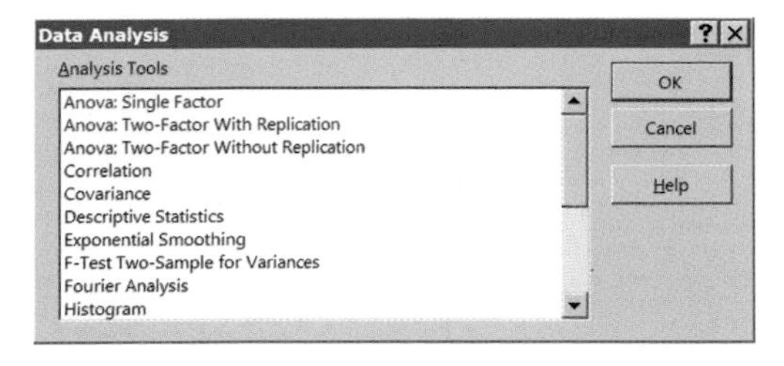

Figure 1.7 The *Analysis ToolPak* provides a range of analytical tools.

Each tool requires the data to be set out in a particular manner; help is available using the Help button.

1.9.2 Other spreadsheets

The Excel spreadsheet that comes as part of the Microsoft Office suite is not the only spreadsheet; of the others available, of particular note are the Open Office and Libre Office programs. These are available from www.openoffice.org and www.libreoffice.org and there are versions available for Windows, Mac and Linux.

Other spreadsheets generally use the same functions as Excel, so it is possible to use another program to produce the same result. Graphics will almost certainly be produced in a different manner to Excel. In this book you will see Excel graphics produced with version 2013 for Windows.

EXERCISES

Answers to exercises can be found in Appendix 1.

1. In a project looking for links between things you would be looking for ____ or ____, depending on the kind of data.
2. Arrange these levels of measurement in increasing order of "sensitivity": *Ordinal, Interval, Categorical.*
3. *Domin, DAFOR* and *Braun-Blanquet* are all examples of ____ scales, whereas *red, blue* and *yellow* would be examples of ____.
4. Which one of the following is not a kind of transect?
   A. Line.
   B. Belt.
   C. Interrupted.
   D. Random.
5. You select the most appropriate collection method to eliminate bias. TRUE or FALSE?

Chapter 1: Summary

Topic	Key points
Scientific method	This is the process by which science operates. In general it involves several processes: planning, recording, analysis and reporting.
Types of project	It is possible to split types of project into several broad categories: differences, correlations and associations. Each type of project requires a different sort of analysis. Studies involving communities of species require slightly different approaches.
Hypotheses	A hypothesis is a special kind of research question, which can be tested using a statistical test. The result of the test determines whether the hypothesis (H1) is supported or should be rejected. In practice you test the null hypothesis (H0: i.e. that there is nothing "interesting"), and rejecting this leads to support of the original hypothesis.
Types of data	Different sorts of data tend to lead you towards a different kind of analysis. The main types of data are: interval, ordinal and categorical.
Sampling effort	You usually collect a sample, i.e. a subset of a larger item (since you cannot measure everything). Sampling effort relates to the way you collect your samples, such as the size of the sampling tools (e.g. quadrat) and how many replicates you collect. You can use a running mean to help determine if you have collected enough data.
Sampling methods	There are different strategies for collecting data; each is suitable for different kinds of situation. The methods are: random, systematic, mixed and haphazard.
The R program	The R program is a powerful programming environment. You can use it to carry out a huge range of statistical analyses and to create graphs. R is open source and available free from the R website.
Using Excel	Excel is great for storing data. You can also use it to create good-quality graphs and carry out a range of statistical analysis. The *Analysis ToolPak* is a useful add-in (Windows only) that can help manage your data.

2. Data recording

The data you write down are of fundamental importance to your ability to make sense of them at a later stage. If you are collecting new data then you are able to work out the recording of the data as part of your initial planning. If you have past data then you may have to spend some time rearranging them before you can do anything useful.

What you will learn in this chapter

» The elements of scientific recording
» Different ways to arrange your data
» How best to arrange your data

2.1 Collecting data – who, what, where, when

It is easy to write down a string of numbers in a notebook. You might even be able to do a variety of analyses on the spot; however, if you simply record a string of numbers and nothing else you will soon forget what the numbers represented. Worse still, nobody else will have a clue what the numbers mean and your carefully collected data will become useless.

All recorded data need to conform to certain standards in order to be useful at a later stage. The minimum you ought to record is:

- *Who*: the name of the person that recorded the data.
- *What*: the species you are dealing with.
- *Where*: the location that the data were collected from.
- *When*: the date that the data were recorded.

There are other items that may be added, depending upon your purpose, as you shall see later.

2.1.1 Biological data and science

Your data are important. In fact they are the most important part of your research. It is therefore essential that you record and store your data in a format that can be used in the future. There are some elements of your data that may not seem immediately important but which nevertheless are essential if future researchers need to make sense of them.

You need to write down your data in a way that makes sense to you at the time and also will make sense to future scientists looking to repeat or verify your work. Table 2.1 2.1 shows some biological data in an appropriate format. Not all the data are shown here (the table would be too big).

Table 2.1 An example of biological data: bat species numbers at various sites around Milton Keynes (only part of the data are shown).

Species	Recorder	Date	Site	GR	Quantity
Pipistrellus pipistrellus	Atherton, M	5-Aug-06	Ouzel Valley	SP880375	21
Myotis daubentonii	Atherton, M	5-Aug-06	Ouzel Valley	SP880375	23
Nyctalus noctula	Atherton, M	5-Aug-06	Ouzel Valley	SP880375	26
Plecotus auritus	Atherton, M	5-Aug-06	Ouzel Valley	SP880375	54
Myotis natteri	Atherton, M	5-Aug-06	Ouzel Valley	SP880375	54
Pipistrellus pipistrellus	Atherton, M	5-Aug-06	Newport	SP874445	43
Myotis daubentonii	Atherton, M	5-Aug-06	Newport	SP874445	11
Nyctalus noctula	Atherton, M	5-Aug-06	Newport	SP874445	9

Every record (a row in your table) always has *who, what, where* and *when*. This is really important for several reasons:

- It allows the data to be used for multiple purposes.
- It ensures that the data you collect can be checked for accuracy.
- It means that you won't forget some important aspect of the data.
- It allows someone else to repeat the exercise exactly.

In the example above, you can see that someone (M Atherton) is trying to ascertain the abundance of various species of bat at sites around Milton Keynes in the UK. It would be easy for him to forget the date because it doesn't seem to matter that much. But if someone tries to repeat his experiment, they need to know what time of year he was surveying at. Alternatively, if environmental conditions change, it will be essential to know what year he did the work.

If you fail to collect complete biological data, or fail to retain and communicate all the details in full, then your work may be rendered unrepeatable and therefore useless as a contribution to science.

Once your biological data are compiled in this format, you can sort them by the various columns, export the grid references to mapping programs and convert the data into tables for further calculations using a spreadsheet. They can also be imported into databases and other computer programs for statistical analysis.

Data collection in the field

When you are in the field and using your field notebook, you may well use shortcuts to record the information required. There seems little point in writing the site name and grid reference more than once for example. You may decide to use separate recording sheets to write down the information. These can be prepared in advance and printed as required. Once again there will be items that do not need to be repeated – a single date at the top of every sheet should be sufficient for example. However, when you transfer the data onto a computer it is a simple matter to copy the date or your name in a column.

In general, you should aim to create a column for each item of data that you collect. If you were looking at species abundance at several sites for example, then you would need at least two columns, one for the abundance data and one for the site. In your field notebook or recording sheet you may keep separate pages for each site and end up with a column of figures for each site. When you return to base and transfer the data to the spreadsheet, you should write your data in the "standard format", i.e. one column for each thing (as in Table 2.1).

Supporting information

As part of your planning process (including maybe a pilot study), you should decide what data you are going to collect. Just because you can collect information on 25 different environmental variables does not mean that you should. The date, location and the name of the person collecting the data are basic items that you always need but there may also be additional information that will help you to understand the biological situation as you process the data later. These things include field sketches and site photographs.

A field sketch can be very helpful because you can record details that may be hard to represent in any other manner. A sketch can also help you to remember where you placed your quadrats; a grid reference is fine but meaningless without a map! Photographs may also be helpful and digital photography enables lots of images to be captured with minimum fuss; however, it is also easy to get carried away and forget what you were there for in the first place. Any supporting information should be just that – support for the main event: your data.

2.2 How to arrange data

As in the example in Table 2.1, it is important to have data arranged in appropriate format. When you enter data into your spreadsheet you ought to start with a few basics. These correspond to the: *who, what, where, when*. There are extra items that may be entered depending on the level of study. These will pretty much correspond to your needs and the level of detail required. If you are collecting data for analysis then it is also important to set out your data in a similar fashion. This makes manipulating the data more straightforward and also maintains the multi-purpose nature of your work. You need to move from planning to recording and on to analysis in a seamless fashion. Having your data organized is really important!

When you collect biological data, enter each record on a separate line and set out your spreadsheet so that each column represents a factor. Table 2.2 shows some data in this layout.

Since the contents of the first four columns are identical, you might consider leaving them out and making an entry at the top (you would certainly do this in your field notebook or recording sheet). This is certainly one option but it is just as easy to fill out the top row and copy the entries down the remaining cells. In this case the data show the abundance of nettles (*Urtica dioica*) at two sites near Preston Montford in Shropshire.

Table 2.2 Data table layout (only part of the data are shown). Showing the most important elements in biological records.

Who	Where	When	What	How many	Other (site)
MG	SJ4314	14-Aug-06	*U. dioica*	12	Pond
MG	SJ4314	14-Aug-06	*U. dioica*	8	Pond
MG	SJ4314	14-Aug-06	*U. dioica*	7	Pond
MG	SJ4314	14-Aug-06	*U. dioica*	32	Wood

You want to see if there are differences between two sites. It would seem easier to simply make two columns, one for the quadrats at the pond site and one for the wood. This is certainly an option but if you stick to this "scientific recording" layout you can easily reproduce the two columns any time you wish (using various tools in the spreadsheet). In fact many analytical programs prefer the layout as it is with the data in one column and the explanatory variable/factor in another. Here is another example.

When you begin to collect your data, you have a variety of ways of writing the results down. For example, imagine that you are looking at the abundance of a plant species in quadrats at two different sites. Your natural instinct would be to write down the abundance of the plant in two columns, one for reach site. If you examined other sites you would create extra columns. Table 2.3 shows the data written in this manner.

Table 2.3 Data table layout. Abundance of *Ranunculus acris* at two sites in Buckinghamshire. Here the abundance is shown in two columns, one for each site.

Upper	Lower
2	4
3	6
4	7
3	8
4	7
6	6
5	5
3	4
4	2

However, you might also write all the abundance data down in a single column, and then use a second column to tell you which site the data were recorded from. Table 2.4 shows this layout.

The first layout (Table 2.3) is perhaps the manner in which you would naturally want to write down your results in your field notebook. For simple projects this is perfectly acceptable. However, when you have a lot of data or more complex situations the second format (Table 2.4) is a better choice. Many computer programs prefer the data to be in this layout and for complex statistical analyses it is essential.

For example, Table 2.5 shows a small part of a complex dataset. Here you have

Table 2.4 Data table layout. Abundance of *Ranunculus acris* at two sites in Buckinghamshire. Here the data are shown in two columns, one for the site and one for the abundance.

Site	Count
Upper	2
Upper	3
Upper	4
Upper	3
Upper	4
Upper	6
Upper	5
Upper	3
Upper	4
Lower	4
Lower	6
Lower	7
Lower	8
Lower	7
Lower	6
Lower	5
Lower	4
Lower	2

recorded the abundance of several butterfly species. You could have recorded the species in several columns, one for each; however, you also have different locations. These locations are themselves further subdivided by management. If you wrote down the information separately you would end up with several smaller tables of data and it would be difficult to carry out any actual analyses. By recording the information in separate columns you can carry out analyses more easily.

The data in Table 2.5 can be split into various subsections using your spreadsheet and the *filter* command (Section 3.2.2). You can also use the *pivot table* function to review the data (Section 3.2.7).

Now you have gone through the planning process. Ideally, you would have worked out a *hypothesis* and know what data you need to collect to support your hypothesis (or to reject it). You will look at hypothesis testing in more detail in Chapter 5. You ought to know at this stage what type of analysis you are going to run on your data (see Table 5.1 in Chapter 5).

Once you have collected data and written it in your spreadsheet in an appropriate format, you are ready to begin to explore the data. This is the subject of the following chapters. Ideally, you should begin by sketching your data graphically. You will see this in Chapters 4 and 6, although there will also be examples throughout the text. After you have a graphical overview of your data, you should summarize it numerically (Chapter 4). Subsequent chapters deal with choosing and carrying out the various analytical methods you might use to support or reject your initial hypothesis. Before you get to

Table 2.5 Data table layout. Complex data are best set out in separate columns. Here butterfly abundance is recorded for four different factors.

Transect	Year	ssp	Man	Count
N	1996	pbf	no	1.15
N	1997	pbf	no	1.54
N	1998	pbf	no	0
N	1996	pbf	yes	1.54
N	1997	pbf	yes	4.62
N	1998	pbf	yes	0
N	1996	pbf	yes	2.78
N	1997	pbf	yes	1.67
N	1998	pbf	yes	0
S	1996	swf	yes	7.11
S	1997	swf	yes	25.53
S	1998	swf	yes	2.37

that, you need to become a bit more familiar with some of the tools that you are going to use, namely Excel and R. Familiarizing yourself with these tools is the subject of Chapter 3.

EXERCISES

Answers to exercises can be found in the Appendix.

1. Which of the following are good reasons to maintain good records (as many as you think are correct):
    A. Allows repeatability.
    B. Allows data to be verified.
    C. So your teacher/supervisor knows where you've been.
    D. So you gain proper credit.
    E. So you don't forget something.
2. There are two kinds of layout that you could use for your data. What are they called?
3. Which of the following is *not* a term you would use to describe a row in your scientific recording dataset?
    A. Record
    B. Replicate
    C. Sample
    D. Observation
4. The *sample* layout allows you the most flexibility in your dataset. TRUE or FALSE?
5. In your dataset each column should represent a separate ____.

Chapter 2: Summary

Topic	Key points
Elements of scientific recording	All scientific data should conform to certain standards to maximize their usefulness. The minimum information should include: who, what, where and when. These elements allow records to be checked and verified (and reproduced if appropriate).
Arranging data	There are alternative ways you may arrange your data. To be most useful each column in your data should be a separate variable. Each row should be a separate record. This layout is known as the recording (or scientific) layout.
Sample layout	In this way of arranging data, each column forms a sample. This is an intuitive way to arrange data, and how Excel best handles data. Sample layout is not the most useful or flexible layout, especially when there are many variables and/or samples.
Recording layout	In this way of arranging data each column is a separate variable. Each row is a single observation (replicate, or record). This arrangement is flexible and extensible. It is the layout used most commonly by R and allows complicated analyses to be carried out with minimal fuss.

3. Beginning data exploration – using software tools

In order to make sense of your data you will need to use some of the tools you have come across already, namely your spreadsheet and the R program. You will see the various specifics as you go along but before you go any further you need to become a bit more familiar with the R program and with some important tools in Excel.

What you will learn in this chapter

» The basics of using R
» How to extend the capabilities of R
» How to manage data using Excel (e.g. sorting, filtering)
» How to make pivot tables using Excel
» How to save data in different file formats
» How to get data from Excel into R

3.1 Beginning to use R

Once you have installed R, you can run it using the regular methods: you may have a shortcut on the desktop or use the *Start* button. Once you have run the program, you will see the main input window and a welcome text message. This will look something like Figure 3.1 if you are using Windows. There is a > and cursor | to show that you can type at that point. In the examples, you'll see the > to indicate where you have typed a command, and lines beginning with anything else are the results of your typing.

The program appearance (GUI) is somewhat sparse compared to most Windows programs. You are expected to type commands in the window. This sounds a bit daunting but is actually not that hard. Once you know a few basics, you can start to explore more and more powerful commands because R has an extensive help system. There are many resources available on the R website and it is worth looking at some of the recommended documents (most are PDF) and working through those. Of course this book itself will provide a good starting point!

After a while you can start to build up a library of commands in a basic text editor; it

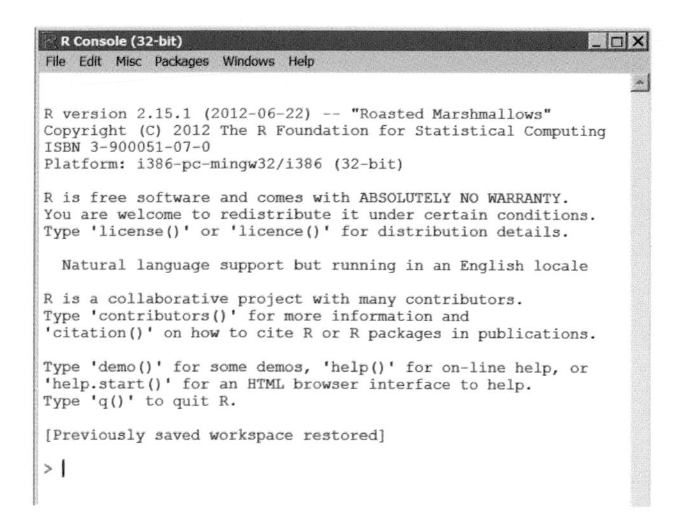

Figure 3.1 The R program interface is a bit sparse compared to most Windows programs.

is easy to copy and paste commands into R. It is also easy to save a snapshot of the work you have been doing for someone else to look over.

3.1.1 Getting help

R has extensive help. If you know the name of a command and want to find out more (there are often additional options), then type one of the following:

```
> help(topic)
> ?topic
```

You replace the word `topic` with the name of the command you want to find out about. The help system that is opened depends on your operating system:

- Windows: opens your default web browser.
- Macintosh: opens a browser-like help window.
- Linux: displays text help in the R console.

You can open the help system in your web browser in any operating system if you type:

```
> help.start()
```

These commands are listed in the opening welcome message (Figure 3.1).

Online help

There are loads of people using R and many of them write about it on the Internet. The R website is a good start but just adding R to a web search may well get you what you need.

3.1.2 Starting with R

R can function like a regular calculator. Start by typing some maths:

```
> 23 + 7
> 14 + 11 + (23*2)
> sqrt(17-1)
> pi*2
```

Notice that R does not mind if you have spaces or not. Your answers look something like the following:

```
[1] 30
[1] 71
[1] 4
[1] 6.283185
```

The `[1]` indicates that the value displayed is the first in a series (and in this case the only value). If you have several lines of answers, then each line begins with an `[n]` label where `n` is a number relating to "how far along" the list the first in that line is. This maths is very useful but you really want to store some of the results so that you can get them back later and use them in longer calculations. To do this you create variable names. For example:

```
> answer1 = 23 + 21 + 17 + (19/2)
> answer2 = sqrt(23)
> answer3 = answer1 + (answer1 * answer2)
```

This time you will notice that you do not see the result of your calculations. R does not assume that just because you created these objects you necessarily want to see them. It is easy enough to view the results; you need to type in the name of the thing you created:

```
> answer1
[1] 70.5
> answer2
[1] 4.795832
> answer3
[1] 408.6061
```

In older versions of R, the = sign was not used and instead a sort of arrow was used:

```
> answer4 <- 93 - 21 + sqrt(41)
> 77/5 + 9 -> answer5
```

This is actually a bit more flexible than the = sign (try replacing the arrows above with = and see what happens), but for most practical purposes = is quicker to type.

Data names

All stuff in R needs a name. In a spreadsheet, each cell has a row and column reference, e.g. B12, which allows you to call up data by specifying the range of cells they occupy. R does not set out data like this so you must assign labels to everything so you can keep track of them. You need to give your data a name; R expects everything to have a name and this allows great flexibility. Think of these names as like the memory on a calculator; with R you can have loads of different things stored at the same time.

Names can be more or less any combination of letters and numbers. Any name you create must start with a letter; other than that, you can use numbers and letters freely. The only other characters allowed are the full stop and the underscore character. R is case sensitive so the following are all different:

```
data1
Data1
DATA1
data.1
data_1
```

It is a good idea to make your names meaningful but short!

3.1.3 Inputting data

The first thing to do is to get some data into R so you can start doing some basic things. There are three main ways to do this:

- The c() command.
- The scan() command.
- The read.csv() command.

The first two methods are outlined next.

Combine c() command

The c() command (short for combine, or concatenate) reads data from the keyboard:

```
> data1 = c(24, 16, 23, 17)
```

Here you make an object called data1 and combine the four numbers in the brackets to make it. Each value has to be followed by a comma. The spaces are optional, R ignores them, but they can be useful to aid clarity and avoid mistakes when typing.

You may easily add more items to your data using the c() command. For example:

```
> data2 = c(data1, 23, 25, 19, 20, 18, 21, 20)
```

Now you have a new item called data2. You could also have overwritten your original item and effectively added to it without creating a new object:

```
> data1 = c(data1, 23, 25, 19, 20, 18, 21, 20)
```

Here you put the old data at the beginning because you wanted to add the new values after, but you could easily place the old data at the other end or even in the middle.

Scan() command

The `scan()` command reads data values from the keyboard but unlike `c()` it does not need to have commas. This makes it easier to enter data:

```
> data3 = scan()
```

R now waits for you to enter data. You can enter a few items and carry on with new lines as you wish but once you are finished, enter a blank line. This signifies to R that you are finished.

```
> data3 = scan()
1: 2 5 76 23 14.4 7 19
8: 21 16.3 112 3
12: 6
13:
Read 12 items
```

In the previous example, data were entered over three lines. Each value is separated with only a space rather than a comma, so it is a little easier to type in.

You can also use the `scan()` function in conjunction with your spreadsheet (or other programs) and the copy and paste functions. If you copy Excel data to the clipboard you can then switch to R and run the `scan()` function. When you paste you will find the data transferred to R. The last thing you need to do is to press *Enter* and tell R you are done. Alternatively you might switch back to Excel and copy another set of values, but you always enter the blank line at the end to tell R to stop scanning. The following examples illustrate the process. You start by typing a name and the `scan()` command.

```
> data.item = scan()
1:
```

The program now waits and a `1:` is shown on the screen to indicate that something is expected. You switch to your spreadsheet and copy a column of cells containing data. Next you switch back and paste (using the keyboard shortcuts or the *Edit* menu).

```
> data.item = scan()
1: 23 17 9 8
```

You see the data entered but the program is not finished. You could paste more data or finish by pressing *Enter* to insert a blank line.

```
> data.item = scan()
1: 23
2: 17
3: 9
```

```
4: 8
5:
Read 4 items
```

The final display shows that the last item entered was a blank (you see a `5:`) and a message tells you how many data items were received (four in this case – item five was the blank).

You'll often have quite extensive datasets and the `scan()` command along with copy and paste can be fairly tedious in these cases. There is a more efficient way to move data from your spreadsheet into R (the `read.csv()` command) but you will return to this later (Section 3.3).

3.1.4 Seeing what data items you have

Once you have a few objects in the memory, you will naturally want to see what you have. Do this by simply typing the name of a variable. For small datasets this is not a problem but if you have lots of data then this might be a bit tedious (R does its best to wrap the text to fit the screen).

```
[1] 24 16 23 17
[1] 24 16 23 17 23 25 19 20 18 21 20
```

The examples above show your `data1` and `data2` items.

List items ls() command

The `ls()` command will show you what objects reside in the memory of R. This will usually comprise the samples that you entered but you may also see other objects that arise from calculations that R performs.

```
> ls()
[1] "beetle.cca" "biol" "biol.cca" "env" "op"
```

The example above shows five objects; some are data typed in directly and others are results from various calculations.

Remove items rm() command

You might wish to remove some clutter and using the `rm()` command allows you to delete objects from R.

```
> rm(data1)
```

This will now get rid of the `data1` object you created earlier.

3.1.5 Previous commands and R history

R stores your commands in an internal list, which you can access using the up arrow. The up and down arrows cycle through recent commands, allowing you to edit a previous command. The left and right arrows move through the command line. Alternatively, you can click the command you want to edit. When you exit R, the previous commands will usually be stored automatically.

Saving history of commands

You can save (or load) a list of previous commands using the GUI. In Windows, the *File* menu gives the option to load or save a history file (Figure 3.2). This may be useful if you have been working on a specific project and wish to recall the commands later (the commands are saved to a text file).

Saving the workspace

You can also save the workspace; this includes all the objects currently in the memory. Imagine you are working on several projects, one might be a bird survey, another a plant database and the third an invertebrate study. It might be useful to keep them all separate. R allows you to save the workspace as a file that you can call up later (the file is in a binary format, readable only by R). You can also send the file to a colleague (or tutor) who can access the objects you were working on (Figure 3.2).

Saving a snapshot of your R session

If you have been working on a series of analyses it may be useful to save an overview of what you have been doing. You can do this using the *File > Save to File* menu option (Figure 3.2). R will save the input window as a plain text file, which you can open in a text editor. You can use this to keep a note of what you did or send it to a colleague (or tutor). You can also use the text file as the basis for your own library of commands; you can copy and paste into R and tweak them as required, R is very flexible.

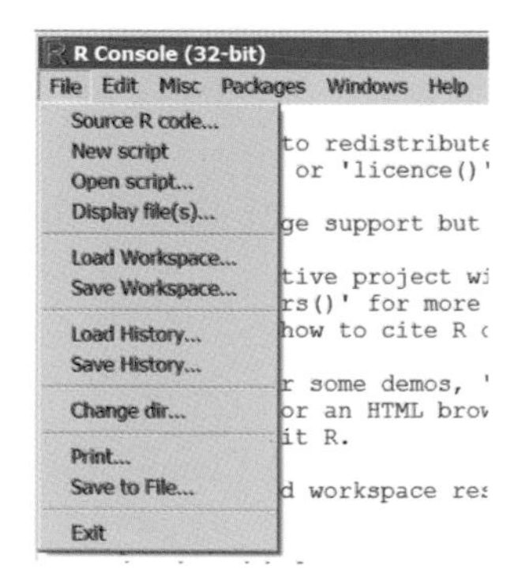

Figure 3.2 The *File* menu in R allows you to save a snapshot of the R working area to a plain text file.

Annotating work

R accepts input as plain text. This means you can paste text from other applications. It is useful to have a library of commands at your disposal. You can annotate these so that you know what they are. R uses the hash character # to indicate that the following text is an annotation and is therefore not executed as a command.

```
> vegemite(vegdata, use= "Domin") # from the vegan package
```

Here the note reminds the user that the *vegan* library (or package) needs to be loaded because this command is not part of the original R program (see Section 3.1.6 for notes about additional libraries of R commands).

3.1.6 Additional analysis packages

The base distribution of R can do a lot but often you will find the need to perform some analysis that it cannot do. Because of the nature of R, there are a lot of people working on it and they have produced a wealth of packages. These are libraries of specialist routines that perform a huge variety of statistical analyses.

It is fairly simple to find and install these extra packages. One method is to search online but the R website has lists of available packages. You may be able to download a zip file. Then you can install the package from within R using one of the menu commands on the toolbar (under the *Packages* menu). Alternatively, if you are connected to the Internet you can use the menu to install packages directly.

You can also install packages using a typed command:

```
> install.packages("package_name")
```

You replace `"package_name"` with the name of the package you want (you need the quotes).

Loading extra analysis routines using the library() command

To load a library of routines into R you use the `library()` command, e.g.

```
> library(vegan)
```

This loads the "vegan" library of analysis tools, which are indispensible for analyses in community ecology (vegan is short for Vegetation Analysis, see Chapter 12). Once they have been loaded, these additional packages may be used like any other. They have their own help files built into R and the syntax for usage follows the standard R format.

3.1.7 Exiting R q() command

When you are ready to end your session in R, you can quit using several methods. You can use the buttons in the GUI like a regular program or use the `quit()` or `q()` command.

```
> q()
```

R will now ask if you want to save the workspace (Figure 3.3). It is a good idea to say yes. This will save the history of commands, which is useful. It will also save any items in memory. When you run R later, you can type `ls()` to list the objects that R has saved. Now you have been introduced to the R program, it is time to look at things you can do in Excel; this is the subject of the following section.

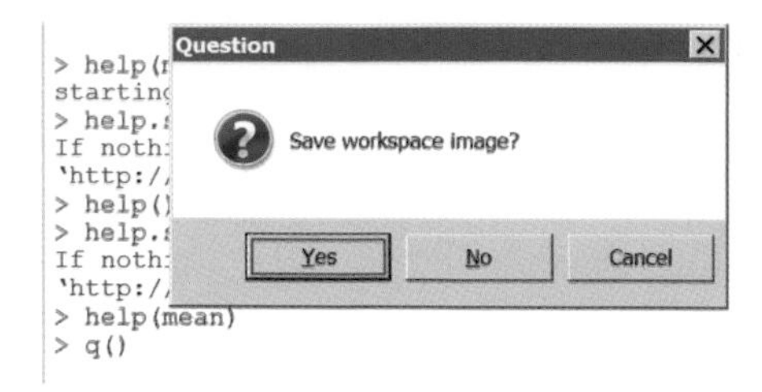

Figure 3.3 When quitting R you are prompted to save the workspace, thus preserving the data in memory.

3.2 Manipulating data in a spreadsheet

Usually you collect data in a field notebook and then transfer them to a spreadsheet. This allows you to tidy up and order things in a way that will allow you to explore the data. It is also a convenient way to store and transfer data. In Section 2.2 you looked at how to arrange your data to begin with. There are many ways you can manage your data subsequently; you'll see some of these now.

3.2.1 Sorting

It is often useful to reorder you data and this is possible by sorting using the *Data > Sort* menu. Figure 3.4 shows the layout for Excel 2013 but the options are similar for all versions.

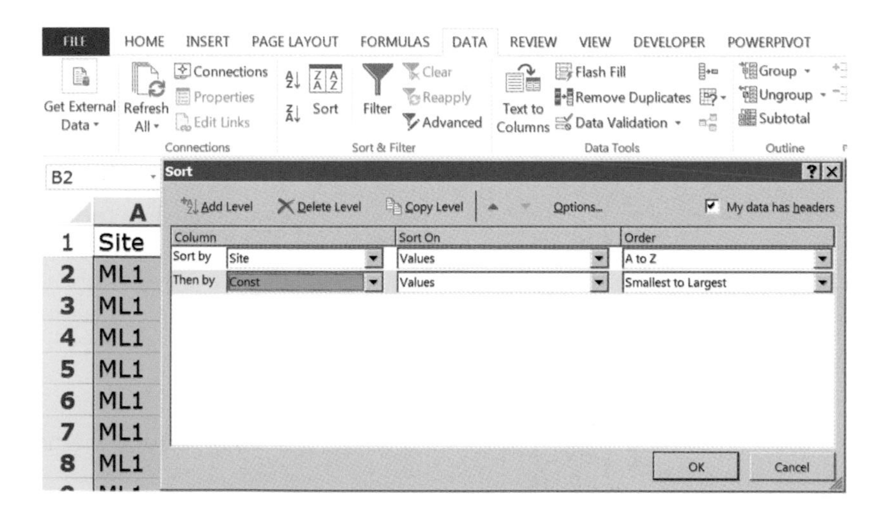

Figure 3.4 Sorting data in Excel, which provides a range of sorting options.

The menu box now allows you to select the columns you wish to sort (and whether you want an ascending or descending order). It is useful to have column names so you recall what the data represent. Ensure the button is ticked where it says *My data has headers*.

3.2.2 Data filtering

Often you'll need to examine only part of your data. For example you may have collected data on the abundance of plant species at several sites. If you wish to look at a single site or a single species (or indeed both) then you can use the filtering ability of Excel. The exact command varies a little according to the version of the program you are using; Figure 3.5 shows Excel 2013 and the filtering option using the *Home* menu (there are also filter tools in the *Data* menu).

Figure 3.5 Using the filter option (in the *Home* menu) to select parts of the data.

You can highlight the columns of data you wish to filter. Excel will select all the columns by default so you do not make an active selection. Once the filter is applied, you see a little arrow in each column heading. It is important to give each column a heading. You can then click the arrow to bring up a list (Figure 3.6).

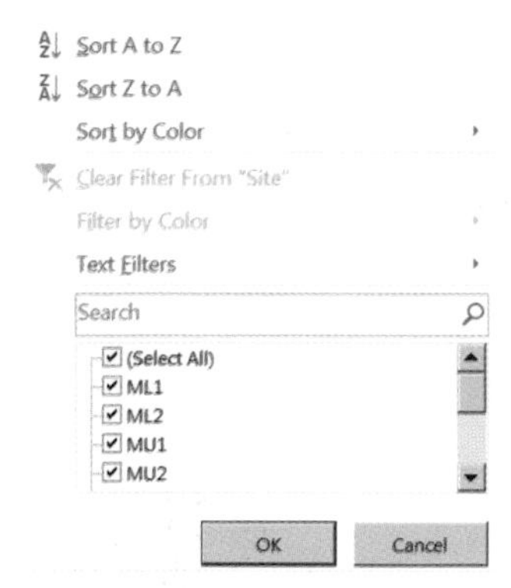

Figure 3.6 Filtering in Excel. Once a filter is applied, it can be invoked using the drop-down menu.

Figure 3.6 shows the filter in action. You can select to display a variety of options from the list available. Here you have site names so could select all, several or just one site to display. By adding a filter to subsequent columns you can narrow the focus.

3.2.3 Paste special

When copying data it is possible to rotate the rows and columns. Some programs require the species to be the rows and the samples to be the columns but other programs need the samples to be the rows. It is simple enough to rotate the data.

1. First of all you highlight the data you want and copy to the clipboard (use the *Home > Copy* button).
2. Next you place the cursor where you want the data to appear (make a new file or worksheet before copying if you want).
3. Now click the lower part of the *Home > Paste* button to bring up various paste options (Figure 3.7). Right-clicking with the mouse also brings up paste options.
4. Choose *Paste Special*, which is usually at the bottom.

Figure 3.7 The paste options are found in the *Home* menu of Excel 2013.

You now have a variety of options (Figure 3.8). If your data are created using formulae then it is advisable to select *Values* from the options at the top. This will not keep the

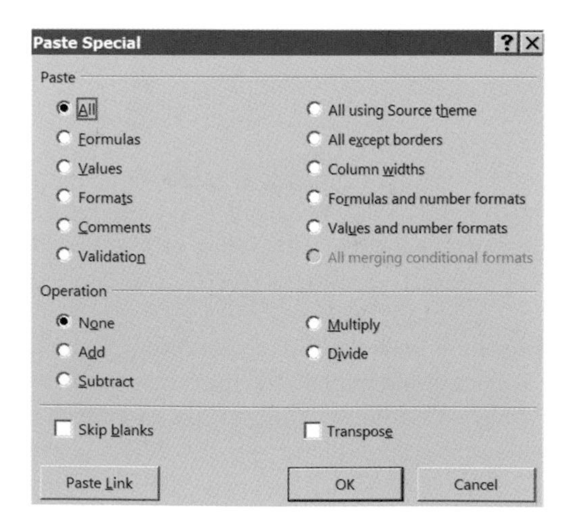

Figure 3.8 Using *Paste Special* in Excel. There are a variety of options for moving blocks of data from one place to another.

formulae but because they would not otherwise be recognized correctly (since you are moving the items) then this is desirable. If you want to preserve formatting (e.g. italic species names) then you can select this option. In this case you also wish to rotate the data so that rows become columns and columns become rows so you select the *Transpose* button. The samples (sites) are now represented as rows and the species are columns.

3.2.4 File formats

You may need to save your data in a format that is not an Excel workbook. Some programs require plain text files, whilst others will accept text where the columns are separated by commas or tab characters. To save files in alternative formats you need to select the worksheet you want to save and click the *File > Save As* button. How you proceed from here depends somewhat on the version of Excel you are using. In Excel 2013:

1. Click the *File* menu.
2. Click the *Save As* option in the sidebar.
3. Click the *Browse* button to select a location to store the file.
4. Click in the *Save as type* box and select the type of file to save (Figure 3.9).
5. Click *Save* to save the file to disk (Figure 3.10).

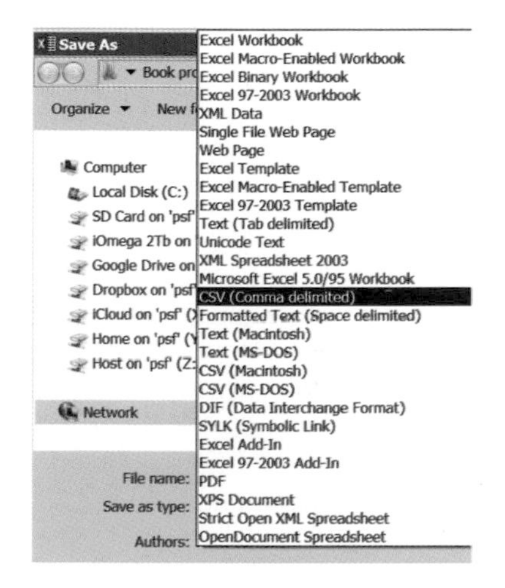

Figure 3.9 Excel 2013 can save files in many formats.

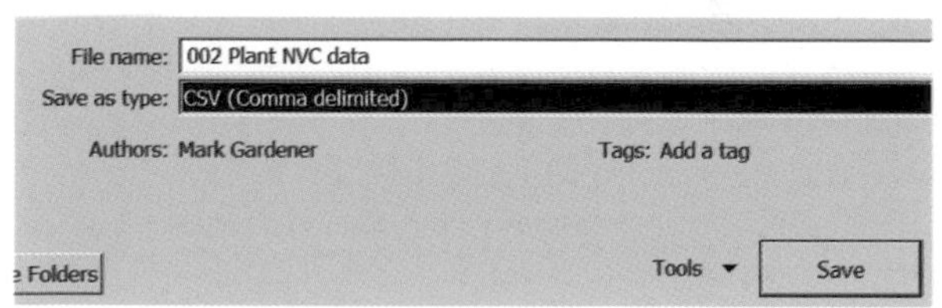

Figure 3.10 The CSV or comma delimited format is used by many other programs.

The comma separated variables (CSV) format in particular is useful as many programs (including R) can read this.

The drop-down menu allows you to select the file type required. CSV is a common format but formatted text and plain text can also be useful.

The CSV format is plain text resulting in the loss of some formatting and you will only be able to save the active tab; multiple worksheets are not allowed. Because of these reasons, Excel will give you a warning message to ensure you know what you are doing. The resulting file is plain text and when you view it in a word processor, you see the data are separated by commas (Figure 3.11). It is a good idea to keep a master XLS (or XLSX) file; you can then save the bits you want as CSV to separate files as required.

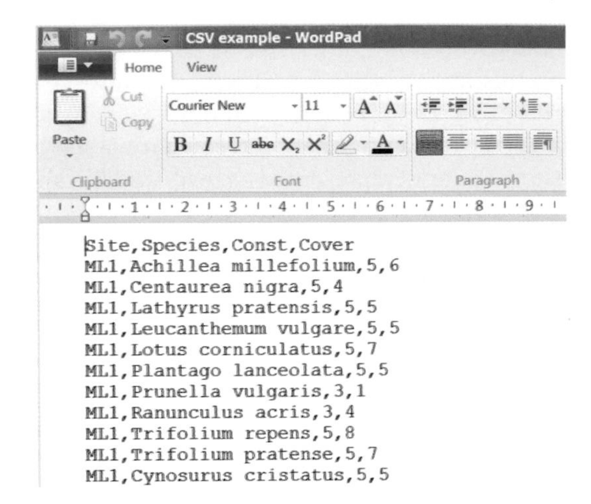

Figure 3.11 CSV format. In this format, commas separate data. When Excel reads the file it will place each item into a separate column when it comes across a comma. Many analytical programs use the CSV format.

In Figure 3.11 you see what a CSV file looks like in a word processor (WordPad). CSV stands for "comma separated variables" and, as this name suggests, you see the data separated by commas. When the CSV data are read by a spreadsheet, a comma causes a jump to a new column.

3.2.5 Opening a CSV file in Excel

When you open a CSV in your spreadsheet, the program will convert the commas into column breaks and you see what you expect from a regular spreadsheet. You can do all the things you might wish in a spreadsheet, such as creating graphs; however, if you want the spreadsheet elements to remain, you must save the file as an Excel file. When you hit the *Save* button you will get a warning message to remind you that the file you opened was not a native Excel file; you then have the option of choosing the format.

3.2.6 Lookup tables

There are occasions when you need to translate your data from one form to another. For example you may have recorded your results using a DAFOR scale, an example of *ordinal* data (recall data types from Section 1.5). When you write your results in your field notebook you record this ordinal scale but later you may wish to present the data numerically (perhaps in order to draw a graph or carry out statistical analysis). In your spreadsheet you can carry out the "translation" using a lookup table.

The `VLOOKUP` function allows you to pick out a value in one column of a table and replace it with a value from another column. In respect to the DAFOR scale you'd replace D with the number 5, A with 4 and so on. The general form of the function is:

```
VLOOKUP(original, table, column, approx._match)
```

In the function, `original` is the reference of the cell you want to replace, `table` points to a table containing the original values and their replacements. The lookup table can contain several columns (the first always gives the original values), so you specify `column`, a number giving the column where the replacements are. Finally you can carry out exact matching by specifying `FALSE`. If you type `TRUE` (or nothing) then approximate matching is carried out.

There is also a `HLOOKUP` function, which does a similar job but by row, rather than column.

In the following exercise you can have a go at using the `VLOOKUP` function.

Note: Fixed cell references

If you precede a cell reference with a $ it is "fixed", and so does not alter if you copy and paste cells. This can be especially useful when using the `VLOOKUP` function. You can use $ before the row or column part of the cell reference (or both).

Have a Go: Use a lookup table to make replacement variables

You do not need any data for this exercise, as you will create a spreadsheet from scratch.

1. Open Excel and start a new workbook. Go to cell A1 and type a heading `Obs`, to top a simple index of observations. Now in cell B1 type a heading `Abund`, to top a column of abundance values, recorded using a DAFOR scale.
2. Type a series of numbers in the *Obs* column, 1–10 will be sufficient. Now go to column B and type some values to represent abundance values in a DAFOR scale (D = dominant, A = abundant, F = frequent, O = occasional, R = rare). Start with `D`, `A`, `F` and so on. Then use lower-case letters so that you end up with ten values.
3. In cell C1 type a heading `Num` for the column that will contain a numeric equivalent for the abundance scale.
4. Before you complete column C, you'll need to make a lookup table to hold the values of the original abundance scale and the corresponding replacement values. So, go to cell E1 and type a heading `Scale`. In the cells underneath type the values of the abundance scale: E2 will contain D, E3 will contain A and so on.
5. In cell F1 type a heading `Ordinal`, to remind you that this column will hold an ordinal value, relating to the abundance scale. Enter the numbers 5 to 1 in cells F2:F6. You should now have a lookup table with one column containing the original DAFOR labels and one containing corresponding numerical values.
6. Now return to cell C2 and type a formula that looks to match an entered abundance in the lookup table and returns a numerical replacement: `=VLOOKUP(B2, $E$2:$F$6, 2)`. Note that you need the $. If you use the mouse to select the cells you can edit the formula afterwards.
7. Copy the formula in cell C2 down the rest of the column. Notice that the

lower-case letters are matched but that the abundance "A" produces an error #N/A for both upper and lower case (Figure 3.12).

C2 =VLOOKUP(B2,$E$2:$F$6,2)

	A	B	C	D	E	F	G	H
1	Obs	Abund	Num		Scale	Ordinal		
2	1	D	5		D	5		
3	2	A	#N/A		A	4		
4	3	F	3		F	3		
5	4	O	2		O	2		
6	5	R	1		R	1		
7	6	d	5					
8	7	a	#N/A					
9	8	f	3					
10	9	o	2					
11	10	r	1					

Figure 3.12 A lookup table must be sorted into alphabetical order if the approximate matching is TRUE (the default).

8. The problem is that the `VLOOKUP` function is trying to find an approximate match. Click on cell C2 and edit the formula: add an extra parameter, `FALSE`, at the end so the function reads: `=VLOOKUP(B2, $E$2:$F$6, 2, FALSE)`. Copy the function down the rest of the column. Your values should now be represented correctly (Figure 3.13).

C2 =VLOOKUP(B2,$E$2:$F$6,2, FALSE)

	A	B	C	D	E	F	G	H	I
1	Obs	Abund	Num		Scale	Ordinal			
2	1	D	5		D	5			
3	2	A	4		A	4			
4	3	F	3		F	3			
5	4	O	2		O	2			
6	5	R	1		R	1			
7	6	d	5						
8	7	a	4						
9	8	f	3						
10	9	o	2						
11	10	r	1						

Figure 3.13 Using FALSE at the end of a VLOOKUP function allows the lookup table to be unsorted.

9. Go back to cell C2 and edit the formula to change `FALSE` to `TRUE`. Copy the formula down the column. Your errors return.
10. Now highlight the values in the lookup table (cells E2:F6), then use the *Home > Sort & Filter > Sort A to Z* button. The lookup table should now be sorted alphabetically and the errors will disappear.
11. The lookup table does not have to be in the same worksheet. Highlight the lookup table (and the headings) in cells E1:F6 and use the *Cut* button, this is on the *Home* menu and looks like a pair of scissors. Now make a new worksheet using the icon at the bottom (or use the *Home > Insert > Insert Worksheet* button). Then paste the cells into the new worksheet using *Home > Paste*.
12. You can rename the worksheet containing the lookup table: right-click the appropriate tab and select *Rename*. If you now return to your original data and

look at cell C2 you'll see the reference to the cells of the lookup table appear with the name of the worksheet and a following exclamation mark!

Having a lookup table in a separate worksheet is often preferable to having the lookup next to the data. You may want more than one lookup table and it is handy to have them together in one place.

The companion website contains a completed version of the spreadsheet from this exercise, called *DAFOR.xlsx*.

Roman numerals

If you want to convert a regular number to a Roman numeral you can simply use the `ROMAN` function, which produces a text value:

```
ROMAN(value, form)
```

The `value` part is usually a cell reference but you can type in a number directly or include a formula that will produce a number. The `form` is a number in the range 0–4, which gives increasingly concise versions of the Roman numerals; this can be helpful for large numbers. You can also specify `TRUE` to get the classic Roman numeral or `FALSE` to get the most concise version.

If you have Excel 2013 or Open Office (or Libre Office) version 4 then you have access to the function `ARABIC`, which allows you to return a number from a Roman numeral:

```
ARABIC(text)
```

You simply give the Roman numeral and the result is an Arabic number. Usually you give a cell reference but you can also specify the text directly (in quotes). The function is not case sensitive, so lower-case letters are fine.

Using the IF() function to ignore blank cells

Sometimes you need to replace values but some spreadsheet cells are blank. If you use the `VLOOKUP` function with blank cells you'll get errors. You can overcome this problem by using the `IF` function.

The `IF` function allows you to take some condition into account; you can do one thing or another according to the result. This allows you to take into account blank cells for example (and many things besides). The basic form of the function is:

```
IF(comparison, do this if it is TRUE, otherwise do this)
```

There are three parts to the `IF` function, which are separated by commas:

1. The first part allows you to decide something; this is usually a formula, which should produce a TRUE or FALSE result.
2. What to do if the decision in part 1 was TRUE.
3. What to do if the result of the decision in part 1 was FALSE.

To look for a blank cell you use a pair of double quotes. This is the first part of your formula `IF(cell = "",)`, meaning that you look to see if the cell is blank. If it is blank then you want to keep it that way so you add a pair of double quotes, which will force the cell to remain blank: `IF(cell = "", "",)`. Finally you insert the `VLOOKUP` part. In practice you would select a non-empty cell and create your `VLOOKUP` formula. Then you could add the `IF` part at the beginning. Once you are happy the formula works on one cell, you can copy and paste over all the data. Doing it this way means you are less likely to get confused and make a mistake.

In the following exercise you can have a go at using the `IF` function in a spreadsheet that contains empty cells.

Have a Go: Use the IF function to take care of empty cells

You'll need the *seashore.xlsx* spreadsheet for this exercise. The file contains two worksheets; one is a completed version so you can check your progress.

Go to the website for support material.

The file shows some data on seaweed abundance on a rocky shore. Data on the abundance of some seaweed species have been collected using an ordinal abundance scale (ACFOR, A = abundant, C = common, etc., in decreasing fashion). Zero abundance is not recorded and the cells are empty. There are 13 sampling points from the high-tide mark down towards the low-tide mark.

Using an abundance scale makes it easy to collect data but is not useful for drawing graphs or carrying out statistics. In this exercise you'll convert the ACFOR scale into a numeric equivalent.

1. Go to the *data* worksheet (the uncompleted version) and you'll see the seaweed data in a table at the top. Highlight all the cells in the table (A1:N7) and copy to the clipboard using the *Home > Copy* button.
2. Click on cell A9 and paste the clipboard to make a copy of the data; use the *Home > Paste* button.
3. Highlight the cells containing the abundance data (B11:N15), then delete them using the *Delete* key on the keyboard (this keeps the formatting).
4. Click in cell P2 and start to make the lookup replacement table. Type the ordinal scale labels into cells P2:P6 (i.e. `A`, `C`, `F`, `O`, `R`).
5. Now click in cell Q2 and type numeric replacements into the cells Q2:Q6 (i.e. 5, 4, 3, 2, 1). You should now have a lookup/replacement table ready for use.
6. Click in cell B11 and type in a formula to replace the ordinal ACFOR value

with a numeric equivalent: `=VLOOKUP(B3,$P$2:$Q$6,2)`, then press *Enter* to complete the operation. This will give an error because cell B3 is empty.

7. Return to cell B11 and click in the formula bar and edit the formula to take into account any empty cells: `=IF(B3="","",VLOOKUP(B3,$P$2:$Q$6,2))`, then press *Enter* to complete the process.
8. Go back to cell B11 and copy it to the clipboard using the *Home > Copy* button.
9. Now paste the cell into the rest of the table (cells B11:N15), using the *Home > Paste* button.

You should now have a copy of the seaweed abundance data but in numeric form (Figure 3.14).

F11 | =IF(F3="","",VLOOKUP(F3,$P$2:$Q$6,2))

	A	B	C	D	E	F	G	H	I	J	K	L	M	N	O	P	Q
1								Transect Station									
2	***Species***	1	2	3	4	5	6	7	8	9	10	11	12	13		A	5
3	*Fucus vesiculosus*					F		F	F	A	O	C	O	C		C	4
4	*Ulva latuca*						F		F	F	O	F	O	F		F	3
5	*Enteromorpha spp.*		F	C			C	A								O	2
6	*Chondrus crispus*								R			R		R		R	1
7	*Corallina officinalis*								O		A	F	F				
8																	
9								Transect Station									
10	***Species***	1	2	3	4	5	6	7	8	9	10	11	12	13			
11	*Fucus vesiculosus*					3		3	3	5	2	4	2	4			
12	*Ulva latuca*						3		3	3	2	3	2	3			
13	*Enteromorpha spp.*		3	4			4	5									
14	*Chondrus crispus*								1			1		1			
15	*Corallina officinalis*								2		5	3	3				

Figure 3.14 Using a lookup table to convert ordinal data as an alphabetic character to a numerical value. Here you have to consider empty cells as well and use an IF statement.

Note that the ACFOR scale is already in alphabetical order. If you use a scale that is not in order (e.g. DAFOR) you must sort your table appropriately (use the *Home > Sort & Filter > Sort A-Z* button).

As you've seen, the `IF` function is very useful at taking into account empty cells.

3.2.7 Pivot tables

If you lay out your data appropriately you can use them in the most flexible manner. Excel (like other spreadsheet programs) has various tools that allow you to explore and visualize your data, helping you to carry out your analyses, check for errors and share your data and results with others.

The key to turning your spreadsheet data into a database is in using a layout in which each column represents a separate variable, whilst each row represents an individual record (i.e. one row for each *observation* or *replicate*).

There are several Excel tools that you can use to help you manage your data; the most important is the PivotTable® tool. A pivot table provides a way of managing and manipulating your data. You can use the PivotTable tool to arrange and rearrange your data in various ways that are useful, but hard to do with the spreadsheet directly. Most often the data are displayed in summary, the sum being the default (you can also show the mean). The more complicated your data, the more useful a pivot table becomes. You can use the PivotTable tool in many ways, such as to:

- Arrange your data in basic sample groups.
- Arrange your data in different groupings.
- View sample means or standard deviation.
- Visualize your samples using graphs.

The process of constructing a pivot table is fairly simple:

1. Click once anywhere in your block of data. There is no need to select any cells, Excel searches around the place you clicked and automatically selects all the occupied cells to form a block.
2. Click on the *Insert* tab in the Excel ribbon.
3. Click the *PivotTable* button on the *Insert* menu; this is usually on the far left.
4. A *Create PivotTable* dialog box opens. This allows you to select data and a location for the completed pivot table. The data are usually selected automatically (see step 1). Once you click OK you move to the next step.
5. From the *PivotTable Field List* task pane you build your pivot table. You drag fields, representing columns in your data, from the list at the top to one of the boxes at the bottom. These boxes represent the areas of the table.

Once you have built a basic pivot table you can alter and customize it quite easily. You can drag fields to new locations and alter the general appearance of the table using the various tools in the *PivotTable Tools* menus. In the following exercise you can have a go at making a pivot table for yourself.

Have a Go: Make a pivot table

You'll need the *butterflies.xlsx* data for this exercise.

Go to the website for support material.

1. Open the *butterflies.xlsx* spreadsheet. You'll see there are three columns: *Count*, the abundance of butterflies; *Habitat*, the location (*grass*, *heath* and *arable*); and *Obs*, the observation (replicate) number (Figure 3.15).
2. Click once anywhere in the block of data. Now click the *Insert* > *PivotTable* button (Figure 3.15).
3. The *Create PivotTable* dialogue box opens (Figure 3.16) and the data should be

selected automatically (if not you can select it now). Choose to place the pivot table in a *New Worksheet* (this is the default) and then click OK.

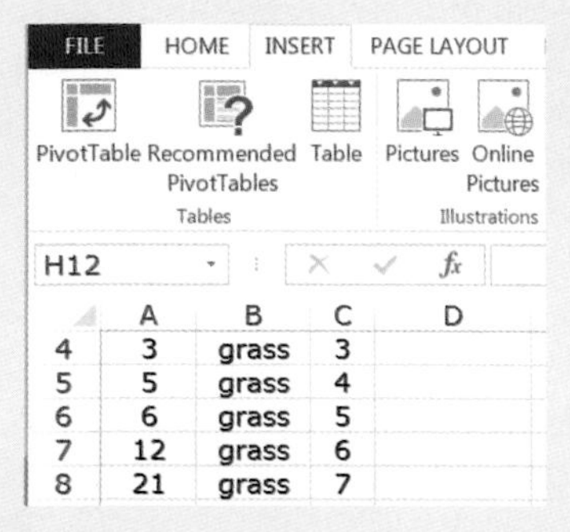

Figure 3.15 The *PivotTable* button is on the Insert menu.

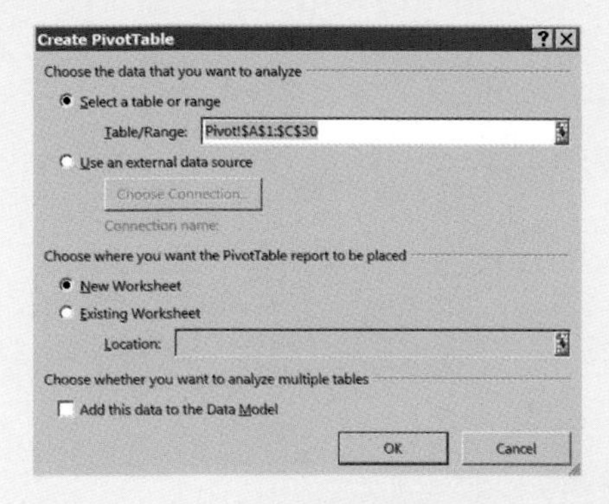

Figure 3.16 The *Create Pivot* dialogue box allows you to choose the destination for your completed pivot table.

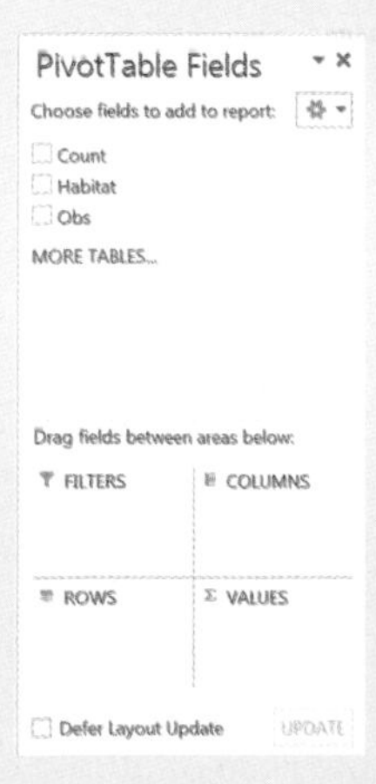

Figure 3.17 The *Field List* allows you to build your pivot table by dragging the fields to the sections of the table.

4. A new worksheet will open and you will see an empty pivot table and the *PivotTable Fields* dialogue box (Figure 3.17). The fields (columns of the original data) are listed at the top of the dialogue and the destinations at the bottom.
5. Drag the fields at the top of the *PivotTable Fields* dialogue box to the appropriate destinations at the bottom as follows: *Obs* to the *Rows* area; *Habitat* to the *Columns* area; *Count* to the *Values* area. Your pivot table should now be complete (Figure 3.18).

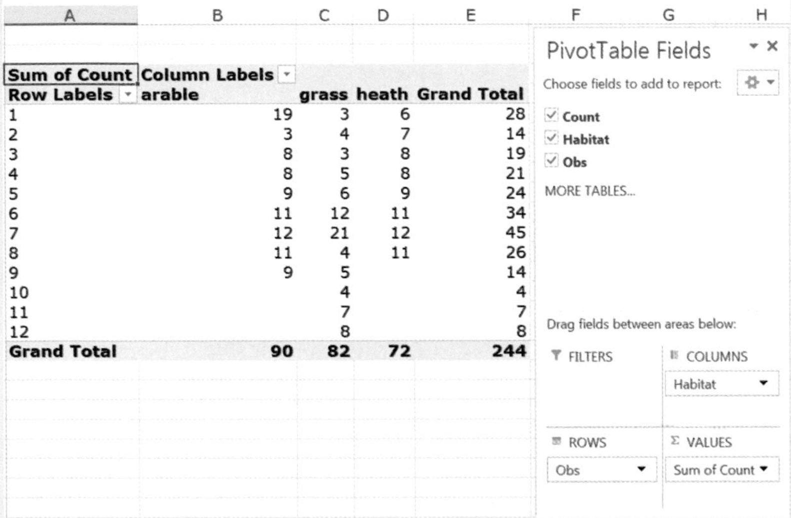

Sum of Count	Column Labels			
Row Labels	arable	grass	heath	Grand Total
1	19	3	6	28
2	3	4	7	14
3	8	3	8	19
4	8	5	8	21
5	9	6	9	24
6	11	12	11	34
7	12	21	12	45
8	11	4	11	26
9	9	5		14
10		4		4
11		7		7
12		8		8
Grand Total	**90**	**82**	**72**	**244**

PivotTable Fields
Choose fields to add to report:
Count
Habitat
Obs
MORE TABLES...
Drag fields between areas below:
FILTERS
COLUMNS
Habitat
ROWS
Obs
VALUES
Sum of Count

Figure 3.18 A completed pivot table.

Click once in the completed pivot table to activate the *PivotTable Tools* menus (they are probably open already). You can now use the various tools to customize your pivot table (the most useful tools are in the *Design* menu).

The PivotTable is a powerful tool, which you can use to help explore and visualize your data.

Customizing and saving pivot tables

When you make a new pivot table the data are usually summarized using the sum. In the previous exercise this was exactly what you wanted. Each cell in your table represents a unique combination of observation and habitat; in other words the original data. If you omitted the *Obs* field item you would arrange your data solely by *Habitat*. Your sum of *Count* items would then represent the sum of all the observations for each site.

One way you can alter the way the data are presented is by clicking the ∑ *Values* box; this brings up a new menu allowing you to alter the *Value Field Settings* (Figure 3.19). This means you can change the kind of summary (to average for example, see Chapter 4).

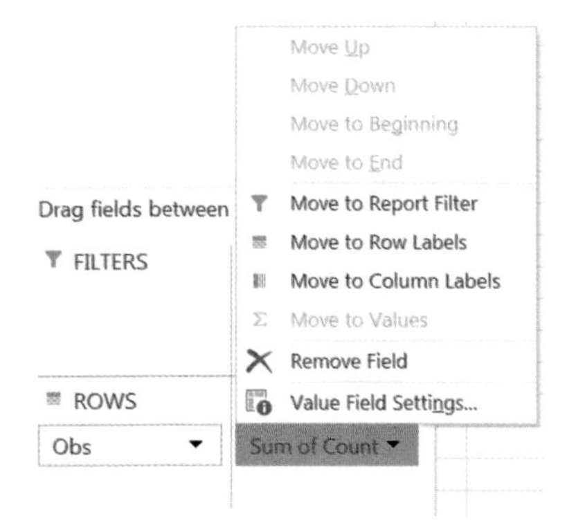

Figure 3.19 Clicking a field item in the Σ *Values* box in the *PivotTable Fields* dialogue box enables a pop-up menu, allowing you to alter the way a field is summarized.

You can also alter the *Field Settings* from the *PivotTable Tools > Analyze* menu. The *PivotTable Tools > Design* menu contains the most useful tools for modifying your table and in the following exercise you can have a go at making a pivot table and familiarizing yourself with some of these tools.

Have a Go: Customize a pivot table

You will need the *birds.xlsx* data for this exercise.

 Go to the website for support material.

Open the *birds.xlsx* spreadsheet. Here you have a survey of some common bird species. Each species has been recorded in several locations, each of which is a different habitat type. The data are in scientific recording format; each column is a separate variable and each row is an observation (the data are shown in Table 3.1).

You can use a pivot table to rearrange these data into a contingency table as a summary.

1. Click once anywhere in the bird data. Now click the *Insert > PivotTable* button. The data should be selected automatically; choose to place the table in a *New Worksheet* and click OK to proceed.
2. You should now see the *PivotTable Fields* dialogue box. Build the pivot table by dragging the fields from the list at the top to the boxes at the bottom like so: *Species* to *Rows* box; *Habitat* to *Columns* box; and *Quantity* to *Values* box (Figure 3.20).
3. Your pivot table is now completed but there is still some work to be done. You'll see that there are empty cells where the *Quantity* is zero (Figure 3.21). It would be useful to replace the empty cells with the number 0.
4. Click once in the pivot table to activate the *PivotTable Tools* menus. Now click

Table 3.1 Bird observation data in scientific recording format. A pivot table can rearrange the data into a contingency table as a summary.

Species	Site	GR	Date	Recorder	Qty	Habitat
Blackbird	Springfield	SP873385	04-Jun-07	Starling, C	47	Garden
Chaffinch	Springfield	SP873385	04-Jun-07	Starling, C	19	Garden
Great Tit	Springfield	SP873385	04-Jun-07	Starling, C	50	Garden
House Sparrow	Springfield	SP873385	04-Jun-07	Starling, C	46	Garden
Robin	Springfield	SP873385	04-Jun-07	Starling, C	9	Garden
Song Thrush	Springfield	SP873385	04-Jun-07	Starling, C	4	Garden
Blackbird	Campbell	SP865395	04-Jun-07	Starling, C	40	Parkland
Chaffinch	Campbell	SP865395	04-Jun-07	Starling, C	5	Parkland
Great Tit	Campbell	SP865395	04-Jun-07	Starling, C	10	Parkland
House Sparrow	Campbell	SP865395	04-Jun-07	Starling, C	8	Parkland
Song Thrush	Campbell	SP865395	04-Jun-07	Starling, C	6	Parkland
Blackbird	Ouzel Valley	SP880375	04-Jun-07	Starling, C	10	Hedgerow
Chaffinch	Ouzel Valley	SP880375	04-Jun-07	Starling, C	3	Hedgerow
House Sparrow	Ouzel Valley	SP880375	04-Jun-07	Starling, C	16	Hedgerow
Robin	Ouzel Valley	SP880375	04-Jun-07	Starling, C	3	Hedgerow
Blackbird	Linford	SP847403	04-Jun-07	Starling, C	2	Woodland
Chaffinch	Linford	SP847403	04-Jun-07	Starling, C	2	Woodland
Robin	Linford	SP847403	04-Jun-07	Starling, C	2	Woodland
Blackbird	Kingston Br	SP915385	04-Jun-07	Starling, C	2	Pasture
Great Tit	Kingston Br	SP915385	04-Jun-07	Starling, C	7	Pasture
House Sparrow	Kingston Br	SP915385	04-Jun-07	Starling, C	4	Pasture

the *Analyze > PivotTable > Options* button (you can also right-click in the pivot table). This opens the *PivotTable Options* dialogue box (Figure 3.22).

Figure 3.20 Pivot table fields from the list can be dragged into the appropriate sections to build a pivot table.

	A	B	C	D	E	F	G
1							
2							
3	**Sum of Quantity**	**Column Labels**					
4	**Row Labels**	**Garden**	**Hedgerow**	**Parkland**	**Pasture**	**Woodland**	**Grand Total**
5	Blackbird	47	10	40	2	2	101
6	Chaffinch	19	3	5		2	29
7	Great Tit	50		10	7		67
8	House Sparrow	46	16	8	4		74
9	Robin	9	3			2	14
10	Song Thrush	4		6			10
11	**Grand Total**	**175**	**32**	**69**	**13**	**6**	**295**

Figure 3.21 Cells with no data are left empty by default. You can alter the settings to replace empty cells with something else.

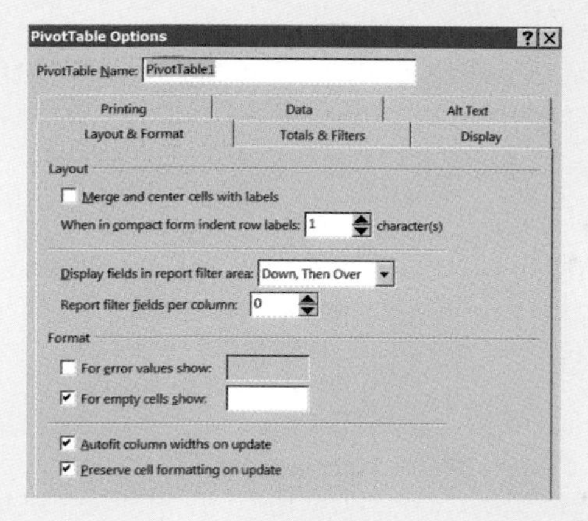

Figure 3.22 The *PivotTable Options* dialogue box allows access to many options that affect the way your pivot tables are displayed.

5. In the *Layout & Format* tab enter 0 (zero) in the box labelled *For empty cells show*, make sure the box is ticked then click OK to return to the pivot table.
6. The empty cells now display a zero. Click once in the pivot table to ensure the *Tools* menus are active. Now click the *Design* menu, which provides some helpful tools (Figure 3.23).

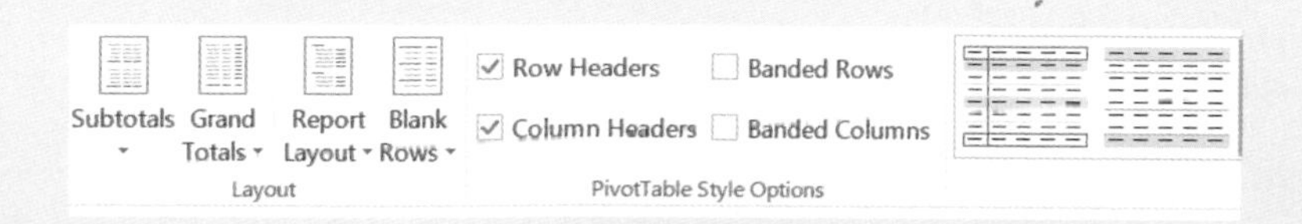

Figure 3.23 The *PivotTable Tools > Design* menu provides useful tools to help format and customize your pivot table.

7. Click the *Grand Totals* button and turn off all totals (i.e. *Off for Rows and Columns*).

8. Click the *Report Layout* button and select *Show in Tabular Form*. Your table is now complete (Figure 3.24).

	A	B	C	D	E	F
1						
2						
3	**Sum of Quantity**	**Habitat**				
4	**Species**	**Garden**	**Hedgerow**	**Parkland**	**Pasture**	**Woodland**
5	Blackbird	47	10	40	2	2
6	Chaffinch	19	3	5	0	2
7	Great Tit	50	0	10	7	0
8	House Sparrow	46	16	8	4	0
9	Robin	9	3	0	0	2
10	Song Thrush	4	0	6	0	0

Figure 3.24 A completed pivot table.

Have a look at some of the other tools; some will have no effect on this table but at least you'll see what sort of things are available.

Now that you have your summary table you can copy the data to a new location. To copy the data you could save the pivot table worksheet to a CSV file or copy and paste to a new file. You will need to use *Paste Special* (Section 3.2.3) otherwise Excel will try to link back to the original data.

There are many other things that you might do with a pivot table; you can have more than one data field in a row or column for example. Whenever you are inserting data into a spreadsheet you should think about how the data will be used and how they may be rearranged using a pivot table. This might lead you to add new columns that can be used for indexing. It is a lot easier to do this at an early stage.

3.3 Getting data from Excel into R

At some point you will wish to transfer data into R. The spreadsheet is a useful tool to set out and manipulate data. You can also carry out a range of analyses and produce graphs (Chapter 6); however, sooner or later you will want to conduct more in-depth analyses and the R program is invaluable for this. In order to import data you will need to learn a few more commands.

The read.csv() command

The `c()` and `scan()` commands are useful for entering small samples of data but you will usually have a lot more. The `read.csv()` command allows you to read CSV files. These may be prepared easily from spreadsheets (Section 3.2.4). By default R expects the first row to contain variable names.

```
> my.data = read.csv(file.choose())
```

As usual when using R you assign a name to store the data. In the above example you call

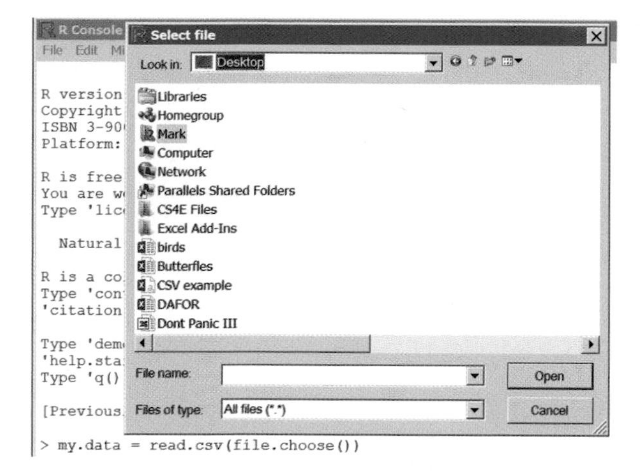

Figure 3.25 Importing data to R. The `file.choose()` command opens an explorer-type window (here shown in Windows 7).

this `my.data`. The `read.csv` part tells R to look for a CSV file. The `file.choose()` part tells R to open an explorer-type window so you can choose the file you want from your computer (Figure 3.25), otherwise you must specify the exact filename yourself (this is what you have to do if you are running R in Linux). There are several variations on this command but since the CSV format is so readily produced it seems unnecessary to learn anything other than the basic form of the command (try looking at the help for this command in R).

Note: Choosing a column to act as row labels

The `read.csv()` command allows you to read a CSV file into R. If you add the parameter `row.names = 1`, you will force R to set the first column as row names instead of regular data. To use a different column use a different value.

Now that you have become a bit more familiar with your computer tools, it is time to look at data and what you can do with them.

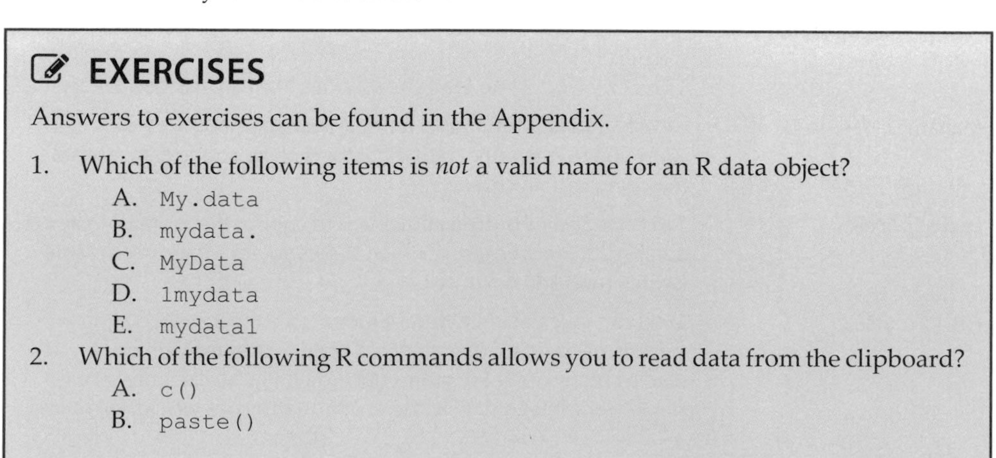

EXERCISES

Answers to exercises can be found in the Appendix.

1. Which of the following items is *not* a valid name for an R data object?
   A. `My.data`
   B. `mydata.`
   C. `MyData`
   D. `1mydata`
   E. `mydata1`
2. Which of the following R commands allows you to read data from the clipboard?
   A. `c()`
   B. `paste()`

   C. `scan()`
   D. `read()`
   E. `read.csv()`
3. If you had data in DAFOR format and needed to convert to a numerical format you could use a ____ table with the ____ or ____ function.
4. A pivot table is a good way to rearrange your data from recording format to sample format. TRUE or FALSE?
5. You can use the ____ command to import data from Excel into R.

Chapter 3: Summary

Topic	Key points
Help using R	From R use `help(topic)` or `?topic` where `topic` is the command you need help on. The command `help.start()` opens the help system in your browser. The Internet is a good source of help in using R.
Maths in R	R uses regular mathematical operators. The * and / are evaluated before + or – so use () as appropriate. Use ^ to raise a value to a power.
Saving results	Make named objects to save items e.g. `mymath = 2 + 2`. The maths to the right of the = is evaluated and stored to the named object, which is created (or overwritten).
Named objects	All stored items have a name. Names are case sensitive and alphanumeric (but cannot start with a number). To use an item, simply type its name. To see the objects in memory use the `ls()` command.
Inputting data	You can input data using `c()`, `scan()` or `read.csv()` commands.
Using the clipboard	The `scan()` command can read the clipboard. Use paste as often as required but press enter on a blank line to end the input process.
Additional packages	You can get extra libraries of R commands, mainly via the R website. Use `install.packages("package_name")`. Use `library(name)` to load the routines from an installed library.
Managing data in Excel	Excel has some useful tools for data management. *Sort* tools allow you to rearrange data. *Filter* tools allow you to display data matching certain criteria.
Paste Special	The *Paste Special* button allows you to control the format of pasted data, including *Transpose*, which allows you to rotate a data table (switch rows and columns).
File formats	Excel can save data in different formats via the *Save As* option in the *File* menu. The CSV format is a plain text format, which can be read by many other programs (including R). You can only save a single worksheet and will lose any formatting (as well as formulae).

Chapter 3: Summary – *continued*

Topic	Key points
Lookup tables	The Excel functions `VLOOKUP` and `HLOOKUP` are useful to "translate" your data from one form to another. You look up a value in the data and replace it with one from a replacement table.
Roman numerals	You can use lookup tables to convert between Roman and regular numbers. The Excel functions `ROMAN` and `ARABIC` can convert between numeric styles.
Empty (blank) cells	The `IF` command can be used to "take care" of blank cells.
Pivot tables	A pivot table is a way to arrange and manage your data. Pivot tables can help you arrange your data into summaries and help prepare data for graphing.
Get data from Excel into R	R can read CSV files. So save data from Excel using CSV format. In R use the `read.csv()` command. Adding `file.choose()` to the command allows you to select the data from a browser-like window. Adding `row.names = 1` sets the first column to be row labels.

4. Exploring data – looking at numbers

Statistics is a subject concerned with numbers but what exactly are statistics? The term has a variety of connotations and not all of them give a favourable impression.

What you will learn in this chapter

» How to summarize data samples
» The different kinds of average
» The different kinds of dispersion (data spread)
» How to use R for summary statistics
» How to explore the distribution (shape) of data
» How to draw histograms and tally plots in R and Excel
» How to test for the normal distribution
» How to transform data to improve normality
» How to calculate running means
» Different statistical symbols

In biological research you often collect a lot of data. Statistics allow you to make sense of these data. Essentially the purpose of statistics is to take a large quantity of information and present it in a simpler and more meaningful manner. A good start would be to produce a graph to summarize your data (Section 4.2). In Chapter 6, you will discover how graphs can help you explore your data and you'll also see graphs used to present results throughout the text (with a summary in Chapter 12). Representing your data graphically is a really important step as it helps you to visualize the situation and inform your approach. You may decide that you need to collect more data or even to revise your statistical method. You should always bear in mind that graphs are useful as part of the general analytical process and not just something to leave until the final report.

Imagine you were interested in the size of water beetles in a local pond. You visit the pond, capture a beetle and measure it. You now know how big the beetles are in the pond right? Unfortunately this is not the case; as with many other biological data, the size of the beetles is variable. A single measurement is not enough and so you collect more beetles and measure each one. The data are shown in Table 4.1.

Go to the website for support material.

Table 4.1 Beetle carapace lengths in millimetres.

Beetle sizes in mm				
17	26	28	27	29
28	25	26	34	32
23	29	24	21	26
31	31	22	26	19
36	23	21	16	30

If you were recording these data into your notebook or transferring the data into a spreadsheet then you would normally put all the values in a single column; however, for space and readability the data are presented in Table 4.1 as five columns.

4.1 Summarizing data

In Table 4.1 you see your beetle sizes in millimetres. Obviously you are excited to have collected all these data and wish to tell your colleagues. How would you go about reporting your data? You could show the table but that might not be impressive and could be quite tedious for the reader. What you need is a way of summarizing your data. One of the most useful ways you can do this is by using an *average*, which is a measure of the "middle" of a set of numbers.

4.1.1 Exploring averages

An *average* is a way of representing the middle of a set of numbers; statisticians talk about the central tendency of a sample. There are several types of average; the most common one used is the *arithmetic mean* or *mean*. To work this out you add up all your numbers and divide by the number of items you have. Written in mathematical notation, this is shown in Figure 4.1:

$$\bar{x} = \frac{\sum x}{n}$$

Figure 4.1 Formula to calculate the arithmetic mean.

The bar over the x signifies a mean, the n represents the number of data items and the capital Greek sigma, $\sum$, tells you to add together all the x values (the observations).

This is not the only average. You could also work out the *median* value. To do this you would arrange the data into numerical order and then select the middle number. In this case you have 25 items so the middle item would be the 13th largest. It is not easy to write this mathematically, the best you can do is shown in Figure 4.2:

$$median = Rank_{\left(\frac{n}{2}\right)+0.5}$$

Figure 4.2 Formula representing the median.

Here you have the rank and n, the number of data items as before. If you had 24 items (an even number) then the median would lie at the 12.5th rank, i.e. between the 12th and 13th values (you take the halfway point).

Finally you could select the most common value in your set of numbers; this is called the *mode*.

Table 4.2 Summary of beetle carapace data.

Summary of data	
Sum	650
n	25
Mean	26
Median	26
Mode	26

Table 4.2 shows the three averages for your set of data. Also shown are the sum and the number of data items. You can see that in this case the three averages are identical (a lucky coincidence!) but this is generally not the case.

Averages in Excel

Calculating averages in the spreadsheet is easy enough if you know how to create simple formulae. You recall that the formula to calculate the mean is the sum of all the values divided by how many items there are (Figure 4.1).

To determine the mean in Excel you need to create a formula for the sum (Figure 4.3), a formula for the count (how many there are) and a formula to work out the mean.

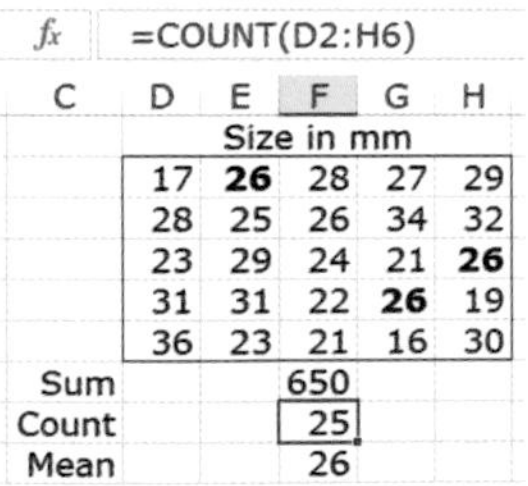

fx =COUNT(D2:H6)

C	D	E	F	G	H
			Size in mm		
	17	**26**	28	27	29
	28	25	26	34	32
	23	29	24	21	**26**
	31	31	22	**26**	19
	36	23	21	16	30
Sum			650		
Count			25		
Mean			26		

Figure 4.3 Calculating the sum using an Excel formula as part of determining the mean.

In Figure 4.3 you can see the beetle size data again. To determine the sum of the values you use the `SUM` function. The values in the brackets simply refer to the rows and columns where the data are to be found. In this case you start at `D2` and continue to `H6`, hence you define the top left and bottom right of the area where the data are found. It would have been better to set out the data in a single column but here the table has been spread over five columns to make it fit on the page easier.

To determine the count you use the `COUNT` function. In this instance you would use the same range of cells. To calculate the mean from the example above you need to divide `F7` by `F8`. Finally you create a formula in cell `F9`. This would simply be `F7/F8`.

This is fine but it would be nice if you could work out the mean in one step; Excel has built-in formulae for working out all three averages. The functions are `AVERAGE`, `MEDIAN` and `MODE` for the mean, median and mode respectively. In your example above you simply replace the name of the function for the one you want and keep the cell range (the `D2:H6` part) the same. It is unfortunate the name of the mean function is `AVERAGE`.

Averages in R

You can use the R program to determine averages. First of all you need your values. In the following example you see the beetle size data from earlier (Table 4.1). The data have been named `bd`, which effectively stores them in the memory and allows you to perform various operations on the numbers. To view the data you simply type the name of the data.

```
> bd
 [1] 17 26 28 27 29 28 25 26 34 32 23 29 24 21 26 31 31 22
[19] 26 19 36 23 21 16 30
```

To get the mean of your beetle data (`bd`) you use the `mean()` command; the result is 26.

```
> mean(bd)
[1] 26
```

To see the median you use the `median()` command. The result is also 26.

```
> median(bd)
[1] 26
```

You might think that the mode would be `mode(bd)` but that produces something unexpected.

```
> mode(bd)
[1] "numeric"
```

You actually have to use a rather daunting looking formula instead:

```
> as.numeric(names(which.max(table(bd))))
[1] 26
```

The good news is that you almost never use the mode! More good news is that once you know what the command is, you can save it in a text file and paste it in when you need it; simply replace the `bd` with the name of your data item.

If you sidetrack briefly and look at the last command, you'll see what R did. The command was:

```
> as.numeric(names(which.max(table(bd))))
```

Take the last part and type that in by itself:

```
> table(bd)
bd
16 17 19 21 22 23 24 25 26 27 28 29 30 31 32 34 36
 1  1  1  2  1  2  1  1  4  1  2  2  1  2  1  1  1
```

This `table()` command has split the data and shows you how many items there are of each value. This is potentially useful, as you shall see soon. You can readily see that the value 26 occurs most (four times) but you need to be a bit more explicit to tell R what is required.

You might have wanted to work out the mean from the sum and the number of items. Try this for yourself:

```
> sum(bd) / length(bd)
```

The `length()` command is like the `COUNT` function in Excel. Here you join the `sum()` and `length()` commands together. Try each one separately to convince yourself that each gives the expected result.

R and data with multiple samples

When you have data that is more complicated than a single sample you cannot simply use the `mean()` and `median()` commands (or others, which you will meet later). These are designed to operate on a single sample. Most often you will import these larger datasets using the `read.csv()` command (recall Section 3.3). The following is a typical example:

```
> hog3 = read.csv(file.choose())
> hog3
  Upper Mid Lower
1     3   4    11
2     4   3    12
3     5   7     9
4     9   9    10
5     8  11    11
6    10  NA    NA
7     9  NA    NA
```

You can see that there are three columns, each being a separate sample. The data represent the abundance of freshwater hoglouse at three different sites. Note that some of the data are not numbers but a special value, `NA`. Unlike Excel, R does not like empty cells and gives them a special assignment. Think of `NA` as standing for *not available*. You'll need to know how to deal with `NA` items; they may be real missing values or (as in this case) they may simply be there to make the lengths of the columns all the same.

Here is another example of data:

```
> hog2 = read.csv(file.choose())
> hog2
   count  site
1      3 Upper
2      4 Upper
3      5 Upper
4      9 Upper
5      8 Upper
6     10 Upper
7      9 Upper
8      4   Mid
9      3   Mid
10     7   Mid
11     9   Mid
12    11   Mid
13    11 Lower
14    12 Lower
15     9 Lower
16    10 Lower
17    11 Lower
```

This time you see the same data but arranged in a different format. This is what could be called *Scientific Recording Format*; each column represents a separate variable, whilst each row is a single observation (replicate). This layout is more powerful than the multi-sample layout and you can carry out more complicated statistics with data laid out like this.

In the example *hog2* data, there are two columns; the one headed *count* is the *response* (or dependent) variable and the one headed *site* is the *predictor* (or independent) variable. Notice that there are no `NA` items in the *hog2* data; you can get missing values with this kind of layout but usually not when you only have two columns (i.e. one predictor variable).

You'll need to know how to deal with both layouts, which you'll see throughout the text. R has several commands that allow you to summarize data that are more complicated than a simple sample; some of these are presented in Table 4.3.

The simplest commands shown in Table 4.3 are `colMeans()` and `colSums()`, with their row equivalents; the former command is illustrated below:

```
> colMeans(hog3)
   Upper       Mid     Lower
6.857143        NA        NA
> colMeans(hog3, na.rm = TRUE)
    Upper       Mid      Lower
 6.857143  6.800000 10.600000
```

Table 4.3 Some summary commands used for complicated data in R.

Command	Details
`colMeans(x)` `rowMeans(x)`	Calculates the mean value for each column (or row) of a data object. To ignore `NA` items use `na.rm = TRUE`.
`colSums(x)` `rowSums(x)`	Calculates the sum for each column (or row) of a data object. To ignore `NA` items use `na.rm = TRUE`.
`apply(x, MARGIN, FUN)`	Allows you to specify any function, `FUN`, which is applied to every row (`MARGIN = 1`) or column (`MARGIN = 2`). To ignore `NA` items use `na.rm = TRUE`.
`tapply(X, INDEX, FUN)`	Allows you to specify any function, `FUN`, which is applied to variable `X`, with a grouping variable `INDEX`. To ignore `NA` items use `na.rm = TRUE`.
`aggregate(y ~ x, data, FUN)`	Allows you to specify any function, `FUN`, by specifying the response and predictor variables `y ~ x`. You can specify the `data` object holding these variables. To ignore `NA` items use `na.rm = TRUE`.

In the first case the `NA` items were not "removed" and you can see the result for samples that contain `NA`. In the second case the `na.rm = TRUE` parameter was used and this resulted in the "correct" mean.

These are useful commands but they only allow you to determine a mean or sum. The other commands in Table 4.3 are a good deal more powerful; the simplest is the `apply()` command:

```
> apply(hog3, MARGIN = 2, FUN = median, na.rm = TRUE)
Upper   Mid Lower
    8     7    11
```

When you specify `MARGIN = 2` you get the function applied to the columns (specify `MARGIN = 1` for rows). Note that the `na.rm = TRUE` parameter can also be used.

The `tapply()` command works on data that are in the scientific recording layout, such as the preceding *hog2* example. In that case there was a response variable (*count*) and a predictor variable (*site*). To use the `tapply()` command you need to specify the response variable and an `INDEX`, in other words a grouping (predictor) variable. However, when you try this on the data you get an error:

```
> tapply(count, INDEX = site, FUN = median)
Error in tapply(count, INDEX = site, FUN = median) :
  object 'site' not found
```

The issue is that the variables *count* and *site* are "contained" inside the *hog2* data object. This means they are not visible to R directly. There are three ways you can overcome this:

- Prepend the data name and a `$` to the variable you want.
- Use the `attach()` command to make the variables within the data object "visible".
- Use the `with()` command to "open" the data object temporarily.

Using the `$` is the simplest way to proceed:

```
> tapply(hog2$count, INDEX = hog2$site, FUN = median)
Lower   Mid Upper
   11     7     8
```

Essentially the `$` subdivides the main data object (*hog2*), allowing the command to access the relevant variables within it.

The `attach()` command works in a different way; it allows R to see inside a data object:

```
> attach(hog2)
> tapply(count, INDEX = site, FUN = median)
Lower   Mid Upper
   11     7     8
> detach(hog2)
```

After the `attach()` command is issued all the variables within the data object are visible to R directly. If you have variables within your data object that have the same names as objects outside the object (i.e. in the main R workspace) then they are temporarily unavailable. This can lead to confusion about which variables are which, so you should use the `detach()` command when you are finished with the data, which "closes" it and returns to "normal".

The `with()` command does a similar thing to `attach()` but is only effective during a single command. Essentially it "tidies up" after itself and you don't need to worry about conflicting data names. You use the command to "wrap around" your main command:

```
> with(hog2, tapply(count, INDEX = site, FUN = median))
Lower   Mid Upper
   11     7     8
```

Note: Separate grouping variables

The `tapply()` command will work with data that are separate from one another, as long as the two items are the same length.

The last of the summary commands from Table 4.3 is called `aggregate()`. This allows you to use a function on a response variable, grouped by a predictor. In this function you give the variables as a sort of formula, `response ~ predictor`. You can also specify where the data are to be found, which is helpful as you do not need to use the `$`, `attach()` or `with()` commands:

```
> aggregate(count ~ site, data = hog2, FUN = median)
   site count
1 Lower    11
2   Mid     7
3 Upper     8
```

Notice that you get a result arranged in a different layout from the other commands; `aggregate()` returns a `data.frame` result, whilst the others produce `matrix`-like results. The different sorts of object are handled by some R commands in different ways. You'll see more about this at times when it is necessary, such as in Chapter 6, when you'll look at graphs.

Slicing up a data object

There are many occasions when you only want to deal with part of a dataset. You saw in the preceding section how to use the `$` to access certain column variables from a data object. There are two main ways to "slice up" a data object:

- Use `data$name`: this only extracts column variables from `data.frame` objects.
- Use `[row, column]`: this permits access to all parts of an object.

The `$` is straightforward to use; you simply use the name of the "holding" data object, a `$` and then the name of the column variable:

```
> names(hog2)
[1] "count" "site"
> count
Error: object 'count' not found
> hog2$count
 [1]  3  4  5  9  8 10  9  4  3  7  9 11 11 12  9 10 11
```

This only works if the data object is in a particular form, the `data.frame`. This is the most likely because that is what you get if you import data files from CSV using the `read.csv()` command.

The square brackets work with more or less any sort of data object. If you have a single sample (a single row of numbers or text is called a `vector`) you specify a single item in the brackets. If you have a two-dimensional object you specify `[rows, columns]`.

Here are some examples that use the beetle size data:

```
> bd
 [1] 17 26 28 27 29 28 25 26 34 32 23 29 24 21 26 31 31 22
[19] 26 19 36 23 21 16 30
```

```
> bd[1]
[1] 17
```

The first element of the *bd* data object.

```
> bd[1:5]
[1] 17 26 28 27 29
```

The first five elements.

```
> bd[18:14]
[1] 22 31 31 26 21
```

Elements 18–14 (in that order).

```
> bd[c(1:3, 12, 9)]
[1] 17 26 28 29 34
```

Elements 1–3, then 12, and then 9. Note the use of the `c()`.

```
> bd[seq(1, 12, 2)]
[1] 17 28 29 25 34 23
```

Elements in a sequence: from 1 to 12, by 2. In other words the 1st, 3rd, 5th, … 11th.

If you have data with rows and columns you need to specify both in the square brackets. If you leave one position empty R assumes "all".

```
> hog3[1,1]
[1] 3
```

The element corresponding to the first row and the first column in the *hog3* data object.

```
> hog3[1:2,]
  Upper Mid Lower
1     3   4    11
2     4   3    12
```

The first two rows and all the columns.

```
> hog3[5:6, "Mid"]
[1] 11 NA
```

Rows 5 and 6 in the column named "Mid". You must give quotes if you use a name.

Being able to split up your data and select parts of a dataset is very useful; you'll see examples throughout the text.

4.1.2 Which average to use

The best average to use is the one that best describes the data! In general the mode is the least useful of the averages and is useful only when the dataset contains a large quantity of numbers (maybe $n > 1000$). The mean and median are much more useful. The mean would be the average to use when the data are fairly symmetrical, with more or less equal values above and below the mean value. The median is a more useful average when there is a preponderance of low values (or high ones).

4.2 Distribution

You wish to use the most appropriate average but how do you determine whether you have a preponderance of low (or high) values or if the data are evenly spread? You need to look at the *distribution* of the data. One easy way to do this is to make a tally plot (e.g. Table 4.4).

In Table 4.4 you see the data separated into categories (often called *bins*); here the lowest bin is for values up to 16, the next covers values > 16 but not above 18 and so on.

Table 4.4 Tally plot to show frequency distribution for beetle carapace lengths. These data appear normally distributed.

Tally	Bin
x	16
x	18
x	20
xxx	22
xxx	24
xxxxx	26
xxx	28
xxx	30
xxx	32
x	34
x	36

For each category you put a cross (or a tick or tally mark) against the appropriate bin. When you are done you see a "picture" of your data. In this instance you can see that the tallies are pretty evenly spread around the mid-point, the plot looks fairly symmetrical about the middle (this is called *normal distribution* or *parametric*). This would be a good situation in which to use the mean as your average. If the plot had appeared like the one below (Table 4.5) then you might make a different choice.

Table 4.5 Tally plot of data that are not normally distributed. There are more values at the low end and the distribution is not symmetrical.

Tally	Bin
x	16
xxxx	18
xxxxx	20
xxxxxx	22
xxx	24
xx	26
xx	28
x	30
x	32
x	34
x	36

In Table 4.5 you see a different situation. You have a *skewed* distribution here and the plot is not symmetrical: you have a long tail at the higher end. This indicates that you have more low numbers than high numbers. In this case you would select the median as your average. This skewed distribution is also known as *non-parametric*.

You can produce a tally plot from your data simply enough using a notebook and pencil; it is the sort of thing you can do as you go along.

Another quick way of looking at the distribution of the data is to use a *stem and leaf plot* (Table 4.6). This is much like a tally plot but you can use actual numbers.

In Table 4.6 the beetle size data from Table 4.1 are written out in two columns with the left column being the 10s and the right column the units.

Table 4.6 Stem and leaf plot of beetle carapace data as an alternative to a tally plot. The advantage over tally plots is that the original data values can be seen.

Stem–leaf plot of data	
1	679
2	112334
2	5666678899
3	1124
3	6

Table 4.6 shows your data as a stem and leaf plot. From this you can see quite easily that your data are fairly evenly distributed. In this instance each block of 10s has been split into two with 0–4 being in one half and 5–9 in the other. This is done to make the plot more readable and to give you more bins. For other sets of numbers it may be okay to use a single block for each division of 10.

You can create a stem and leaf plot simply in a notebook using a pencil, which makes it a useful tool for when you are out in the field. In the examples shown here the values are in order, which makes it easier to read but is not essential. You can generate a stem plot using R quite simply using the `stem()` command.

```
> stem(your.data)
```

You simply replace the `your.data` part with the name of the data you are using. There is no easy method of making a stem and leaf plot in Excel (but see Chapter 6).

4.2.1 Histograms

The tally plot (e.g. Table 4.4) and stem and leaf plots (Table 4.6) are both useful ways of looking at the distribution of the data. Both can be done in the field as you are collecting data. Another way of presenting the distribution is to create a *histogram* or bar plot of the data (Figure 4.4). In this case you have a series of bars, each relating to a bin category, and their height is related to the frequency of the data in that bin.

Figure 4.4 shows a histogram of the beetle data in Table 4.1. The x-axis shows the size classes (i.e. the bin categories) and the y-axis shows the frequency. Here the horizontal gridlines have been preserved (as dashed lines) but often these will be omitted. In a true histogram the x-axis represents a continuous variable (although divided into parts) and the bars should be touching; here they have been separated by a short interval for clarity.

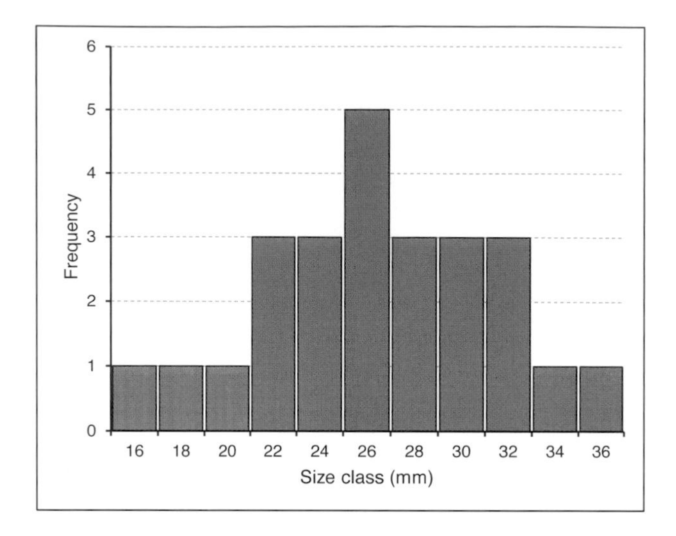

Figure 4.4 Histogram of beetle carapace data. This plot was produced in Excel 2013.

Histograms in Excel

In Excel you can make a histogram using a simple bar (column) chart (see Chapter 6 for more about bar charts); you need two sets of information:

- The *bins* (the size classes).
- The frequencies (how many observations there are in each *bin*).

The frequency data are compiled using the `FREQUENCY` formula, which works as follows:

```
FREQUENCY(sample_data, bins)
```

In the `FREQUENCY` formula the first parameter (`sample_data`) points to the data, the second parameter (`bins`) points to the range of bins. This type of formula is called an array formula in Excel and is entered simultaneously in all the cells where you want the results. You can do this by highlighting all the cells and entering the formula. Instead of pressing `ENTER`, you press `CTRL+SHIFT+ENTER` (in Windows); see the Excel help for details (on a Mac press `cmd+Shift+ENTER`).

Each bin is written in the spreadsheet as a single numerical value. In reality each bin represents a range of values. In Figure 4.4 for example the first bin (16) represents values ranging up to 16. The next bin (18) represents values greater than the previous bin but only up to 18.

In order to create a histogram you follow these steps:

1. Determine how many bins you'll need. This is something of a "black art" but 7–11 is usually about right. Work out the lowest value you need and the interval between bins to give you the required number of bins.
2. Use the mouse to select some blank cells next to the bins (you want one empty cell for each bin). This is where your frequency results will appear.
3. Start typing the formula `=FREQUENCY(`

4. Now either type the range of cells corresponding to the data or use the mouse to select them.
5. Type a comma and then select the range of cells containing the bins (or type the cell range directly.
6. Type a closing) but do *not* press *Enter*.
7. Press `Ctrl+Shift+Enter` to complete the formula (on a Mac use `cmd+Shift+Enter`).

You should now have a set of values representing the frequencies corresponding to the bins (Figure 4.5).

fx {=FREQUENCY(A2:A26,E2:E12)}

C	D	E	F	G
		Bins	Freq	
max	36	16	1	
min	16	18	1	
bins	11	20	1	
int	2	22	3	
		24	3	
		26	5	
		28	3	
		30	3	
		32	3	
		34	1	
		36	1	

Figure 4.5 Using the FREQUENCY function to assign frequencies to bins (size classes) in the preparation of a histogram.

Now you have the frequency data you can draw the histogram itself. The graphs are all found in the *Charts* section of the *Insert* ribbon menu (Figure 4.6).

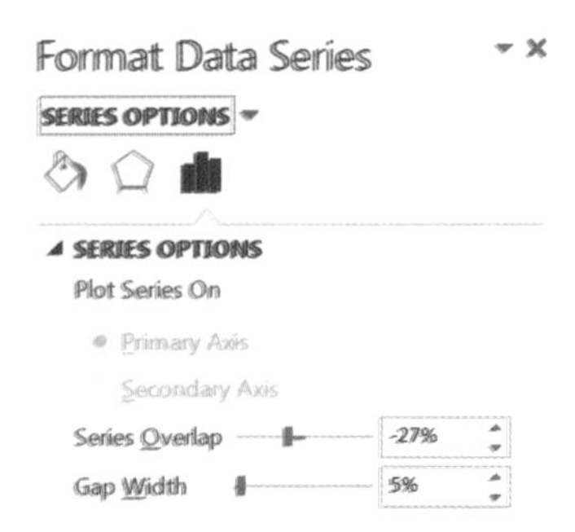

Figure 4.6 The *Charts* section of the *Insert* menu is the place to find the chart tools in Excel 2013.

To make a histogram you'll need the following steps:

1. Click once in your worksheet; click an empty cell that is *not* adjacent to any data. Excel searches for data and tries to be helpful; generally it fails so it is best to be in control yourself.
2. Click the *Insert > Column Chart* button and choose a basic *2-D Column* chart (Figure 4.7).

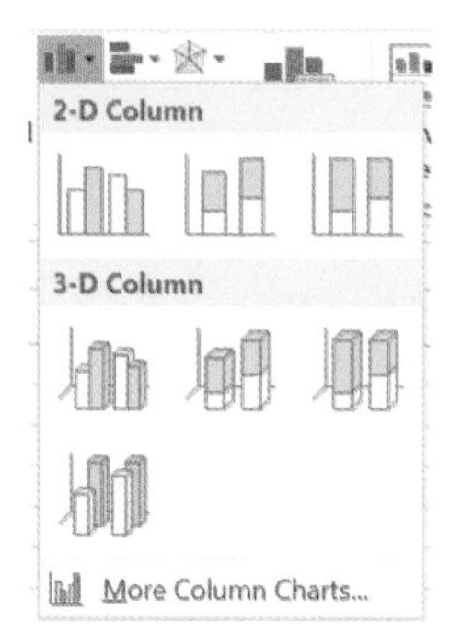

Figure 4.7 A basic 2-D Column chart is used for a histogram.

3. You should now have an empty chart (you'll see the outline only) and the *Chart Tools* menus will be visible. Click the *Design* > *Select Data* button.
4. Click the *Add* button in the *Legend Entries (Series)* section; then choose the cells that contain the frequency data. You can also choose a cell that contains a name if you like. Click OK once you are done.
5. Now click the *Edit* button in the *Horizontal (Category) Axis Labels* section. then select the data corresponding to the bins. Click OK once you've done that.
6. Click OK once again and your chart should now be completed. There are various ways you can alter the formatting and style of the chart. The tools you need appear in the *Chart Tools* menus once you've clicked on the chart. You can also right-click on the chart and bring up a menu that allows access to many formatting options (Figure 4.8).

Figure 4.8 Right-clicking an Excel 2013 chart brings up menus allowing access to many formatting tools.

Your chart will still need a bit of work to get it into an appropriate format but the basics are there from the start. For your histogram one useful option is to reduce the gap between the bars:

1. Click on the chart to activate the *Chart Tools* menus.
2. Now click the *Format* menu and use the dropdown box on the far left (in the *Current Selection* section). Choose the data series (*Series 1* is the default name but you may have chosen a name).

3. Now click the *Format Selection* button, which opens a dialogue box enabling you to alter settings for the bars (Figure 4.9). Alter the *Gap Width* to a small value (5% is about right).

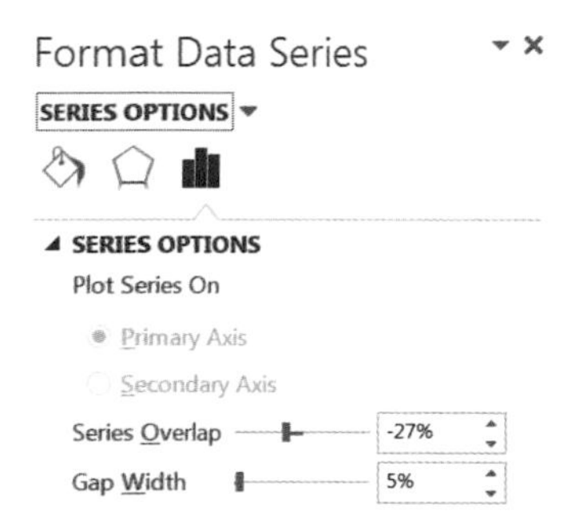

Figure 4.9 Altering the options for the data series: reducing the gap width makes the histogram look more like a classic histogram.

There are many other chart elements that can be tweaked; have a go at using the various formatting tools and see the effects on your chart.

You can use the *Analysis ToolPak* (Section 1.9.1) to generate the frequencies that you will need. You still need to create a list of bins but the *ToolPak* removes the need for you to grapple with the `FREQUENCY` formula.

You run the histogram routine by clicking *Data Analysis* on the *Data* ribbon menu. You then select *Histogram* from the options and select your data. In the following exercise you can have a go at using the *Analysis ToolPak* for yourself.

Have a Go: Use the *Analysis ToolPak* to make a histogram

You will need the *beetles.xlsx* data file for this exercise.

 Go to the website for support material.

1. Open the *beetles.xlsx* spreadsheet. Go to the *beetle sizes* worksheet, where you will see a simple column of values. You will use these and determine the frequencies to use in a histogram.
2. Click in cell C1 and type a heading for the bins; `Bins` will do nicely.
3. Go to cell C2 and enter the number `16`. In cell C3 enter `18`. Fill in the rest of the values down the column until you reach the value `36`. You should now have 11 values from 16 to 36 in steps of 2.
4. Now click the *Data* > *Data Analysis* button, which will open the *Data Analysis* dialogue box. Scroll down and select the *Histogram* option; click OK to move on.
3. The *Histogram* dialogue box allows you to select the data cells and the bins. You can also choose to have a histogram drawn for you (Figure 4.10).

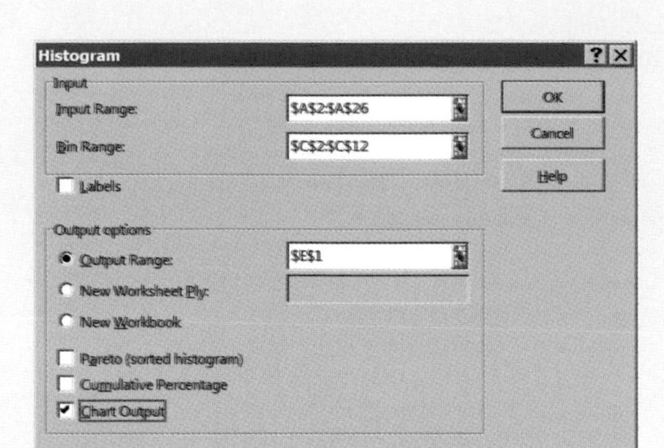

Figure 4.10 The *Histogram* dialogue box from the *Analysis ToolPak* makes drawing histograms an easy process.

4. Click in the box labelled *Input Range* and select the range of cells for the data, which is cells A2:A26.
5. Click in the box labelled *Bin Range* and select the range of cells for the bins you just made, which is cells C2:C12.
6. You can select the location to place the output in the *Output options* section. Click the button then click in the box beside it and click in cell E1; the result will go there shortly.
7. Now check the box labelled *Chart Output* so that Excel will draw the chart for you. Click OK once you are ready and the results will be produced immediately (Figure 4.11).

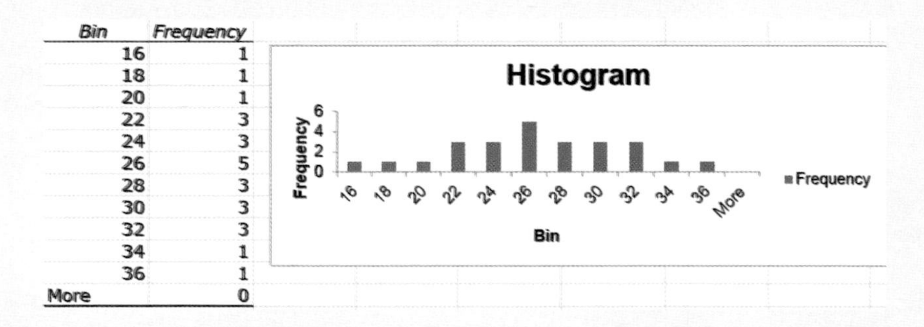

Bin	Frequency
16	1
18	1
20	1
22	3
24	3
26	5
28	3
30	3
32	3
34	1
36	1
More	0

Figure 4.11 The *Analysis ToolPak* can produce frequency results and a histogram, which will require some editing.

You can now edit the histogram using the regular *Chart Tools*. There are additional worksheets showing completed histograms using the *Analysis ToolPak* and "longhand".

Note: The *Analysis ToolPak* and bins

The *Analysis ToolPak* tends to add an extra bin, labelled *More,* at the end. This is also evident in the histogram. You can either leave it in place or simply edit the chart so that the data range does not extend to this bin (use the *Chart Tools > Design > Select Data* button).

The *Analysis ToolPak* is a convenient way to make a histogram in Excel, and can save you a lot of time.

Histograms using R

In R you can create histograms using the `hist()` function. You simply give the name of the data and the command does the rest.

```
> hist(bd)
```

Here the beetle data from Table 4.1 were used to make the histogram (Figure 4.12).

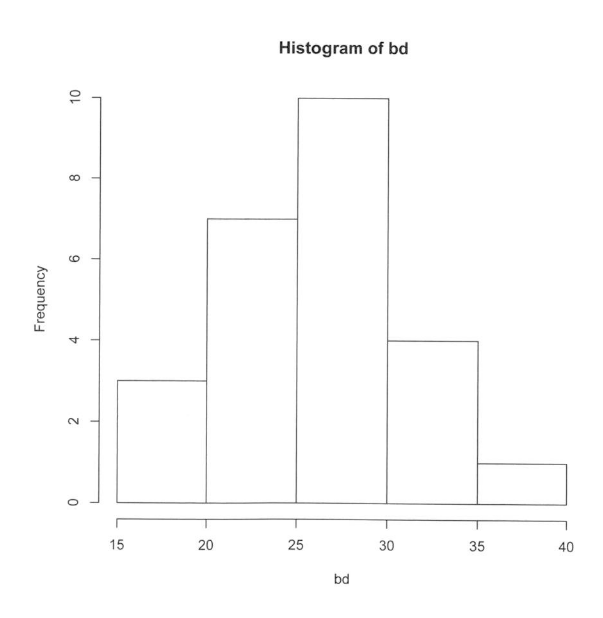

Figure 4.12 Histogram of beetle size data created using R and the hist() command.

It is possible to specify the exact bins in R but in practice it is generally best to let the program work out the best arrangement for itself. Notice how R arranges the values on the category axis (the *x*-axis). The values are placed at the edges of the bars, highlighting the fact that the axis represents a continuous scale rather than discrete categories.

It is also possible to use points (joined with smooth lines) instead of bars but this generally does not look good unless you have a lot of data. In the R program there is a way to produce a density plot, which is a similar chart. The `density()` command generates the data required to make the plot.

A histogram by definition splits the sample into blocks (usually called bins) and how the breakpoints are defined may well alter the picture. The `density()` function provides an alternative, which can help you interpret the data. For example:

```
> plot(density(bd))
```

It is possible to create a regular histogram and add the density lines afterwards. In the following exercise you can have a go at this for yourself.

Have a Go: Use R to make a histogram and density plot

You will need the *beetles.xlsx* data file for this exercise.

Go to the website for support material.

1. Open the *beetles.xlsx* spreadsheet. Go to the *beetle sizes* worksheet where you will see a simple column of values. You will transfer these to R and make your plots.
2. Open R then return to the spreadsheet and use the mouse to highlight the column of beetle sizes. Copy to the clipboard using the *Home* > *Copy* button.
3. Now switch to the R program and type the following command:

```
bd = scan()
```

4. R will now wait for you to enter some data so copy the contents of the clipboard; use *Edit* > *Paste* from the R menu or *Ctrl+V*. You'll see the data in the window but R is waiting (you can add more data if you need). Press *Enter* on the keyboard and the data entry is completed. R will give a message saying *Read 25 items*.
5. Now the data are stored in the item called *bd*. Make a basic histogram by typing:

```
> hist(bd)
```

6. Notice that the *y*-axis is labelled *Frequency*. Redraw the histogram but set the *y*-axis to sum to unity rather than to use frequency. For good measure fill the bars with a bit of colour:

```
> hist(bd, freq = FALSE, col = "lightblue")
```

7. Notice that the *y*-axis is now labelled *Density*. Add the density plot line using the following command:

```
> lines(density(bd))
```

The `lines()` command adds lines to an existing plot and the `density()` command calculates the co-ordinates from the data. Your plot is now complete and should resemble Figure 4.13.

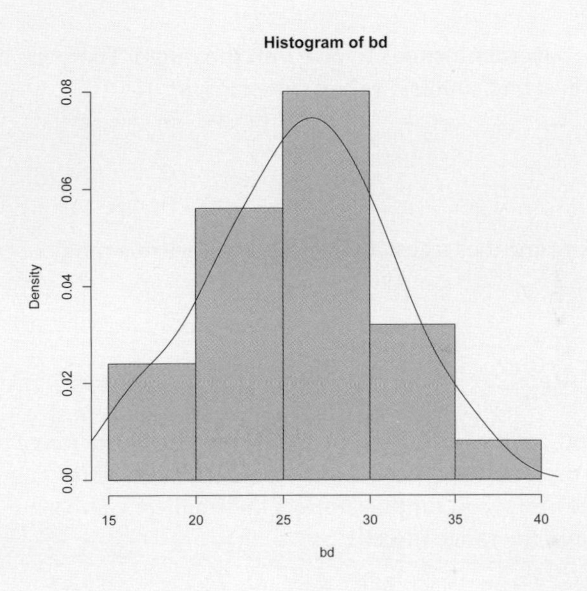

Figure 4.13 Density plot of beetle carapace lengths superimposed on a histogram of the same data using R commands.

The histogram (with or without density lines) is a useful diagnostic tool, which can help you to decide if your data are normally distributed or not. This decision is important, as it will affect the kind of statistical test you can employ (see Chapter 5) as well as the summary statistics (see Table 4.10).

4.3 A numerical value for the distribution

So far you have seen how to put a value to the middle, or central tendency, of your data (the average) and how to describe the distribution graphically. It would be nice if you could use a number to describe your distribution to go with the number that you have for the average.

4.3.1 Range

The simplest thing would be to present the range of values, i.e. the smallest and largest. For the beetle data (Table 4.1) you would have a range of 36 – 16 = 20. This is all very well but not all that informative. You could have two sets of data with the same range but very different distributions (see the previous tally plots Table 4.3 and Table 4.4 for example); however, you can gain some insight into the distribution if you use the median as well. In the first example, the median will lie more or less half way between the ends of the range. In the second example, the median will lie closer to the low end.

Range in Excel

In Excel you can use several functions to give you the range. There are built-in functions to determine the largest and smallest values for example:

```
MAX(range)
MIN(range)
```

These functions determine the largest and smallest values in a range of cells. You can also use:

```
LARGE(range, n)
SMALL(range, n)
```

In these formulae you get the *n*th largest or smallest value in the range of cells. So, to get the first largest you use `1`, for the second smallest use `2` and so on.

To determine the range you simply subtract the smallest value from the largest; there is no function to obtain the range directly.

Range in R

You can do something similar using R; the largest value (the maximum) is found using the `max()` function whilst the minimum is found using `min()`. In the following example you'll see the beetle data once again:

```
> bd
 [1] 17 26 28 27 29 28 25 26 34 32 23 29 24 21 26 31 31 22
[19] 26 19 36 23 21 16 30
> max(bd)
[1] 36
> min(bd)
[1] 16
> max(bd) - min(bd)
[1] 20
```

To get the range you obviously take the smaller value away from the larger one.

You can also get the range as two values in one step using the `range()` command:

```
> range(bd)
[1] 16 36
```

The range is not the most illuminating summary statistic but it can be useful, especially when combined with other values, as you'll see now.

4.3.2 Quartiles

Now you have the end points of your distribution (the range) and the middle (median or mean). You can also divide the data into further halves (so you end up with four chunks) and the results are the *quartiles*. If you take the middle value between the lowest value and the median, you get the lower quartile. If you take the middle value between the highest point and the median, you get the upper quartile. These two values are also known as the first and third quartiles respectively. The median is the second quartile and the endpoints are the zero and fourth (lowest end and highest end). For the beetle data the values are shown in Table 4.7.

Table 4.7 Quartile values for beetle (*Haliplus lineatocollis*) carapace lengths. The median is the middle value and the inter-quartiles (Q1, Q3) are halfway between the median and each end.

Min	Q1	Median	Q3	Max
16	23	26	29	36

Quartiles in Excel

You can get these quartile values with Excel using the `QUARTILE` function:

```
QUARTILE(range, quartile)
```

The function has two parts: the first is the range of cells containing the data you want to summarize. The second part is a number from 0 to 4. Zero would represent the lowest quartile (i.e. the minimum value), 1 is the lower quartile, 2 is the second quartile (i.e. the median), 3 is the upper or third quartile and 4 gives the fourth quartile or maximum value.

Quartiles in R

In R you can produce the quartiles using similar commands; there are two to choose from:

```
> summary(bd)
   Min. 1st Qu.  Median    Mean 3rd Qu.    Max.
     16      23      26      26      29      36
> quantile(bd)
  0%  25%  50%  75% 100%
  16   23   26   29   36
```

The first command, `summary()`, is a general command and will produce a different report depending on what your data are (as you shall see later). Here you get all the quartiles and the mean as well. There is a more specialized command, `quantile()`, for producing quartiles, and you see this presented after the `summary()` command.

As well as producing all the quartiles, the command can produce a single value for you. You simply need to tell R which quartile you want. You can see that you have five to choose from. To get the 25% quartile you type:

```
> quantile(bd, 0.25)
```

To get the 75% you type:

```
> quantile(bd, 0.75)
```

Of course if your data are called something other than `bd` you need to change the command to reflect this.

Note: quantile() command

The R command is called `quantile` and *not* quartile, like in Excel. Strictly speaking quartiles split things into quarters but a quantile can be any proportion (try using 0.2 in the command for example.

Now you have a better description of your data using fewer numbers than the original dataset. You can produce a plot of these five values to give a visual representation of your data. The graph is called a box–whisker plot.

4.3.3 Boxplot

It is always a good idea to represent your data graphically before you do anything else and the box–whisker plot is a good place to start. You will look at different types of graph in Chapter 6 (including details of how to make them) but it is useful to see the boxplot now in brief.

Figure 4.14 shows a boxplot (also known as a box–whisker plot) of the beetle data from Table 4.1. This boxplot shows the five descriptive values: the median is shown as a stripe. The box shows the inter-quartile range, i.e. the distance between the first and third quartiles. The whiskers extend out to the full range, i.e. the maximum and minimum. This sort of plot is possible to do in Excel but it is not straightforward and the program needs a lot of coercing (you need to use one of the stock graphs, see Chapter 6). It is, however, simple to do in R using the `boxplot()` command.

To create the boxplot you put the name of your data (you'll use the `bd` beetle data from before) in the brackets:

```
> boxplot(bd)
```

This is a bit sparse but you'll see later how to add better axis titles and other items to make your plot look better. For the time being you can gain an impression of your data just

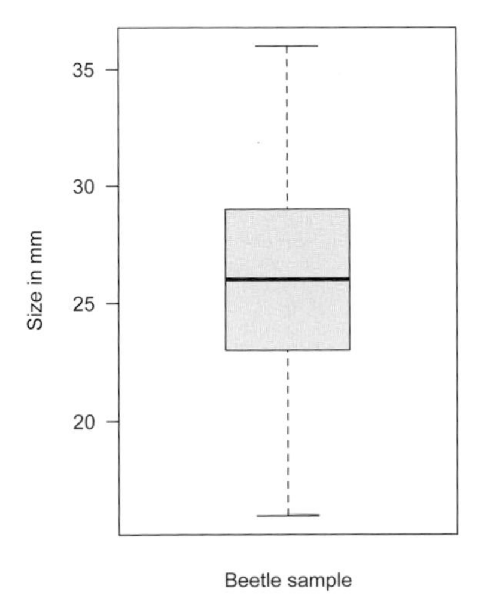

Figure 4.14 Box–whisker plot of beetle (*Haliplus lineatocollis*) carapace lengths. The stripe shows the median, the whiskers show the maximum and minimum values and the box shows the inter-quartiles.

from looking at the plot. If the data are normally distributed, then the middle stripe will be more or less in the middle and the box–whiskers will be symmetrically arranged either side. If the data are skewed, then the stripe and the box–whiskers will not be symmetrical. It might be helpful to think of the boxplot as a kind of contour map showing your sample of data as a kind of hill. Just like reading a map you can visualize the shape of the ground by looking at the contours. You'll see more about boxplots in Chapter 6.

4.3.4 Standard deviation

The range and quartiles provide useful measures of the distribution of the data. However, there are other measures that may be used, especially if the data appear to be symmetrically arranged (as in the case of the beetle data, Table 4.1). One such measure is called *standard deviation*. To calculate it you use the formula shown in Figure 4.15.

$$s = \sqrt{\frac{\sum (x - \bar{x})^2}{n - 1}}$$

Figure 4.15 Formula to calculate standard deviation of a sample.

The steps you require to calculate standard deviation are as follows:

1. Determine how many items there are in your sample, call this value n.
2. Calculate the mean for the sample: sum the values and divide by n.
3. Subtract the mean from each item in the sample ($x - \bar{x}$). You'll get positive and

negative values because some of the original data are larger than the mean and some are smaller.

4. Sum the differences; you will find that you get zero, which is not helpful. What you must do is to square each of the differences $(x - \bar{x})^2$, which will get rid of the negative signs (this is a common "trick" in statistics).
5. Sum the squared differences; the result is called the *sums of squares*.
6. Divide the sums of squares by $n - 1$. The result is called the *variance* (denoted by the symbol s^2).
7. Take the square root of the variance from step 6; this is the *standard deviation* (denoted by the symbol s). Because you squared the differences in step 5 this brings the result back into the same order of magnitude.

You can see an example of the calculation in Table 4.8. In this table the column labelled x shows the data values. At the bottom the row labelled $\sum$ gives the sum for each column. The row labelled n simply gives the number of items in the sample (8). The row labelled $\bar{x}$ shows the mean ($44 \div 8 = 5.5$). The column labelled $x - \bar{x}$ shows the mean (5.5) subtracted from each of the data values; these sum to zero (0). The last column $(x - \bar{x})^2$ shows the previous column squared (which removes negative signs). The final two rows show steps 6 and 7; the variance is $18 \div (8 - 1) = 2.57$ and the square root of this is the standard deviation (1.60).

Table 4.8 Calculating the standard deviation of a sample.

	x	$x - \bar{x}$	$(x - \bar{x})^2$
	4	–1.5	2.25
	3	–2.5	6.25
	5	–0.5	0.25
	6	0.5	0.25
	8	2.5	6.25
	6	0.5	0.25
	5	–0.5	0.25
	7	1.5	2.25
$\sum$	44	0	18
n	8		
$\bar{x}$	5.5		
s^2	2.57		
s	1.60		

What you have done essentially is to take the average deviation from the mid-point (the mean).

Standard deviation in Excel

The calculation steps required to compute standard deviation are simple enough to calculate using Excel (and many calculators). In fact there is a function to do it directly: `STDEV`, which does all the calculations in one go:

```
STDEV(sample_range)
```

You simply provide the cell range for your sample in the function.
The variance (standard deviation squared) can also be calculated directly in Excel using the `VAR` function.

Standard deviation in R

In R, the command to work out standard deviation is `sd()`. Below you can see the command in action on the beetle data from Table 4.1, which are saved to an item called `bd`.

```
> bd
 [1] 17 26 28 27 29 28 25 26 34 32 23 29 24 21 26 31 31 22
[19] 26 19 36 23 21 16 30
> sd(bd)
[1] 5.049752
```

The variance can be calculated using the `var()` command.

Properties of the standard deviation

Figure 4.16 shows two sets of data. They both have the same mean (40) and number of items ($n = 25$) but they have quite different standard deviations.

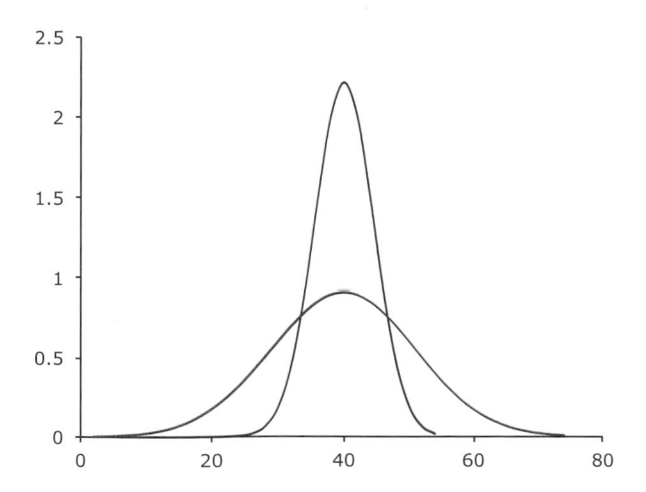

Figure 4.16 Frequency plots of two samples. Both have the same mean but have differing dispersions (standard deviation for the narrow curve = 4.5, standard deviation for the wider curve = 11).

In Figure 4.16 you see how the data with the largest standard deviation has the wider, flatter spread on your frequency graph. The thinner plot has a standard deviation of 4.5 whilst the fatter plot has a standard deviation of 11.

Standard deviation has an important property. If you plot your frequency graph and draw vertical lines at 1 standard deviation above and below the mean you will encompass (between the lines) 67% of all the data. If you move your lines out to 2 standard deviations (plus and minus) you encompass 95% of all the data (Figure 4.17).

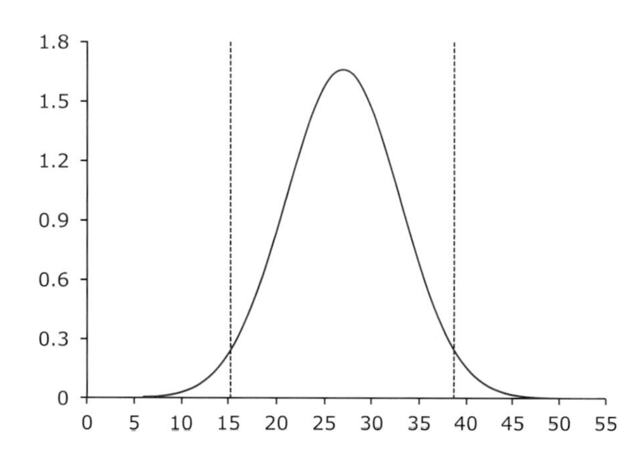

Figure 4.17 Properties of the normal distribution curve. Dashed lines are shown at +2 and –2 standard deviations from the mean. Only 5% of the data lies outside this range.

Figure 4.17 shows a sample of data with a mean of 27 and standard deviation of 6. The dotted lines are at +2 and –2 standard deviations and the space between them is occupied by 95% of the data. Put another way, only 5% of the data lie further than 2 standard deviations from the mean. This property becomes important in statistical testing (Chapter 5).

Why use (n – 1)?

You may have wondered why you divided your formula by $(n - 1)$ rather than n. The reason is that you are working on a sample. In most cases you will be collecting data from only some of the available items. For example, your beetles from Table 4.1 represent some of the individuals that were in the pond. You might spend a long time and try to capture and measure every single beetle. What you do instead is to collect a representative sample of the population. You are therefore working out estimates of the mean and standard deviation (if you worked out median and range, they would also be estimates). This is taken into consideration by using $n - 1$. Think of it as a kind of correction factor. If n is quite small then subtracting 1 might make quite a large difference. If n is large then subtracting 1 makes a smaller difference. In this case as n approaches the true size of the population (beetles in this instance) then the correction factor gets smaller and smaller. Let's illustrate using a new example.

You are asked to determine how large the leaves on a shrub actually are. This seems simple enough so you rip off a bunch and get out a ruler. It turns out that your handful was ten leaves. Because you didn't measure all the leaves, you divide by $n - 1$ instead of n when you do your calculation of standard deviation. What you have done is to use nine instead of ten as the divisor in your calculation, a 10% difference. In a fit of zeal you go out and get nine more handfuls and measure a further 90 leaves, making 100 in total. Now you divide by 99 (which is $n - 1 = 100 - 1$) and get a new estimate. Now your divisor is only 1% different. As you collect more and more leaves, your correction factor (the –1 part) is getting smaller and smaller (Table 4.9).

Table 4.9 Effects of using *n* – 1 in calculation of standard deviation.

n	*n* – 1	% diff.
10	9	10
100	99	1
1000	999	0.1
10000	9999	0.01

Of course the final value of the standard deviation is not altered by the same degree because the formula for standard deviation is a little more complicated. So, how much difference does it actually make? Here is an example that shows the correction factor in action (Table 4.10). The readings are for the leaves from the shrub in the current example. All the numbers are the maximum length in millimetres. Each handful contains leaves of exactly the same size, which is not very realistic but helps illustrate the effect of the correction factor.

Table 4.10 Leaf size data (max length in mm) to illustrate the effect of *n* – 1 in standard deviation calculation.

	31	31	31	31	31	31	31	31	31	31
	33	33	33	33	33	33	33	33	33	33
	31	31	31	31	31	31	31	31	31	31
	36	36	36	36	36	36	36	36	36	36
	29	29	29	29	29	29	29	29	29	29
	32	32	32	32	32	32	32	32	32	32
	35	35	35	35	35	35	35	35	35	35
	34	34	34	34	34	34	34	34	34	34
	38	38	38	38	38	38	38	38	38	38
	27	27	27	27	27	27	27	27	27	27
n	10				50					100
$\bar{x}$	32.6				32.6					32.6
s_n	3.14				3.14					3.14
s_{n-1}	3.31				3.17					3.15

The first column in Table 4.10 contains the readings from the first ten leaves. At the bottom are some of the summary statistics. You can see that the population standard deviation is

3.14 whilst the sample standard deviation is 3.31. Your correction factor has made quite a difference. In the middle you see summary statistics for the first 50 leaves. The mean is of course the same because each handful has leaves of identical size. The population standard deviation is also the same because it's based on n, which you haven't corrected. The sample standard deviation is now much closer. You have used a divisor of ($n - 1 = 49$) which is only a 2% "correction" (compared to the 10% you used for the first 10 leaves). The final column shows the summary for all 100 leaves. The two measures of standard deviation are now very close (the correction factor is only 1% of the divisor).

4.4 Statistical tests for normal distribution

In many cases it is sufficient to look at your histograms or stem plots (Section 4.2) in order to determine if your data are normally distributed; however, there are ways to examine the distribution of the data mathematically. The properties of the normal distribution are known quite explicitly. If you know the mean and standard deviation, you can plot a graph. The formula to work out the normal distribution is shown in Figure 4.18.

$$y = \frac{1}{\sigma\sqrt{2\pi}} e^{-\frac{(x-\mu)^2}{2\sigma^2}}$$

Figure 4.18 Formula to determine normal distribution.

The formula looks quite daunting and you might be glad to know that you shall not delve into it in any great detail. There are only two quantities that need to be known: standard deviation (σ) and mean (μ). All you really need to know is that it is possible to use this to determine if a sample of data follows the normal distribution. The simplest way to do this is to use the Shapiro–Wilk test in the R program.

4.4.1 Shapiro–Wilk test for normality in R

Once you have some data, you can perform a test for normality quite simply using the `shapiro.test()` command.

```
> shapiro.test(your.data)
```

Here you replace the `your.data` part with the name of your sample of values. The example below shows the Shapiro–Wilk test being carried out on the beetle data you met earlier (Table 4.1).

```
> bd
 [1] 17 26 28 27 29 28 25 26 34 32 23 29 24 21 26 31 31 22
[19] 26 19 36 23 21 16 30
> shapiro.test(bd)
    Shapiro-Wilk normality test
data:  bd
W = 0.9882, p-value = 0.9885
```

You can see that the result is not significant ($p = 0.9885$). In other words, the data (bd) are not significantly different from a normal distribution. You also have some data from beetles collected at a different time of year; these data are called Mar and the result is shown below:

```
> Mar
 [1] 18 21 18 51 46 47 50 16 19 15 17 16 52 17 15 16 17 49
[19] 48 16 18 16 17 15 20
> shapiro.test(Mar)
    Shapiro-Wilk normality test
data:  Mar
W = 0.6662, p-value = 2.644e-06
```

In this case you see that the *p*-value is significant (here *p* is a very small number), which indicates that the sample (Mar) is skewed, i.e. is significantly different from a normal distribution.

4.5 Distribution type

When your data are symmetrical, as in the beetle data (see the tally plot, Table 4.3), the distribution is called normal distribution or *parametric*. In this case the mean makes the best average and the standard deviation the best measure of the spread (also called *dispersion*). When your data are not normally distributed, i.e. non-parametric or *skewed* (see the tally plot, Table 4.4), the median makes the best average and the dispersion is best modelled using the range and quartiles. The most appropriate measures of average and dispersion are summarized in Table 4.11.

Table 4.11 Measures of centrality and dispersion (spread) to use according to distribution of data.

	Parametric	**Skewed**
Average:	Mean	Median
Dispersion:	Standard deviation	Range and quartiles

4.5.1 Standard error

When you collect data you are usually sampling from a larger population. If you think back to the beetle data in Table 4.1, you collected individuals from a pond and measured them. You did not get every beetle from the pond. What you have is a subset of the pond, a sample. When you work out the mean (or the median) you are really estimating the population mean. As you collect more individuals, your estimate should get closer to the real true mean of the population.

Imagine that you took a sample from a population. You can determine the mean in the usual way. Now you will return to the pond and collect another sample. You can work out the mean again. The two means are unlikely to be exactly the same. This situation is represented in Figure 4.19. The larger curve represents the complete population and shows the frequency distribution (normally distributed in this case). You also see three smaller curves. These represent three samples taken from the larger population (you went to the pond three times).

If you look at the means of the samples, you see that some are smaller than the overall population mean and some are larger than the population mean. You could carry on taking samples and calculating the means and find out that the means would be normally distributed, just like the population data. If you took these sample means, you could work out how far away they were from the original population mean. You can also work out the standard deviation of these sample means from the population mean.

This deviation of samples from the population mean is called standard error.

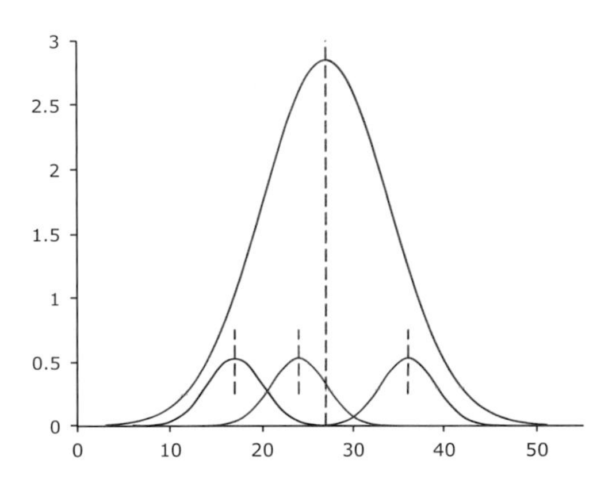

Figure 4.19 Standard error. The large curve represents a complete population. The small curves represent separate smaller samples taken from this population. Standard error is the standard deviation of these sample means from the population mean.

If you look at Figure 4.19, you can see that if the overall population had a very large standard deviation, the large curve would be wide and fat. You would expect your samples to be a bit more widespread and this would make your standard error larger. If the original population had a small standard deviation, the large curve would be tall and thin. Your samples would probably be closer together and result in a smaller standard error.

Standard error can be thought of as a measure of how good an estimate of the actual mean your effort is.

An example will make a good illustration of standard error. Previously you looked at some leaf sizes and the data were shown in Table 4.9. In this instance it was remarkable that each handful of ten leaves was identical in size. It would of course be a lot more realistic if each handful of leaves were different sizes. In Table 4.12 you see more leaf data. This time each handful (your sample) contains ten leaves but they are all different sizes.

Each sample of ten leaves has a mean calculated ($\bar{x}$). If you take a mean of all 100 leaves you find that it is 32.6. Here you see that the mean of each sample of ten leaves is slightly different. The mean of each sample varies; in some cases it is lower than the true value and in others it is higher. You can think about it like this. You have a total population; in this case you have 100 leaves and their size is normally distributed around a mean value (32.6 mm). Each time you take a sample you have a smaller set of numbers,

Table 4.12 Leaf sizes in millimetres. Each sample of ten leaves has an associated mean value calculated. The overall mean (μ) is 32.6.

	34	35	33	26	35	35	35	32	38	29
	32	34	33	29	37	37	31	37	31	28
	30	38	34	33	33	32	29	35	29	30
	36	32	31	33	35	35	31	34	29	29
	36	34	37	28	27	32	28	32	36	32
	36	31	33	38	33	32	36	33	34	31
	41	34	30	27	34	34	29	27	37	31
	35	34	27	35	28	31	31	36	32	32
	30	33	36	34	29	33	28	34	37	34
	34	33	31	30	36	26	29	35	34	35
$\bar{x}$	34.4	33.8	32.5	31.3	32.7	32.7	30.7	33.5	33.7	31.1

also distributed normally around a mean. This new mean is an estimate of the true mean of the bigger population.

Note: Leaf data

The data in Table 4.12 are available in a spreadsheet, *leaves.xlsx*, for you to practice with.

Go to the website for support material.

If you look at the means of the samples you can demonstrate that they are normally distributed using a stem and leaf plot (Table 4.13).

Table 4.13 Stem and leaf plot of means of ten samples of leaves (sizes in millimetres).

30	8
31	13
32	558
33	777
34	2

You can determine the standard deviation of these ten means. If you do this you get 1.22, which is a lot smaller than the standard deviation of the 100 leaves (which is 3.11). This new measure of standard deviation is of course the standard error.

You have seen how you could work out standard error in theory but this requires you to know the actual population mean and the whole point of sampling is to avoid having to collect absolutely everything. What you need is a way to estimate it (Figure 4.20).

$$SE = \frac{s}{\sqrt{n}}$$

Figure 4.20 Formula to estimate standard error.

The formula (Figure 4.20) shows how you estimate standard error. You use the standard deviation and divide by the square root of the sample size.

You can do a similar thing even if the data are not normally distributed. You could take medians of samples from a larger population and determine the differences from the overall median; however, the medians of the samples are not likely to be normally distributed, so it is not very useful! What you do instead is to use the quartiles in lieu of the standard error. When you summarize some non-parametric data, you give the median as your central value and the inter-quartile range, i.e. the difference between the third and first quartile (you looked at quartiles in Section 4.3.2).

The effect of both standard error and the inter-quartile range is to smooth out the effects of extreme values in your sample.

Standard error is a useful measure and is commonly used to show sample variability on graphs (e.g. as error bars). It is also used in statistical testing. These tests rely on the properties of the normal distribution. Recall from Figure 4.18 how most of the data lie between the mean and 2 standard deviations. What standard error means in practice is that the more data items you have, the more tightly clustered around the true mean the estimated samples means will be. You can be more confident that you are near the real average when you have more data items. You will come on to the idea of confidence interval next.

4.5.2 Confidence intervals

The *confidence interval* is another kind of measure of the variability of a sample. A confidence interval defines a range of values within which you can be confident that the true mean sits. In fact you put a value to your confidence; 95% is the general level but you can have other levels (higher or lower).

You need the mean and the standard error to work out a confidence interval. From Figure 4.18 you know that 95% of the data will be within 2 standard deviations either side of the mean in a normally distributed population. You are dealing with a sample rather than the complete population so you use the standard error to do something similar. You can estimate the confidence interval as 2 × standard error. For your beetles you get CI = 2 × 1.01 = 2.02. This means that 95% of the beetles in your sample ought to be between 23.98 and 28.02 mm in size. You can write this as: 26 ± 2.02 mm (n = 25, CI 0.95), indicating to your reader mean, spread and sample size like before.

Different levels of confidence interval

Your confidence interval calculated above was an estimate. It so happens that you can determine confidence intervals for 95%, 99% and 99.9%. The properties of the normally distributed curve have been studied intensively and statisticians have determined that 95% of the data will lie between ±1.96 standard deviations (rather than the 2 mentioned previously). Table 4.14 shows how many standard deviations you expect for different levels of probability. You can use these values to determine confidence intervals. These values are part of a family of numbers relating to the normal distribution and you'll come across the related t statistic later (Section 7.1) when you look at differences between two samples.

Table 4.14 Values used for calculating confidence intervals.

Number of standard deviations	p level
1.96	0.05
2.58	0.01
3.29	0.001

The p level is the level of probability. You have been using the percentage of the data that lies within the confidence interval (e.g. 95% of the data will lie within the interval) but you can also think of this in another way. You can turn this around and say that 5% of the data will lie outside of the confidence interval. In other words there is a probability of 0.05 that the data will lie outside the confidence interval. If you use 1.96 instead of 2 to work out your beetle confidence interval (at 95%, $p = 0.05$) you get 26 ± 1.98 mm, meaning that 95% of all the beetles ought to be between 24.02 and 27.98 mm in size. Alternatively only 5% will be smaller than 24.02 mm or bigger than 27.98 mm in size.

Note: Confidence intervals

The values reported in Table 4.14 for determining the confidence interval are estimates based on large sample sizes. For small samples the t-distribution should be used (see Table 7.1 for critical values).

Confidence intervals in statistical testing

You can use these confidence intervals to say something meaningful about two (or more) samples. For example, let's return to the beetle example. You have a sample from a population and have determined the mean and 95% confidence interval. If you took beetles from another pond you could do the same for the new sample.

If the two confidence intervals do not overlap you can be pretty sure that there is a real difference between the sizes of beetles from the two ponds. In fact you can be 95% certain because your confidence interval tells you that only 5% of the data are likely to lie outside this range.

This idea is behind statistical tests that look at differences between samples and where the data are normally distributed. This is why the properties of the normal distribution are so important. The idea of the confidence interval also leads you to think about things in terms of probability or likelihood. There are few occasions when you can be totally certain of your results. This is because biological data is variable. Rather than look for absolute certainty you strive to put a value to the probability that your result is not down to chance. When you set up your hypothesis (see Sections 1.4 and 5.1) you are looking to put a value on how likely it is that your hypothesis is correct.

4.5.3 Reporting the variability of data

When you report the results of a simple sampling exercise you need to be succinct. The three important things you need to report are: average, spread and sample size.

If your data appear to be normally distributed then you use the mean and standard deviation as the average and spread (see Table 4.10). The median and quartiles would be the way to summarize your data if the sample was not normally distributed.

If you examine the beetle data in Table 4.1 you can work out the mean and standard deviation. You get $\bar{x} = 26$ and $s = 5.05$. You already know $n = 25$. To report this you could write:

> "The mean beetle size was 26 mm ($s = 5.05$, $n = 25$)."

The reader can see the situation instantly. There were 25 beetles captured and measured. The average size was 26 mm and the standard deviation was 5.05. You may decide to use the standard error instead of the standard deviation. You work that out to be 1.01 and would write:

> "The mean beetle size was 26 mm (SE = 1.01, $n = 25$)."

You can also report the confidence intervals:

> "The mean beetle size was 26 mm ($CI_{0.95} = 1.98$, $n = 25$)."

Note that you should say at what level your confidence interval was calculated for (in this case a subscript was used).

When you draw graphs you should represent the spread of data as well as the average (using error bars, which you will see in Chapter 6). You can represent the spread using any appropriate measure; which one you select depends somewhat on the purpose of your graph (see Chapters 6 and 13).

4.6 Transforming data

The properties of the *normal distribution* are so well understood that it underpins quite a few statistical analyses. It is so important that wherever possible you strive to collect normally distributed data. Some data you would not expect to be normally distributed. For example, collecting invertebrates often results in skewed data with a lot of low counts followed suddenly by a large one! At other times you may expect to get a normal distribution, e.g. measurements of size or weight, but do not. You can sometimes get over this by collecting more data and filling in the gaps.

If you end up with data that is not normally distributed, there are things that you might do to the numbers to make the distribution more like your ideal. There is no reason to suppose that a regular scale of numbers is inevitably the best to use for your data. For example, you may collect data that varies in orders of magnitude. This may well be a logarithmic scale. Log scales are reasonably common in the natural world (pH being a prime example).

In order to improve the fit to a normal distribution, you simply perform a mathematical operation on your data. Then you examine the new data and see if the distribution is more like the parametric ideal. There are several commonly used mathematical operations:

- Logarithm.
- Square root.
- Arcsine (also called angular).
- Reciprocal.

Essentially what you do is take your original data and apply the mathematical operation to all the values. Thus you create a new set of values based on the mathematical operation; hopefully this produces a more normally distributed sample.

4.6.1 4.6.1 Logarithmic transformation

The log transformation is potentially useful when you have data that range across one or more orders of magnitude. These data may arise when dealing with organisms of very different sizes for example. You apply the transformation by using standard log to the base 10 in most cases but there may be situations when the natural log (or indeed any base) seems more appropriate. There is a potential problem with this, as any zero values will give an error (as $\log_{10} 0 = \infty$). To get over this possible problem, it is usual to add +1 to all your data, like so:

$$x_t = \log_{10}(x + 1)$$

You then carry on with your data analysis using the new values. You can work out logs using Excel or R quite simply and the commands are very similar. In Excel you use `LOG(data, base)` and in R you use `log(data, base)`. In both cases you replace the `base` part with a numerical value that corresponds to the base you want (commonly 10). If you do need to add 1 to take care of zero values you simply modify the command, e.g. `log(data + 1, base)`. In R there is even a special command `log1p()`, which adds 1 to all data before taking the (natural) logarithm.

In Excel the default is to use base 10 and if you omit the `base` part of the command that is what you get. In R the default is to use the natural log and that is what you get if you omit the base part of the command. To get natural logs in Excel you use a different command `LN(data)`.

Note: Logs and negative values

Negative values are a problem with logarithms, since you cannot get a log of a value less than zero. There are two main approaches: add a constant value to all your data (such that the minimum becomes 1) or simply ignore the sign, take the logarithm and reinstate the sign afterwards.

4.6.2 Square root transformation

The square root transformation can be useful in certain cases. It often seems to work in cases where you are counting things, the abundance of an invertebrate in kick samples for example. Applying the transformation is simple enough but usually you add 0.5 to each value before the square root to improve the move to normal distribution:

$$x_t = \sqrt{(x + 0.5)}$$

In Excel the `SQRT` function will do the trick. In R you use the `sqrt()` command.

4.6.3 Arcsine transformation (angular transformation)

When you are dealing with items that have some natural boundary, you often find that because the ends of the distribution are constrained you do not get normal distribution. Examples would include percentages and proportions, where you usually have boundaries of 0–100 and 0–1 respectively. In such cases the arcsine transformation can improve the distribution. To carry out the transformation, you take the arcsine of the square root of the proportion:

$$x_t = \text{arcsine} \sqrt{p}$$

In the formula, p is a proportion. If you have percentage data, you divide by 100 to create the proportion and then apply the square root and take the arcsine of that. The result is an angle and will range in value from 0° to 90°. Table 4.15 shows some percentages that have been transformed using this method.

Table 4.15 Examples of percentage values transformed using the arcsine method.

Percentage	Arcsine $\sqrt{p}$ (°)
0	0.00
10	18.43
25	30.00
33	35.06
50	45.00
75	60.00
90	71.57
100	90.00

Once you have your new values you can perform your various analyses. Of course if you need to report the original means then you convert the values back to their proportions:

$$p = [\sin (x_t)]^2$$

Obviously if you are dealing with percentages then the final step is to multiply by 100.

Both Excel and R use radians by default; to convert to degrees you need to multiply by $\pi/180$. In Excel π is given using `PI()` and in R it is simply `pi`. For example $\sin 45° = \sqrt{2} \div 2 = 0.707$. To do this in R you use the `sin()` command:

```
> sin(45 * pi/180)
```

In Excel you use:

```
SIN(45 * PI()/180).
```

To go from degrees to radians you multiply by $180/\pi$. To carry out the angular transformation of 10%, for example, you would use the `asin()` command in R:

```
> asin(sqrt(0.1))*180/pi
```

In Excel you type:

```
ASIN(SQRT(0.1))*180/PI()
```

In both Excel and R you can convert a range of values and you would replace the 0.1 with the range of cells (in Excel) or the data name (in R) that you want. You must not forget that percentage values need to be divided by 100 first.

4.6.4 Reciprocal transformation

The reciprocal transformation might prove useful in certain cases where you have various dilutions for example. In these cases the original values are simply transformed thus:

$$x_t = 1/x$$

4.6.5 Reporting results from transformed variables

When you use transformations on your analyses you are improving your chances of getting a realistic result but making it harder to understand the resulting numbers. What you must do at some point is to untransform the data to report mean values for example; however, the act of transforming the data is generally asymmetric and measurements of standard error would be misleading. What you need to do is to determine the confidence limits using the transformed values and report these in their untransformed form.

4.7 When to stop collecting data? The running average

You have seen how to summarize data using averages and spread (e.g. standard deviation, quartiles). In Section 1.6.3 you thought about how much data you ought to collect and introduced the idea of the *running mean*. However, you did not go into any detail at that point because you had not even considered dealing with averages.

The idea behind the running mean is to give you an idea of how close to the true mean you might be. You start with a single value, your first measurement. Then as you add more data you calculate the mean as you go along, recalculating each time you obtain a new measurement. You can plot these means in a simple scatter graph as you proceed.

Generally speaking the more data you collect the less "wobbly" your running mean will become. In Figure 4.21 you can see a running mean taken using ten measurements.

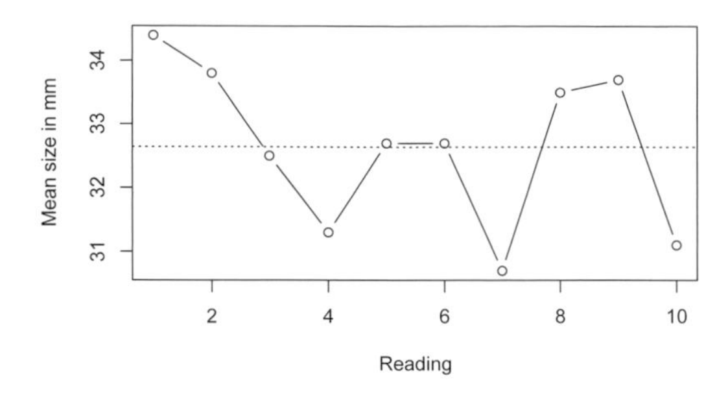

Figure 4.21 A running mean taken over ten replicates. The dashed line shows the overall mean.

In Figure 4.21 you can see how the running mean changes at each sampling event. If you have more sampling events (i.e. more replicates) you are likely to see the running mean "settle down" and the curve flatten off close to the true mean (Figure 4.22).

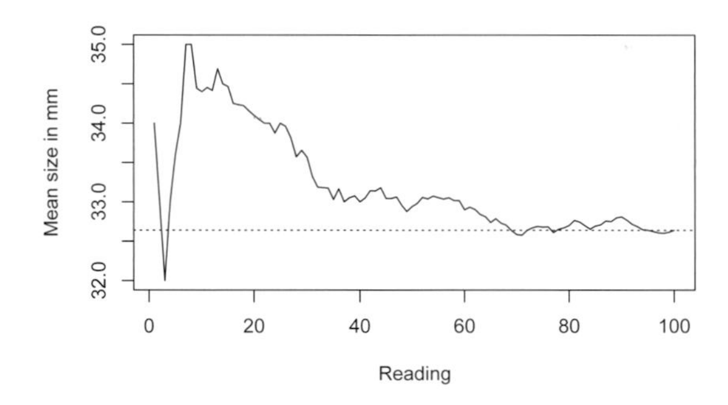

Figure 4.22 A running mean taken over 100 replicates. The dashed line shows the overall mean value.

You can do this sort of graph in your field notebooks as you go along. At other times you are likely to want to check that you have collected enough data by doing the running means after you've returned from the field (you assume that you could go out again and collect more data another time). If you are creating a report or simply exploring the data you will probably want to do this on a computer. You can use your spreadsheet or the R program to work out and display running means quite easily.

4.7.1 Running means in Excel

You can calculate running means easily using a spreadsheet. Excel provides you with the `AVERAGE` function for example and this can be pressed into service quite simply. You need to use a range of cells that incorporates each successive data item, which you can do with only a tiny tweak. The key is to use a $ as part of the cell reference; the $ "fixes" a location so that it is not altered when a cell is copied and pasted. If your data were in cells A1:A10 you would use `AVERAGE(A$1:A1)` as the formula to work out the mean of the first item. When you copy the formula down the column the `$1` part remains fixed so the next item would become `AVERAGE(A$1:A2)`. As you copy further down the column the first part remains fixed and the second part alters until the last one would read `AVERAGE(A$1:A10)`.

In the following exercise you can have a go at working out a running mean and plotting a graph for yourself.

Have a Go: Calculate and plot a running mean in Excel

You do not need any data for this exercise, as you can type in values directly to a new spreadsheet. However, the data are available in the spreadsheet *leaves.xlsx*, which is on the support website.

Go to the website for support material.

1. Open your spreadsheet and make a new workbook. Navigate to cell A1 and type in a heading for the data; `Size` will do nicely.
2. Now go to cell A2 and type in some values down the column (cells A2:A11); use the first column of data from Table 4.11: `34, 32, 30, 36, 36, 41, 35, 30, 34`.
3. Go to cell B1 and type a heading for the running mean column; `RunMean` will do nicely.
4. In cell B2 type a formula to calculate the mean: `=AVERAGE(A$2:A2)`; note that you only need the $ in front of the first number 2.
5. Copy cell B2 to the clipboard; use the *Home* > *Copy* button.
6. Highlight the cells B3:B11 and paste the clipboard using the *Home* > *Paste* button. You will see the running mean values appear immediately. Look at the formulae in the cells and see how the first cell has been "fixed".
7. Now you should make a chart to visualize your running mean; you will use a line chart. Use the mouse to select the cells B1:B11, which hold the values for the running mean and the heading. Now click the *Insert* > *Line Chart* button (it is in the *Charts* section). There are several options; select a chart with line markers, since there are not many data items (Figure 4.23).

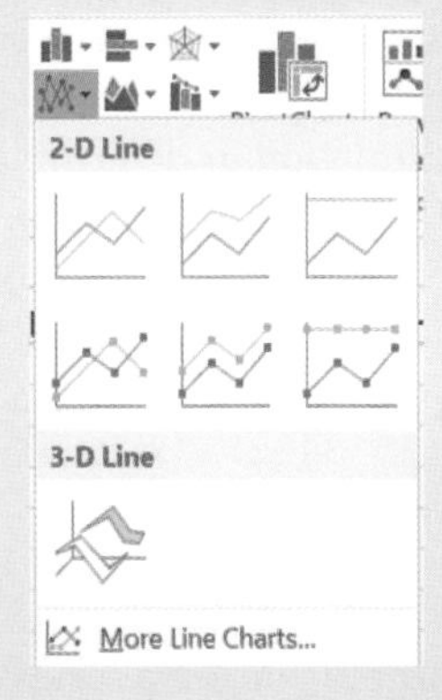

Figure 4.23 Line chart options in Excel 2013 are in the *Charts* section of the *Insert* menu.

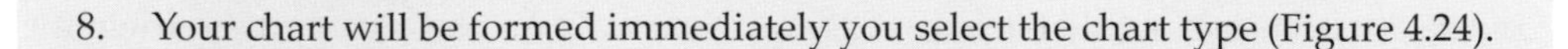

8. Your chart will be formed immediately you select the chart type (Figure 4.24).

Figure 4.24 A line chart used to visualize a running mean in Excel 2013.

Your basic line chart (Figure 4.24) will require some editing if you want to share it (see Chapters 6 and 13), which you can do using the tools on the *Chart Tools* menus as well as by right-clicking on the chart itself.

The Excel graph you produced in the preceding exercise was a line chart (Figure 4.24). You only had one set of figures so Excel generated an index for the x-axis, which was fine for the purpose. You'll see more about producing charts in Chapter 6.

4.7.2 Running means using R

Once you have your data in R you can use a variety of commands to explore and investigate. You have already met a few such as the `mean()`, `sd()` and `hist()` commands. There is not a built-in function to determine the running mean; however, there are several functions that relate to cumulative data. The one you will use is `cumsum()`. This determines the cumulative sum of a vector of values.

```
> lf
 [1] 35 34 38 32 34 31 34 34 33 33
> cumsum(lf)
 [1]  35  69 107 139 173 204 238 272 305 338
```

In the example you see a series of values from the leaf size data that you met in Table 4.11 (column 2) when discussing standard error (Section 4.5.1). Here you have ten sizes (in mm) that represent a single sample. You have called the sample of ten leaves `lf`, and you apply the `cumsum()` command by putting its name in the brackets. You see the result as a new list of values; each one is the sum of all the values up to that point. So, the third value is 107, which is the sum of the first three values (35, 34 and 38).

To get a running mean you need to divide each of your cumulative values by how far along you are. For example you would divide the first value (35) by one, as it is the first. You divide the second value (69) by 2 because it is the second (giving 34.5). R provides a command allowing you to deal with sequences, `seq()`. There are several ways it can be used but the one you want here allows you to step along a series of values.

```
> seq(along = lf)
[1] 1 2 3 4 5 6 7 8 9 10
```

To generate a running mean you need to divide the cumulative sum by how far along the series of numbers you are. The following shows what you get with your leaf (`lf`) data:

```
> seq(along = lf)
 [1]  1  2  3  4  5  6  7  8  9 10
> cumsum(lf) / seq(along = lf)
 [1] 35.00000 34.50000 35.66667 34.75000 34.60000 34.00000
 [7] 34.00000 34.00000 33.88889 33.80000
```

Your running mean consists of ten values. Just by examining the numbers you can see that the last few seem to be evening out; however, it might be useful to draw this as a graph. You can make a primitive plot using the following command (your data are the `lf` part):

```
> plot(cumsum(lf)/seq(along= lf), type= "b")
```

The `plot()` command makes a simple scatter graph. The *y*-data are the means and the *x*-data are a simple numerical index. You could plot just the points but it is more useful to see them joined up, so you add the `type = "b"` to tell R to create both lines and points.

In the following exercise you can have a go at making some running mean plots for yourself.

Have a Go: Use R to calculate and plot running means

You'll need the *leaves.xlxs* spreadsheet data for this exercise. The spreadsheet contains the leaf sizes from Table 4.11.

Go to the website for support material.

1. Open the *leaves.xlsx* spreadsheet. You'll see ten columns of leaf size data.
2. Open R and begin the process of transferring in some data from Excel. Type:

```
lf = scan()
```

3. R will now display a `1:` and wait for you to enter some data. You can type numbers (separated with spaces) or use the clipboard. Return to Excel.

4. In Excel select a sample of data, for example cells B3:B12, which represent the first sample of ten leaf sizes. Copy the data to the clipboard using the *Home* > *Copy* button.
5. Switch to the R program. Paste in the values from the clipboard use *Edit* > *Paste* or use Ctrl+V from the keyboard. R will display `11:` and wait for more data, but none will be forthcoming so press *Enter* to complete the data input (R will display *Read 10 items*).
6. Now display the running mean by typing the following command:

```
> cumsum(lf) / seq(along = lf)
```

7. Make a plot using the following commands:

```
> plot(cumsum(lf)/seq(along = lf), type = "b",
ylab = "Running Mean")
> abline(h = mean(lf), lty = 2, col = "blue")
```

You should now have a plot that resembles Figure 4.25.

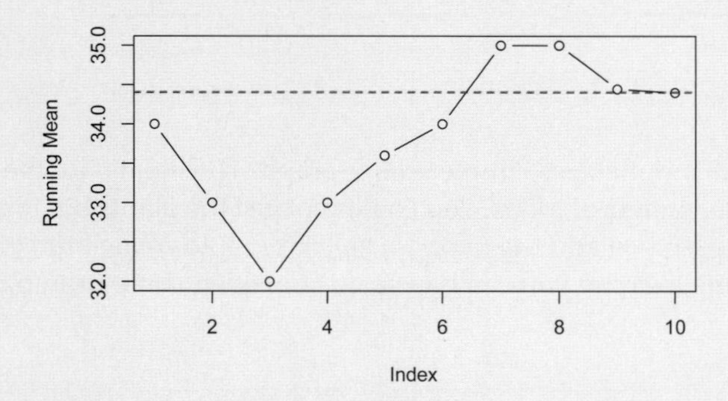

Figure 4.25 Plot of running means drawn using R. The dashed line represents the overall mean.

Try some other data samples. Remember that you can recall previous commands using the up and down arrows, so you do not have to type the commands in full every time.

The plot you drew in the preceding exercise (Figure 4.25) has an additional line representing the mean as well as axis titles; there are other ways you can customize your R plots; you'll deal with these commands later (Chapter 6).

4.8 Statistical symbols

You often need to use symbols in scientific writing. These symbols are used to convey some important information in a concise manner. In statistics there are a number of commonly used symbols and some of these you have already met.

In Table 4.16 you can see some of the most commonly used symbols and abbreviations. The mean is usually written with an overbar, e.g. $\bar{x}$, but you may also see the Greek letter μ (mu). Strictly speaking, overbar is used for your estimate and μ is used for the absolute mean; in most life sciences you are dealing with estimates. The standard deviation is given a simple letter *s* but you may also see the Greek letter σ (sigma). Generally *s* is used to show your estimate of standard deviation whilst σ is used to represent the actual standard deviation.

Table 4.16 Some commonly used symbols in maths and statistics.

Symbol	Meaning
$\bar{x}$ or μ	Arithmetic mean
σ or *s*	Standard deviation
Σ	Sum of
n	Number in sample
s^2	Variance
t	Result of a *t*-test
U	Result of a *U*-test
z	Result of a *z*-test
χ^2	Result of a chi-squared test
r_s	Spearman rank correlation coefficient
r	Pearson's product moment
$\|x\|$	Modulus, the absolute value of *x*
x^2	*x* to the power of 2
*x*E+3	*x* times 10 to the power of +3 (1000)
$p < 0.05$	Probability is less than 5%

The uppercase Greek letter sigma, Σ, is used to mean "sum of", i.e. everything added together. The number of items in your data list (sample) is given the designation *n*. A commonly used measure of dispersion in statistics is *variance* and this is written as s^2, which is useful as it really is the standard deviation squared.

Different stats tests usually have their own letter, especially the fairly simple ones: *t* for the *t*-test and *U* for the *U*-test. The two major types of correlation have similar letters so Pearson's coefficient is *r* and Spearman's rank coefficient is r_s. The chi-squared test uses the lowercase Greek letter chi, χ.

The level of significance of a stats result is of great importance so the probability of this result happening by chance is given the letter *p*. Generally it is best to quote $p < 0.05$, $p < 0.01$ rather than the exact value (this is perceived as showing off so don't put $p = 0.0000023$); if not significant you either write $p > 0.05$ or n.s.

In some cases you are not interested in the sign of a difference but merely its size, so $x_1 - x_2$ might give a positive or negative result but you only want the magnitude. You indicate this using the modulus, a pair of upright braces, so you write $|x_1 - x_2|$ to indicate this. This is sometimes called the absolute value and in Excel and R you obtain the modulus with similar commands, `ABS` and `abs()` respectively.

If you only have plain text and cannot use superscripts or subscripts then you may want to indicate more complex numbers using a basic notation. For example 10^2 equals 100 but how to write it without the superscript? Common notation uses a "hat" (properly called a caret): 10^2 = 100.

Really large or really small numbers can be given scientific notation so 10000 (which ought more properly to be written 10 000) becomes 1×10^4 and 347,348 becomes 3.47×10^5. On the other hand 0.0000035 would be 3.5×10^{-6}. Without superscripts you can write the right-hand part (the exponent) as an "e". So you get 1e+4, 3.47e+5 and 3.5e–6. Excel tends to use uppercase `E`, and R a lowercase `e` (but both accept either). As long as you are consistent it does not matter.

Logarithms are not shown in the table but $\log_{10}$ or $\log_2$ (log by itself implies base 10) are usual ways to write logs. The natural log ($\log_e$) is often given its own symbol ln (or Ln).

EXERCISES

Answers to exercises can be found in the Appendix.

1. The ____ is the best measure of central tendency (average) to use when your data are non-parametric.
2. Which of the following combinations are suitable measures of dispersion to use when your data are Gaussian (normally distributed)?
    A. Standard deviation, inter-quartile range
    B. Standard error, range
    C. Inter-quartile range, confidence interval
    D. Variance, standard error
    E. Confidence interval, mean
3. The ____ and ____ plot are forms of chart that allow you to visualize data distribution. Both use the ____ of data in size classes called ____.
4. If your data vary by orders of magnitude the arcsine transformation is a good choice to help make the data more normally distributed. TRUE or FALSE?
5. Here are some data (height of *Agrostis capillaris* plants in cm): 28, 26, 28, 27, 29, 28, 24, 26, 28, 32. Are these data normally distributed?

Chapter 4: Summary

Topic	Key points
Summarizing data	You need three key items to summarize a numerical sample: an average (central tendency), a measure of dispersion (how spread out around the middle are the data) and the sample size (the replication).
Averages	There are three kinds of average: mean, median and mode. The mean is used for describing normally distributed data (symmetrical around the middle value). The median is used to describe skewed (non-parametric) data (the middle value when data are arranged in order). The mode is the most frequent value. In Excel the `AVERAGE` function works out the mean. The `MEDIAN` function computes the median and the `QUARTILE` function the quartiles. In R the `mean()`, `median()` and `quantile()` functions are used.
Dispersion	You can measure the spread of data in several ways. If data are normally distributed then standard deviation, variance, standard error and confidence intervals are all useful measures. For non-parametric data the range or inter-quartiles are more useful. Calculate standard deviation in Excel using `STDEV`; in R use the `sd()` command.
Data distribution	In normal (parametric or Gaussian) distribution the data are symmetrically arranged around the mean. If the data are skewed (non-parametric) the median is a better measure of the middle. The shape of the data affects the kind of summary statistics and other analyses you can use.
Visualize data distribution	The tally plot is a simple way to visualize the shape of the data. Use simple tallies to show the frequency of data in data size classes (bins). The histogram is a form of bar chart used to visualize data distribution. In Excel compute data frequency using `FREQUENCY` (an array function), then draw a bar chart. You can also use the *Analysis ToolPak*. In R use the `hist()` command to draw histograms.
Testing for normal distribution	You can use the `shapiro.test()` command in R to carry out the Shapiro–Wilk test for normality.
Transforming data	You can sometimes improve the normality of data by applying a mathematical transformation. Commonly used transformations are: logarithmic, square root, arcsine and reciprocal.
Running means	A running mean can be helpful in determining if your samples are large enough. You take a mean after each successive observation and plot them on a chart. When the curve flattens out you have sufficient replication.
Statistical symbols	Symbols are used as a quick aide-memoire for various statistical quantities.

5. Exploring data – which test is right?

Different sorts of data require different analytical approaches. An important part of the planning process is working out what sort of project you are going to carry out, as this will inform how you collect data and what you do with them afterwards.

What you will learn in this chapter

» How to distinguish different types of project
» How to formulate a hypothesis and null hypothesis
» How to choose the most appropriate statistical test

5.1 Types of project

It can be helpful to think about the type of project you are undertaking, as this can help guide you towards the most appropriate method(s) of analysis. You'd do this as part of the planning process because it also helps to determine the sampling methods and a host of other minor (yet important) details. You looked briefly at types of project in Section 1.2. Not all projects will require a hypothesis test, but they will all require some kind of statistical summary. Some of the statistics you will require are cogent to only one sort of project, such as measures of diversity (see Chapter 12); others are more general (e.g. averages).

You can split projects into four main types:

- *Descriptive*: In a descriptive project you do not carry out a hypothesis test but rather use various statistical measures to summarize a situation. For example, you may wish to look at the community of plants in an area. You can do this using various measures of diversity, which you will meet in Chapter 12.
- *Differences*: In a project involving differences you are looking to create two (or more) samples of data. You will use summary statistics to help describe the samples but your main aim is to compare the samples to see if they are different. There are innumerable things you could measure and thus compare; these may be biological or environmental variables. Your general aim is to set up and test a hypothesis about

the possible differences (see Chapters 7 and 10). However, there are some kinds of difference that are less suitable for classic hypothesis tests. Community studies often require a special statistical approach; in Chapter 12 you'll see examples of how to compare communities using similarity (or dissimilarity).

- *Correlation and regression*: In a project involving correlation or regression the emphasis is on linking variables, rather than splitting them apart (as in a differences project). Correlation is the simpler of the two; here you look for the strength and direction of the link between two variables (see Chapter 8). Often one is a biological variable and the other environmental. Regression is an extension of correlation where you seek a mathematical relationship to the link between variables. In projects involving regression you may have more than two variables (see Chapter 11).
- *Association*: In a project involving association the idea is to look for a link between things that are categorical (see Section 1.5). These projects of association often involve counting things that fall into a certain category. You'll see more about association analysis in Chapter 9. In ecology the categories could be different species; in Chapter 12 you'll see how you can use association to explore the similarity between assemblages of species.

All your data will need to be visualized in some way, regardless of the sort of project. You've seen some graphs (e.g. histograms) used already in Chapter 4. You'll see more about graphs and data visualization in Chapter 6.

5.2 Hypothesis testing

As part of the planning process one of the first things you will need is an idea of what you are trying to show/determine. Examples might be:

> "There are more buttercups (*Ranunculus repens*) in the south field compared to the north field."

or:

> "There is a positive link between stream flow and abundance of stonefly (*Perla bipunctata*) larvae."

In broad terms, a hypothesis should be something that you can put to the test, something you can prove or disprove. In reality you can rarely be 100% certain so you talk about supporting or rejecting your hypothesis.

In scientific hypothesis testing, you will have biological data; these will be variable in nature and so your conclusions are generally based on the likelihood that you are correct rather than absolute certainty. In science you will accept your conclusions only if there is less than a 5% chance that the result could have happened due to chance. This is where the statistical tests come in; they allow you to put a value on this likelihood and so give you a tool to decide if you are above or below this 5% threshold.

In practice, it is a lot easier to disprove something than to prove it; this concept is carried into biological hypothesis testing. You switch the emphasis around and create a null hypothesis. In the case of the examples above your null hypotheses would be:

> "There is no difference in buttercup (*Ranunculus repens*) abundance between the north and south fields."

and:

> "There is no link between stream flow and stonefly (*Perla bipunctata*) larvae abundance."

So, the null hypothesis is not just the opposite of the original idea but a new idea in its own right: the null hypothesis is what is actually tested by a statistical test. Once you have your null hypothesis (often abbreviated to H0 or H_0) you can move on to choosing the correct test to apply to your data. Ideally you do this before you collect any data at all as the type of test might influence the data collection. Thus the original hypothesis (often called the alternative hypothesis, H1 or H_1), the H0 and the selection of the stats test become part of the planning process. Once you know what you are aiming for, this will help you to decide which statistical test is right for your situation.

Note: Directional hypotheses

Your null hypothesis is always that nothing is happening. However, you ought to make your alternative more predictive. Rather than just state that there *will be a difference between two samples*, give a prediction. This shows that you have done some background research and made a prediction based on prior evidence. This does not affect the H0. Remember that the H0 is what you are really testing; your original H1 is simply your prediction.

Not all projects are suitable for classic hypothesis testing; these tend to be ones that are purely descriptive, but studies of ecological communities, for example, also fall into this group. The kinds of approach required for the study of community ecology are dealt with in detail in the companion volume to this work (*Community Ecology: Analytical Methods Using R and Excel*, Gardener 2014); however, in this book you'll see two important elements, diversity and similarity (Chapter 12).

5.3 Choosing the correct test

Choosing the correct test should be part of your project planning process. Different tests have different requirements, e.g. some need a minimum amount of data, so you need to know what you are aiming for. In simple terms, there are two broad categories of stats tests: those that split data and those that link data. If you are looking for differences in buttercup abundance between two fields/sites (as in the example above) then you are looking to split the data. In the second example you are looking for a link (between flow and abundance).

Some books and computer programs provide a decision tree to enable you to select the correct test; Figure 5.1 shows the decision chart from the *Don't Panic III* spreadsheet (see: www.gardenersown.co.uk/Education/dontpanic.htm).

The decision tree in Figure 5.1 provides a number of questions and you branch off accordingly. One question that crops up in several places is "are your data normally

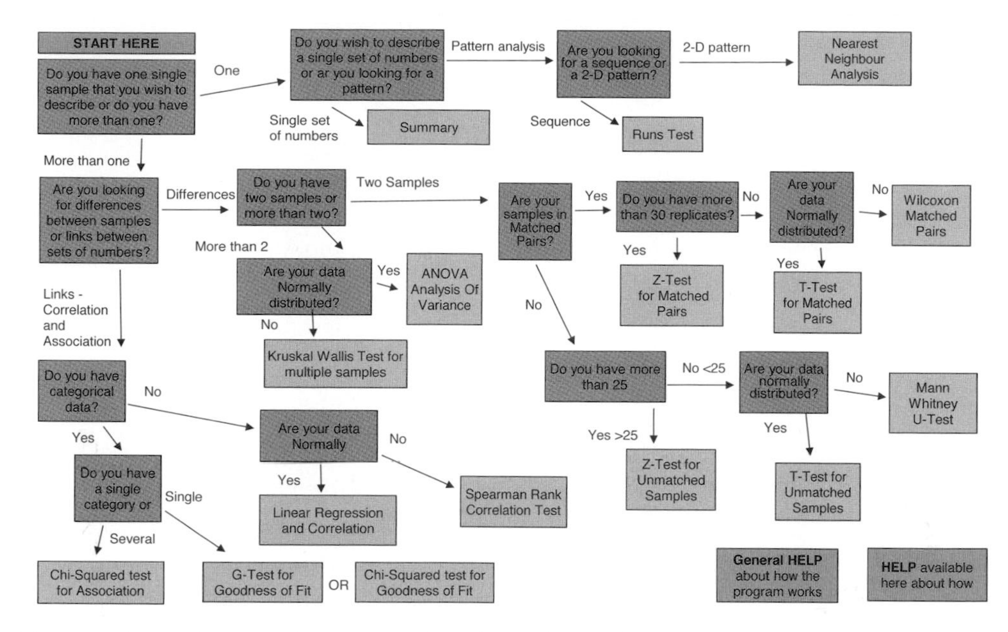

Figure 5.1 Decision tree for statistical analysis. This guides you towards the correct test for your data/situation (reproduced from *Don't Panic III*: www.gardenersown.co.uk/Education/dontpanic.htm).

distributed?" This is a very important consideration because the distribution (Section 4.2) of the data affects the statistical approach quite profoundly. You can also see that in some cases you need a lot more data to proceed (25–30 replicates). Once you have made your initial choice you should then determine what the requirements and limitations of your test are because this will influence your sampling strategy (you may of course alter your strategy in order to use a more suitable stats test). After looking at ways to explore your data graphically (Chapter 6), the next few sections will deal with specific types of statistical test. There are various kinds of test and each has a particular speciality.

In Table 5.1 you can see a formalized method to help you decide which test is the right one. You start at number 1 and compare the pair of statements (each has *a* and *b* options). Then you select the most appropriate and read the option at the end of the line. Eventually you come to a final decision. This is somewhat simplified, as you will see as you go along, but it gives you a good framework and a place to start thinking about your analyses. The decision chart does not cover studies of communities or descriptive projects.

You should carry out this process ideally as part of your original planning. Once you have determined the approach you need to take, you can begin to plan the data collection. Eventually you collect your data and can set about analysing it. A good start would be to draw graphs, histograms and box–whisker plots of the individual samples as well as comparing samples, perhaps with bar charts or scatter plots. You looked at distribution graphs like histograms and stem–leaf plots in Section 4.2. You will see graphs in Chapter 6 where you will start to think about which types of graph can be useful in which circumstances. After the chapter on graphs, you will focus on the mechanics of the various statistical tests. As you progress, you should bear in mind that producing a graphical summary of the situation is always a good idea. This will be picked up again in Chapter 13, where you'll see ways to present and share your results.

Table 5.1 Aid to choosing the correct statistical test.

1a	Looking for... differences between items/samples.	Goto 2
1b	Looking for... links between items.	Goto 5
2a	Data are normally distributed (parametric).	Goto 3
2b	Data not normally distributed (i.e. skewed).	Goto 4
3a	There are only two samples to compare.	*t*-test (Section 7.1)
3b	There are more than two samples.	ANOVA (Section 10.2)
4a	There are only two samples to compare.	*U*-test (Section 7.2)
4b	There are more than two samples.	Kruskal–Wallis (Section 10.3)
5a	The data are categorical.	Goto 6
5b	The data are continuous variables.	Goto 7
6a	There are two sets of measured categories.	Chi-squared association (Section 9.1)
6b	There is one set of categories.	Goodness of fit (Section 9.2)
7a	Data are normally distributed (parametric).	Goto 8
7b	Data not normally distributed (i.e. skewed).	Goto 9
8a	There is one dependent variable and one independent.	Pearson product moment (Section 8.2)
8b	One dependent variable and several independent.	Multiple regression (Section 11)
9a	There is one dependent variable and one independent.	Spearman's rank correlation (Section 8.1)
9b	One dependent variable and several independent.	This is tricky and beyond the scope of this book

EXERCISES

Answers to exercises can be found in the Appendix.

1. There are two forms of project that explore links between things; these are ____ and ____. The difference between these projects is the kind of data.
2. In a project involving correlation you always have a biological variable and an environmental variable. TRUE or FALSE?
3. Find two matching hypotheses (i.e. a main hypothesis H1 and a matching null hypothesis H0) from the following:
   A. There is a correlation between mayfly abundance and water speed
   B. There are more mayflies in slow-moving pools than in fast-moving riffles
   C. There is a positive correlation between mayfly abundance and water speed
   D. There are more mayflies in fast-moving riffles than in slow-moving pools
   E. There is no correlation between mayfly abundance and water speed
   F. There is no difference in mayfly abundance between slow-moving pools and fast-moving riffles
   G. There is a difference in mayfly abundance between slow-moving pools and fast-moving riffles
   H. There is a negative correlation between mayfly abundance and water speed
4. The *t*-test is a test for differences between several samples that are normally distributed. TRUE or FALSE?
5. Choose the most appropriate test from the following choices for this scenario: You are looking at several different species of bee. You have recorded the number of visits of these bee species to several kinds of flower.
   A. Analysis of variance
   B. Regression
   C. Association
   D. Spearman rank correlation
   E. Goodness of fit

Chapter 5: Summary

Topic	Key points
Types of project	Different projects require different analytical approaches: Descriptive: You simply describe the situation, often with summary statistics (e.g. mean, standard deviation). Species diversity would be an example of a descriptive project (although it is possible to compare samples, see Chapter 12). Differences: Where you are looking for differences between two or more samples. In community ecology this would include analysis of sample similarity (see Chapter 12). Correlation and regression: Where you are looking for links between variables, usually a biological variable and one or more environmental variables. Association: Where you are looking for links between the frequency of items that are in categories (i.e. they cannot be ranked), such as species and habitat.
Hypotheses	The hypothesis is a special kind of research question. Your statistical test will support or reject your hypothesis. In practice you create a null hypothesis (H0), which is that there is no difference (or correlation or association). The H0 is then tested. If this is rejected, your original hypothesis (H1) is supported. Note that the H0 is not simply the opposite of the H1. Think of the H0 as the *dull* hypothesis.
Choosing the correct test	Different analytical approaches are required for different situations. Differences: *t*-test, *U*-test, analysis of variance, Kruskal–Wallis. Correlation and regression: Spearman rank, Pearson, multiple regression, curvilinear regression, logistic regression. Association: chi-squared, goodness of fit tests. Community: diversity, similarity.

6. Exploring data – using graphs

Visualizing your data is very important. A graph can help you decide how to tackle a problem and also help others understand your results.

What you will learn in this chapter

» Different sorts of graph and their uses
» Choosing the appropriate type of graph
» How to make charts and use the chart tools in Excel
» How to make graphs using R

6.1 Introduction to data visualization

Graphs are useful for several reasons. They can help you to visualize the data and decide which statistical test is the best. You may spot patterns in the data and gain a better understanding of what you are dealing with. Graphs are also useful for summarizing your final results, especially when you present your findings to other people (see Chapter 13).

6.1.1 Types of graph and their uses

There are many sorts of graph (or chart) and each has its own particular merits. Some charts can be pressed into service for several uses.

- *Tally plots and stem–leaf plots*: Used for showing sample distribution (see Section 4.2).
- *Density plots*: Used for showing sample distribution (see Section 4.2).
- *Histograms*: Used for showing sample distribution (see Section 4.2).
- *Bar charts*: Used to show differences between samples. Each bar shows the value of something in a category. You can display multiple categories by grouping bars together or by stacking them upon one another (see Table 6.1).
- *Box–whisker plots*: Used to show differences between samples. You can also use them to summarize the shape of samples (see Section 4.3.3 and Table 6.1).

- *Scatter plots*: Used to show the relationship between two variables. You can also use a scatter plot to help show the result of a paired differences test (see Chapter 7).
- *Line plots*: Used to show how a variable changes over time. You can display more than one variable. You can also use a line plot to show a running mean (see Section 4.7). The line plot looks like a scatter plot with joined-up dots but is more like a bar chart as the x-axis usually has fixed intervals.
- *Pie charts*: Used to display compositional data. The data in a sample is displayed as a series of slices. The whole pie adds up to 360° and each slice shows the proportion of the total.

In Table 6.1 you can see examples of these graphs along with a summary of the uses.

Table 6.1 Types of chart and their various uses.

<table>
<tr><th>Type</th><th>Illustration</th><th>Purpose</th></tr>
<tr><td>Tally plot</td><td>16 | I
18 | I
20 | I
22 | III
24 | III
26 | IIIII
28 | III
30 | III
32 | III
34 | I
36 | I</td><td>Sample distribution. The tally plot is a simple histogram showing the frequency of items in bins (Sections 4.2 and 6.2).</td></tr>
<tr><td>Stem–leaf plot</td><td>1 | 679
2 | 112334
2 | 5666678899
3 | 01124
3 | 6</td><td>Sample distribution. The stem-leaf plot is similar to a tally plot but shows numbers instead of tallies, each digit is equivalent to a tally mark (Sections 4.2 and 6.2).</td></tr>
<tr><td>Histogram</td><td> 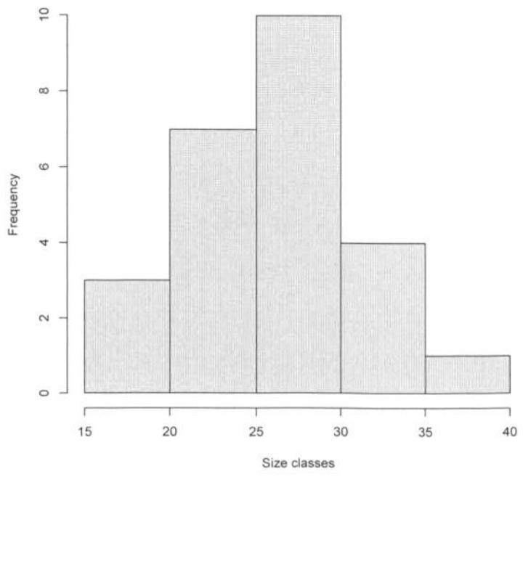Frequency: 0, 2, 4, 6, 8, 10
Size classes: 15, 20, 25, 30, 35, 40</td><td>Sample distribution. In a classic histogram the x-axis is a continuous variable. The bars show frequency of items in a bin (Sections 4.2 and 6.2).</td></tr>
</table>

Type	Illustration	Purpose
Density plot	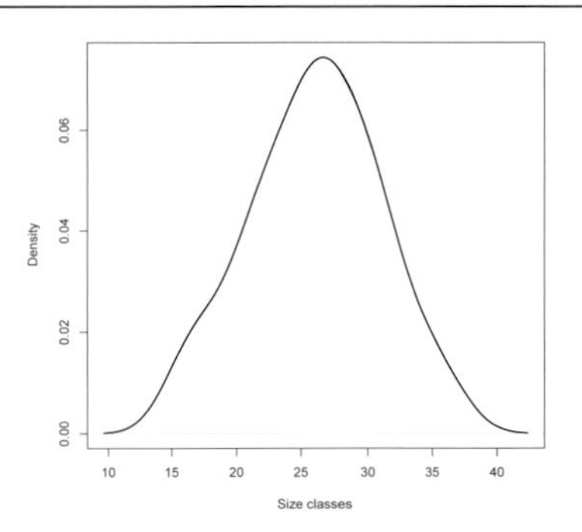	Sample distribution. The density plot is like a histogram but the area under the curve sums to unity (Section 6.2).
Box–whisker plot	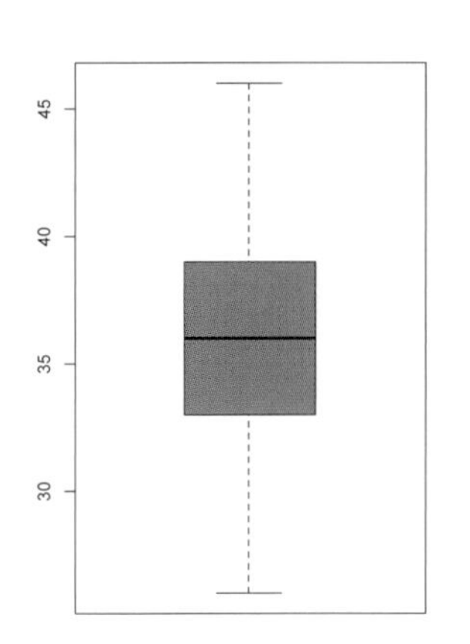	Sample distribution. The boxplot shows five important summary statistics for a sample: the median, max, min, upper and lower quartiles. These can help to visualize the distribution (Section 4.3 and 6.3.2).
Bar chart	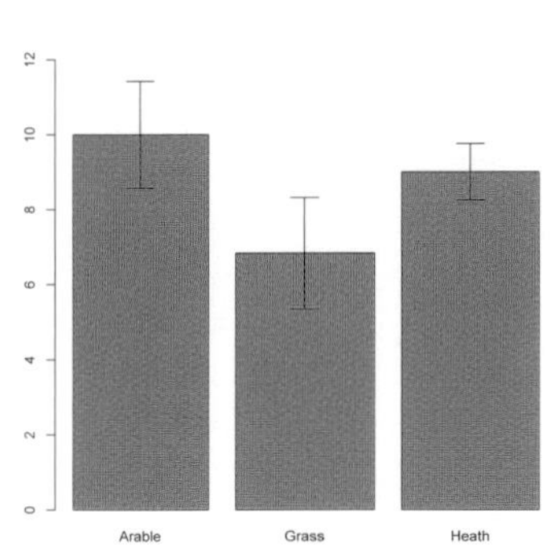	Differences between samples. Each bar shows the value of a single category. Error bars can show variability. The bars can be drawn horizontally (Section 6.3.1).
Box–whisker plot	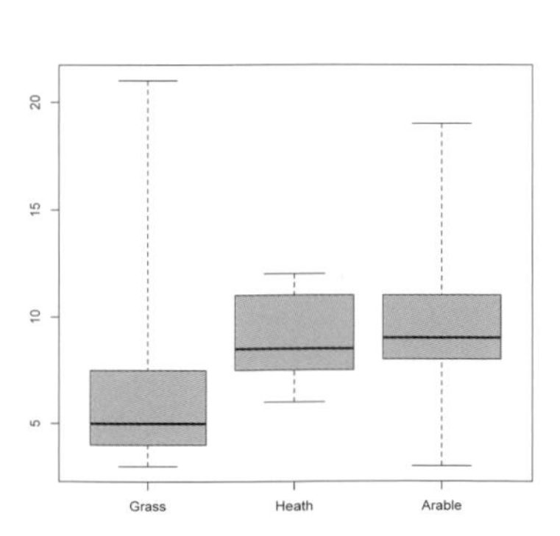	Differences between samples. A boxplot is akin to a bar chart but shows more information. They are especially useful because the *y*-axis does not have to begin at zero (Section 6.3.2).

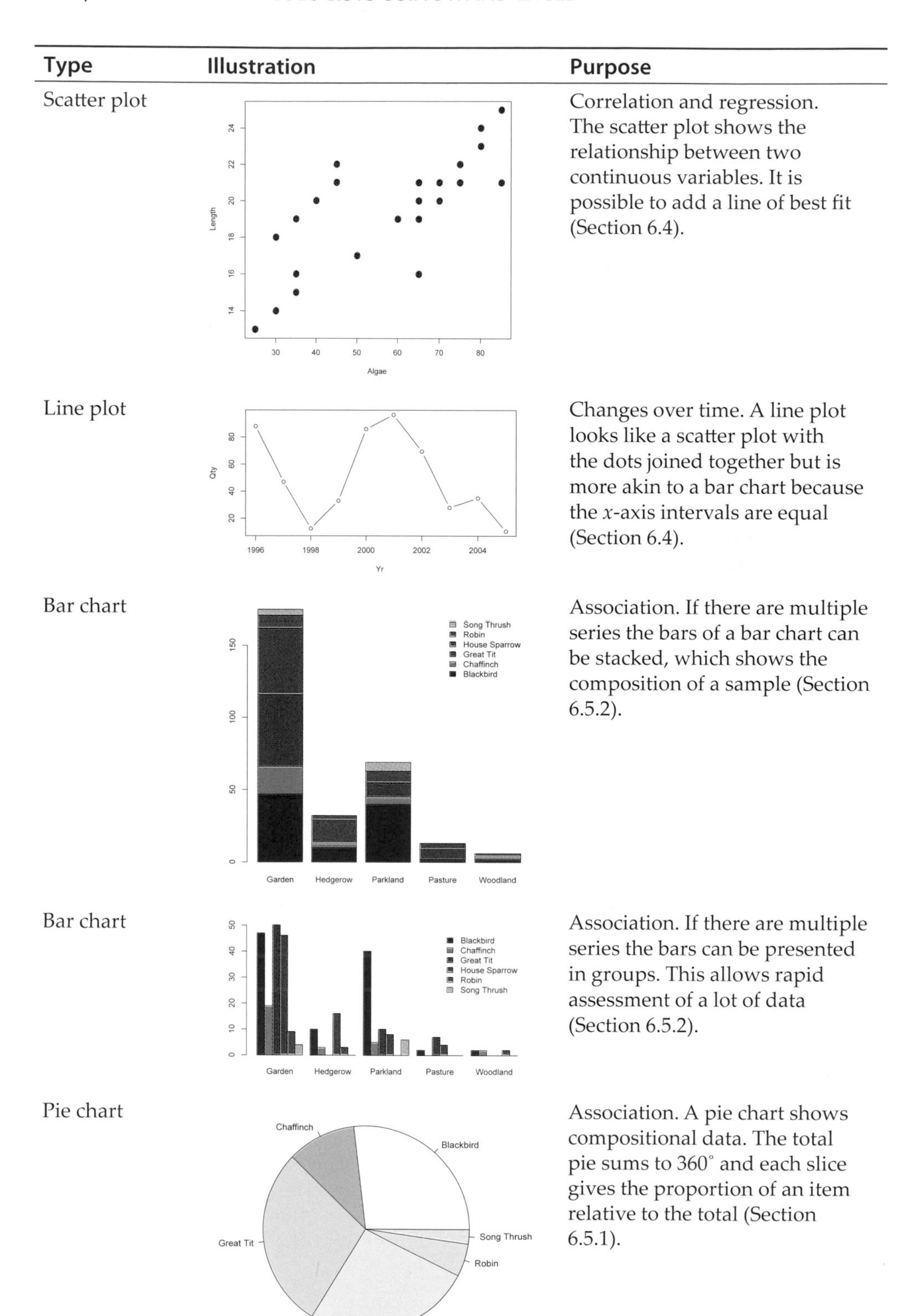

Type	Illustration	Purpose
Scatter plot		Correlation and regression. The scatter plot shows the relationship between two continuous variables. It is possible to add a line of best fit (Section 6.4).
Line plot		Changes over time. A line plot looks like a scatter plot with the dots joined together but is more akin to a bar chart because the x-axis intervals are equal (Section 6.4).
Bar chart		Association. If there are multiple series the bars of a bar chart can be stacked, which shows the composition of a sample (Section 6.5.2).
Bar chart		Association. If there are multiple series the bars can be presented in groups. This allows rapid assessment of a lot of data (Section 6.5.2).
Pie chart		Association. A pie chart shows compositional data. The total pie sums to 360° and each slice gives the proportion of an item relative to the total (Section 6.5.1).

Type	Illustration	Purpose
Bar chart	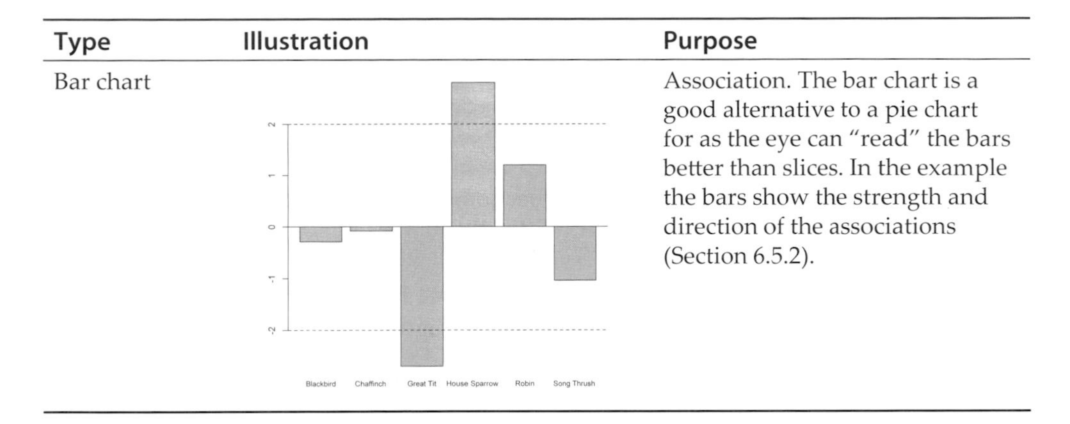	Association. The bar chart is a good alternative to a pie chart for as the eye can "read" the bars better than slices. In the example the bars show the strength and direction of the associations (Section 6.5.2).

The graphs shown in Table 6.1 should cover most of your requirements although there are some other chart types that can be useful (e.g. the dendrogram, which you'll see later in this chapter and in Chapter 12).

It can be helpful to think about the purpose to which you want to use your graph, and then you can choose the most appropriate chart. In Table 6.2 you can see a summary of the main chart types, grouped by purpose.

Table 6.2 Summary of graph types to use for different purposes.

Purpose	Types of graph
Illustrating distribution	Stem–leaf plot, tally plot, histogram, density chart, box–whisker plot (Sections 4.2 and 6.2).
Illustrating differences between samples	Bar chart, box–whisker plot (Section 6.3).
Illustrating correlations	Scatter plot (Section 6.4).
Illustrating associations	Pie charts, bar charts (Section 6.5).
Illustrating sample composition	Pie charts, bar charts (Section 6.5).
Illustrating sample sizes	Line plot of running average, e.g. mean or median (Section 4.7).
Illustrating changes over time	Line plot (Section 6.4).

Table 6.2 shows the various uses to which you can put your graphs (Excel calls graphs charts so I use the terms interchangeably). It is always useful to draw your data before you carry out any statistical testing. A visual assessment of the situation can help you to make better choices about what to do with your data. You can tweak your charts later on, when you want to share your results with others (see Chapter 13).

Both R and Excel have wide-ranging graphical capabilities. However, they produce charts in quite different ways, which is the subject of the following sections.

6.1.2 Using Excel for graphs

In Excel the process of constructing a chart begins with the *Insert* menu on the ribbon menu (Figure 6.1).

6.1.3 Using R for graphs

R has very extensive graphical capabilities but its approach is quite different to that of Excel. In R you build your graphs using commands typed from the keyboard. If you do not like a particular colour, for example, then you must redraw the chart; you cannot edit it like you can in Excel. At first this seems somewhat cumbersome but in reality you can proceed quite rapidly. The keyboard up arrow recalls previous R commands, which you can edit. This means you can redraw a chart quite rapidly.

There are four sorts of graphics commands in R:

- Commands that draw a plot; such as to draw a bar chart or a scatter plot.
- Instructions (parameters) that you add to a plotting command to modify or embellish the plot; such as to add axis titles or alter the plot colour.
- Commands that add additional items to an existing plot; such as additional data points, lines or text.
- Commands that alter the general graphical output, which you type separately from the plotting commands. Examples include altering plot margins and setting explicit plot dimensions.

The basic plotting commands produce a general graph; there are commands for bar charts, scatter plots and so on (Table 6.4).

Table 6.4 Basic chart commands in R.

Type of plot	Command	Details
Stem–leaf plot	`stem(x)`	Makes a stem–leaf plot in the main R console. This is not really a graphic but plain text.
Histogram	`hist(x)`	Makes a histogram of the selected sample. You can alter the breakpoints and so alter the number of bars that are produced.
Bar chart	`barplot(x)`	Makes a bar chart. You can draw the bars horizontally. If there are multiple series you can choose to have the bars stacked or beside one another.
Box–whisker plot	`boxplot(x)`	Makes box–whisker plots. You can draw the plot horizontally. The length of the whiskers can be controlled (default is 2/3 IQR).
Scatter plot	`plot(x,y)`	Makes a scatter plot of two variables. You can plot points only, lines only or both.
Line plot	`plot(x)`	Makes a line plot if you specify `type = "l"`.

The basic commands shown in Table 6.4 produce graphs with set defaults. You can specify additional parameters to these commands to alter the appearance of your plots (Table 6.5).

You'll see examples of how these graphical parameters operate in action as you see examples of producing the individual plots.

Table 6.5 Commonly used graphical parameters/instructions added to R plotting commands.

Parameter type	Command	Details
Colours	`col`	Alters the default colours on the plot. There are variants that specify the colours of certain components (e.g. `col.lab`, `col.axis` for labels and axis).
Plotting symbol	`pch`	Alters the plotting symbol used in scatter plots. You generally specify a number but characters can be given in quotes.
Size	`cex`	Alters the "magnification" of plot elements. Values >1 make items bigger (<1 smaller). There are variants that specify the size of certain components (e.g. `cex.lab`, `cex.axis`).

Titles	`main`
	`xlab`
	`ylab`
		Allows you to specify titles (there are often default titles). Titles should be in quotes. Note that `xlab` always refers to the bottom axis and `ylab` to the left.

Line width	`lwd`	Alters the width of plotted lines. Specify a number (>1 wider, <1 narrow).
Line style	`lty`	Alters the style of plotted lines: 1 = solid, 2 = dashed, 3 = dotted, 4 = dotdash.
Axis scale	`xlim` `ylim`	Alters the extremes of the axes. You specify two values, the start and the end e.g. `xlim = c(0, 10)`. Note that `xlim` always refers to the bottom axis and `ylim` to the left.
Axis label orientation	`las`	Alters the way axis labels are oriented: 0 = all parallel to axis (the default), 1 = all horizontal, 2 = all normal to axis, 3 = all vertical.

Most of the general plotting commands will obliterate any existing chart window. However, some commands allow you to add additional items to an existing plot. These commands allow you to add additional data, text and other items (Table 6.6). You've already met one example of adding to an existing plot in Chapter 4, when you made a histogram and added a density plot over the bars.

Most of the graphical parameters have default settings and you can alter these by using the `par()` command. Type `help(par)` in R to see the full range of graphical parameters you can set.

You can create a new (blank/empty) plot window and set it to a particular size. The command is specific to your operating system:

- Windows: `windows()`
- Macintosh: `quartz()`
- Linux: `X11()`

Table 6.6 Commands that add to existing R charts.

Addition type	Command	Details
Titles	`title()`	This command allows you to add titles to an existing chart (see Table 6.5). If there is an existing title it will be overplotted and not rewritten.
Additional data points	`points()`	This command allows you to add points to an existing plot; you specify the *x,y* co-ordinates of the points as well as any other graphical parameters (e.g. `pch`, `cex`).
Additional data lines	`lines()`	This command allows you to add lines to an existing plot; you specify the *x,y* co-ordinates of the data points (which are then joined by the line segments) as well as any other graphical parameters (e.g. `pch`, `cex`).
Text	`text()`	This command allows you to add text to the plot window. You specify the co-ordinates and the text. You can also alter the colour, size, font and rotation of the text.
Marginal text	`mtext()`	This adds text to the margin area of the plot.
Sections of line	`segments()`	This adds sections of line to a plot; you specify the *x,y* co-ordinates of the start and end points. You can use this command to make error bars for example.
Arrows	`arrows()`	This is similar to `segments` but allows you to add arrowheads (use flat ones to make the hats of error bars).
Straight lines	`abline()`	This adds straight lines to a plot. You can specify a horizontal or vertical position (e.g. to make gridlines). You can also give *a* and *b* components (intercept and slope), to draw a line of best fit.

Note: R plot windows

The "plot window" commands are operating system specific but the `X11()` command will work on any operating system. In Windows it will open a regular graphics window, on a Mac it will open a window using the X11 system (which is not the same as the quartz system).

Typing the command will create a new plot window with the default size but you can override it by giving new `width` and `height` settings, e.g.:

```
> windows(width = 6, height = 4)
```

The previous command would create a new empty plot window with a width of 6 inches and a height of 4 inches.

In the following sections you'll see examples of different types of plot.

6.2 Exploratory graphs

An exploratory graph is one that you produce to help you see what you've got before you launch into any statistical tests. If you were looking at a correlation you'd want to have a sneak preview of the potential relationship. You'll see specific scenario-based graphs later; in this section you'll focus on those graphs that tell you something about the sample data you have, in other words:

- Tally plots.
- Stem–leaf plots.
- Histograms.
- Density plots.
- Running means.

You've already seen how to produce some exploratory graphs in Chapter 4, where you looked at all the graph types in the preceding list. This section will therefore be short and act more as a review.

6.2.1 Exploratory graphs using Excel

Excel is best suited to producing bar charts (for histograms, Section 4.2.1) and line plots (for running means, Section 4.7.1).

Tally plots in Excel

It is possible to coerce Excel into making a tally plot. In order to do that you'll need to work out the bins you want to split the data into. Then you'll need to work out the frequency of items in each bin; you can use the `FREQUENCY` function to do this (recall Section 4.2.1).

Now you have the frequencies you need to draw the tally marks. The easiest way to do this is to use the `REPT` function. This function repeats a piece of text however many times you specify:

```
REPT(Text, Repeats)
```

You can set the `Text` part to `"I"` or `"X"` and the `Repeats` part to the cell containing the frequency. Copy the formula and you have your tally plot.

Note: Tally plots in Excel

On the support website you can find some additional notes about histograms and making tally plots in Excel.

Histograms in Excel

Before you can make a histogram you'll need the frequency data for the various bins. This means that you'll need to determine the appropriate values for the bins, and then use the `FREQUENCY` function to scan the data and work out the frequencies in the various bins. Remember that the `FREQUENCY` function is an array function so you need to highlight the cells where the result is to be placed, type the formula then press `Ctrl+Shift+Enter`.

Once you have the frequency data you can prepare a bar chart (Excel does not do a "proper" histogram). Use the *Insert > Column Chart* button to make your chart. The frequencies will form the main data and the bins will be the *x*-axis categories. You'll need to make the bars wider, so double-click on the bars and reset the *Options > Gap Width* to a small value (5% is about right).

You can also use the *Analysis ToolPak* to make a histogram (column chart) for you. You will need to work out the bin values but there is no need to compute the frequencies. Click the *Data > Data Analysis* button, and then choose *Histogram*. You'll still need to edit the resulting chart.

Running means in Excel

Firstly you'll need to calculate the running mean; this is easily achieved with the `AVERAGE` function. Once you have the numerical values you can plot them as a line chart (not a scatter plot). Select the data then use the *Insert > Line Plot* button. You can choose a plot with point markers or with a line only. Once you have your chart you can use the *Chart Tools* menus to edit it as you require.

6.2.2 Exploratory graphs using R

R can produce the exploratory charts you'll need:

- Stem–leaf plot: `stem()`
- Histogram: `hist()`
- Running mean: `plot()`

It is a bit trickier to make a tally plot but it is possible to cobble together something using the `hist()` command to work out the breaks and frequencies for you.

Note: Tally plots in R

On the support website you can find some additional notes about making a dot chart as an alternative to the histogram using the `hist()` command as a base.

Stem–leaf plots in R

The `stem()` command makes a stem–leaf plot from some numerical data. You simply provide the name of your data. The command attempts to display the results neatly:

```
 bd
 [1] 17 26 28 27 29 28 25 26 34 32 23 29 24 21 26 31 31 22
[19] 26 19 36 23 21 16 30
> stem(bd)
  The decimal point is 1 digit(s) to the right of the |
  1 | 679
  2 | 112334
  2 | 5666678899
  3 | 01124
  3 | 6
```

You can alter the number of rows using the `scale` parameter:

```
> stem(bd, scale = 3)
  The decimal point is at the |
  16 | 00
  18 | 0
  20 | 00
  22 | 000
  24 | 00
  26 | 00000
  28 | 0000
  30 | 000
  32 | 0
  34 | 0
  36 | 0
```

Anything to the right of the vertical bars shows you a tally (frequency), whilst the position of the vertical bars indicates something about the decimal place.

You've essentially got two significant figures to play with:

```
> x
  [1] 0.262 0.774 0.595 0.095 0.158 0.701 0.269 0.297 0.563
[10] 0.345 0.996 0.215 0.591 0.409 0.540 0.810 0.011 0.675
[19] 0.880 0.408
> stem(x)
  The decimal point is 1 digit(s) to the left of the |
  0 | 106
  2 | 16705
  4 | 114699
  6 | 707
  8 | 18
 10 | 0
```

Generally it is best to leave the default `scale` setting.

Histograms in R

The `hist()` command produces true histograms; unlike Excel the bars form a continuous *x*-axis:

```
> hist(variable)
```

Here is a vector of numbers saved as the variable `test.data`:

```
21 26 27 32 41 43 52 51 48 18 14 25 27 31 26 28
```

To create a histogram you type:

```
> hist(test.data)
```

> **Note: Histogram test data**
>
> If you want to mimic the text and copy the histograms you can enter the data yourself. Type the following:
>
> ```
> test.data = scan()
> ```
>
> Then enter the values shown in the text, separate the numbers with spaces. When you are done press *Enter* twice and the data will now be stored in R.

To plot the density rather than the actual frequency you need to add the parameter `freq = FALSE` like so (Figure 6.3):

```
> hist(test.data, freq= FALSE, las = 1)
```

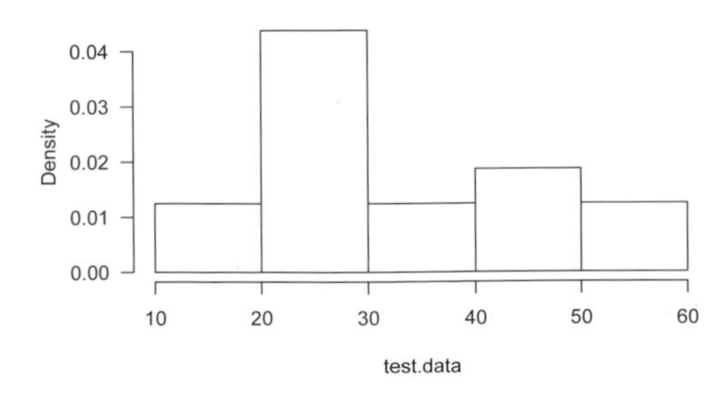

Figure 6.3 Histogram drawn using density instead of frequency. The area under the bars sums to unity.

The histogram in Figure 6.3 shows density instead of frequency, the area under the bars sums to unity. Note also that the axis labels are all horizontal (`las = 1`), which makes them easier to read. This is useful but the plot is a bit basic and boring. You can change axis labels and the main title using `main`, `xlab` and `ylab` parameters. Here is a new plot with a few enhancements (Figure 6.4):

```
> hist(test.data, col = "cornsilk", xlab = "Data range", ylab =
"Frequency of data", main = "Histogram", font.main = 4)
```

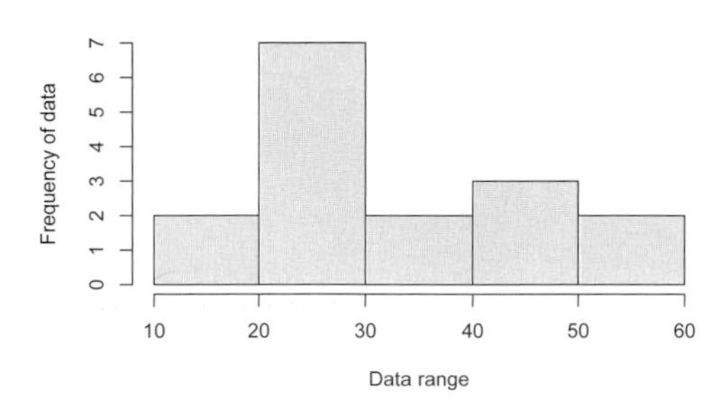

Figure 6.4 Histogram with coloured bars and custom titles.

These commands are largely self-explanatory. The 4 in the `font.main` parameter sets the font to italic (try some other values).

By default R works out where to insert the breaks between the bars. You can change the number of breaks by adding a simple parameter (Figure 6.5):

```
> hist(test.data, breaks = 10, col = "lightblue") # 10 breaks
```

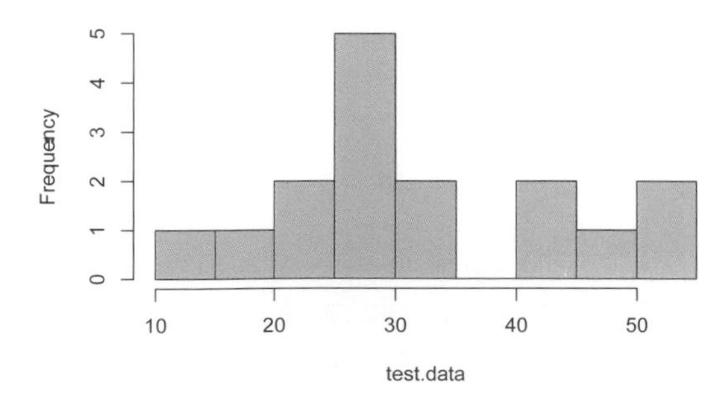

Figure 6.5 Histogram with specified number of breakpoints.

The # tells R that what follows is a comment, useful for creating your own library of commands.

Alternatively you can be more specific and set the breaks exactly. In the following example (Figure 6.6) the range of the *x*-axis is also specified:

```
> hist(test.data, breaks = c(10, 15, 20, 30, 40, 60),
       col = "pink", xlim = c(0, 70)
```

Notice how the exact break points are specified in the `c(x1, x2, x3)` format and the `xlim` parameter in a similar manner, `c(lower, upper)`.

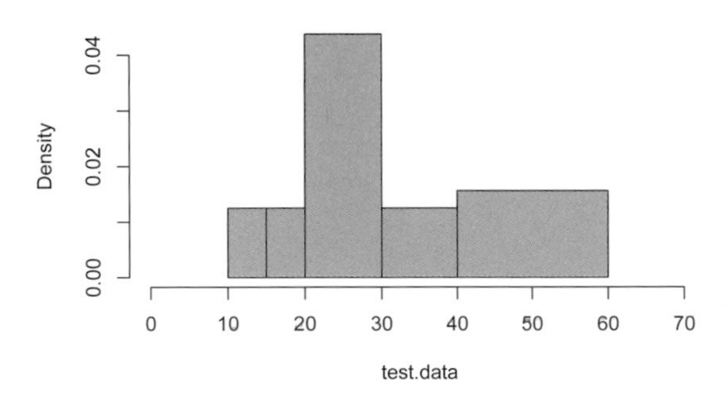

Figure 6.6 Histogram drawn with customized break points and *x*-axis limits.

The `xlim()` and `ylim()` commands are useful if you wish to prepare several histograms and want them all to have the same scale for comparison.

Running means in R

You already saw how to produce running means using R in Section 4.7.2. First of all you need the running mean values; then you simply use the `plot()` command. You need to specify `type = "l"` or `type = "b"` so that you get a line.

The following commands produce a plot that resembles Figure 6.7:

```
> lf
 [1] 33 33 34 31 37 33 30 27 36 31
> cumsum(lf)/seq(along = lf)
 [1] 33.00000 33.00000 33.33333 32.75000 33.60000 33.50000
 [7] 33.00000 32.25000 32.66667 32.50000
> plot(cumsum(lf)/seq(along = lf), pch = 16, type = "b",
      ylab = "Mean size mm")
> abline(h = mean(lf), lty = 3, col = "blue")
```

In the preceding commands you really did not need to display the running mean values; they were shown so you can see how the plot is constructed more clearly.

The main point of the example is to show you that you only need to specify the

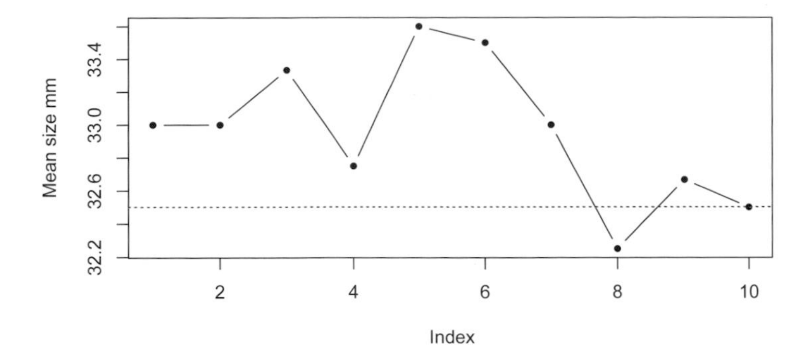

Figure 6.7 A running mean line plot. The dotted line shows the overall mean.

y-values. The `plot()` command makes a simple index for the x-values. Note that the plotting symbol was changed using the `pch` parameter. The last line is a command that adds something to an existing plot (see Table 6.6), in this case a straight (horizontal) line representing the overall mean.

6.3 Graphs to illustrate differences

There are two main types of graph used to illustrate differences between things:

- Bar charts, called column charts in Excel.
- Box–whisker plots, also called boxplots.

In both cases each bar (or box) represents a sample. Your chart therefore has one axis that is categorical (the samples, usually the x-axis), and one axis that shows values (usually the y-axis).

The y-axis shows the "value" of each sample; usually you'll present the data as an average (a mean or median, see Chapter 4), but sometimes other values are used (such as the sum). The y-axis represents the *response* variable (or *dependent* variable), whilst the x-axis represents the *predictor* (or *independent*) variable.

6.3.1 Bar charts to illustrate differences

Bar charts are widely used and easily recognizable graphs. Since each bar represents a single sample they are a useful way to visualize differences between samples.

Bar charts in Excel

Bar charts are available from the *Charts* section of the *Insert* menu in Excel (see Figures 4.6 and 6.1). Excel calls a chart with vertical bars a *column chart* and one with horizontal bars a *bar chart*. You'll see the terms bar chart and column chart used interchangeably and it is a matter of preference if you have the bars vertical or horizontal. Generally vertical bars are most easily interpreted but there are occasions when horizontal bars are useful; when there are a lot of bars and you want to fit them on one page it is easier to have a tall layout, which also makes it easier to read long axis labels (Figure 6.8).

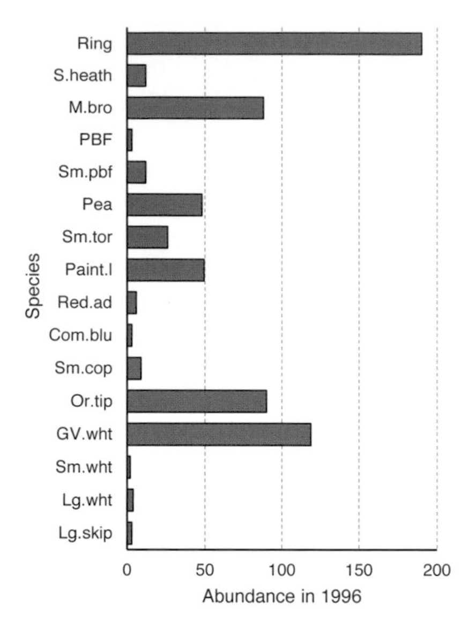

Figure 6.8 A bar chart with horizontal bars can make for a better presentation. Axis labels are more readable and a large number of bars can fit the display more easily.

The simplest situation is where you have values representing two or more samples; the data in Table 6.7 provide a typical example.

Table 6.7 Abundance of freshwater hoglouse at three sampling sites.

Obs	Upper	Mid	Lower
1	3	4	11
2	4	3	12
3	5	7	9
4	9	9	10
5	8	11	11
6	10		
7	9		

Before you can chart the data in Table 6.7 you will need to produce the appropriate summary data (see Chapter 4). In this case you deem that a median is the most appropriate average so use the `MEDIAN` function to produce something like Table 6.8.

Table 6.8 Median abundance of samples of freshwater hoglouse at three sampling sites. These data can be used to make a column chart in Excel.

	Upper	Mid	Lower
Median	8	7	11

Once you have your data you can make your column chart. You can proceed in two ways:

- Select the data (the values and the sample headings).
- Select no data (click in an empty cell).

In any event you need to click the *Insert > Column Chart* button and choose a basic layout. If you selected the data then Excel will form the chart directly. If you did not select any data you must use the *Chart Tools > Design > Select Data* button to choose the cells that contain the values and the axis labels.

Adding error bars to Excel column charts

When you summarize a sample of data you ought to include a measure of the variability (Section 4.3). You can incorporate this measure into your bar charts by using error bars. Excel draws error bars that run from the top of the bar up and/or down by an amount you specify. You can add error bars using the *Chart Tools > Design > Add Chart Element > Error Bars* button (Figure 6.9).

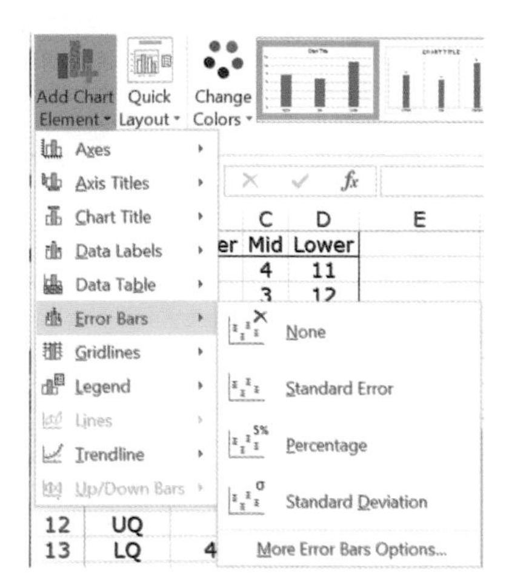

Figure 6.9 The *Error Bars* button allows you to add error bars to charts (Excel 2013).

There are several standard options for adding error bars, but none of them work! You want to select the option that says *More Error Bars Options*. This allows you to *Specify Value*, where you can choose spreadsheet cells that hold the appropriate values.

The value you specify will determine the length of the error bar so if you have calculated quartiles you will need to work out the difference between these and the median (Table 6.9). In Table 6.9 you can see the median and the two inter-quartiles (UQ and LQ). The lengths of the error bars to be plotted on the chart are in the last two rows. When you have non-parametric data (as in the current example) the error bars may not be symmetrical. The bar chart for the data in Table 6.8 and the error bars (Table 6.9) produce a plot that looks something like Figure 6.10.

Once you have your error bars you can use the *Chart Tools* menus to edit and alter them like any other chart element.

Table 6.9 Summary statistics for three samples of freshwater hoglouse. Error bar lengths are determined by the difference between the quartiles and the median.

Statistic	Upper	Mid	Lower
Median	8	7	11
UQ	9	9	11
LQ	4.5	4	10
Up	1	2	0
Dn	3.5	3	1

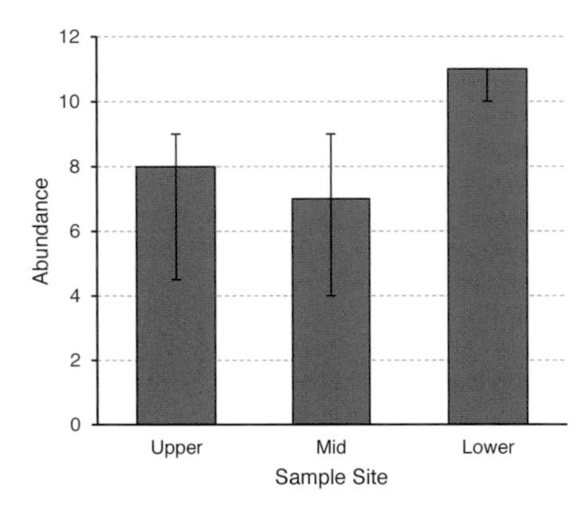

Figure 6.10 Abundance of freshwater hoglouse at three sites. The bars show the medians and error bars inter-quartile range. Note that the error bars are not symmetrical.

Note: Example data

The data from Table 6.8 are available on the support website. The file is called *Hoglouse.xlsx* and contains the data, summary statistics and an example column chart (there are two worksheets, one with just the data and the other with the summary and chart).

Download available from the support website.

Multiple data series and Excel bar charts

When you have several data series you have a wider range of options. Look at the data in Table 6.10, which show the abundance of some butterfly species over several years.

You can use the data in Table 6.10 to make bar charts in several different layouts:

- Single species with bars for different years.
- Single year with bars for different species.

Table 6.10 Abundance of butterfly species for several years.

	1996	1997	1998	1999	2000
M.bro	88	47	13	33	86
Or.tip	90	14	36	24	47
Paint.l	50	0	0	0	4
Pea	48	110	85	54	65
Red.ad	6	3	8	10	15
Ring	190	80	96	179	145

- Multiple species with bars for different years.
- Multiple years with bars for different species.

The kind of chart you get depends on which data you select. If you select the top row (the years) and one or more rows of butterfly data you will get a chart with a cluster of bars for each year. Each of the clusters will have a different coloured bar for each species.

If you select the first column (the species names) and one or more columns you will get a cluster of bars for each species, with the individual bars being a different year.

Note: Selecting non-adjacent cells

If you want to select non-adjacent cells hold down the *Control* key whilst using the mouse.

The *Column Chart* button gives a variety of choices for the overall style of column chart (see Figure 4.7); the first option is the *Clustered Column* option (the name appears when you hover over it with the mouse).

There are other options, such as stacked bar charts. You'll see these later (Section 6.3.2) when you look at charts for illustrating associations (although that is not the only use for stacked column charts). In the following exercise you can have a go at making some column charts for yourself.

Have a Go: Use Excel to make column and bar charts

You'll need the data from Table 6.10 for this exercise; the data are in a file called *butterfly table.xlsx*. There is also a CSV version, which will open in Excel but is intended for import into R.

Download available from the support website.

1. Open the *butterfly table.xlsx* file. The data show the abundance of some butterfly species over several years.

2. Highlight all the data cells A1:F7, and then click the *Insert > Column Chart* button.
3. Choose the *Clustered Column* option, which is the first one. Click the icon and the chart will be produced immediately. If you hover over the button you will see a preview in Excel 2013.
4. Click once on the chart to make sure the *Chart Tools* menus are activated. Now click the *Design > Quick Layout* button. Hover over the various options and see the effects. Choose layout 9, which places a legend on the right and makes spaces for the axis titles.
5. Click the box containing the blank *x*-axis title (it will say *Axis Title*); then type a new name, `Butterfly Species` will do nicely.
6. Now alter the *y*-axis title in a similar manner, `Abundance` would be a sensible title here.
7. Delete the main title by clicking it then using the *Delete* button on the keyboard.
8. Try a few other adjustments by using the tools on the *Format* menu; you can also double-click the chart in Excel 2013 to activate the *Format* dialogue box. You can then select chart elements and use the dialogue box to help format those elements. You can look to Figure 6.11 as a guideline.

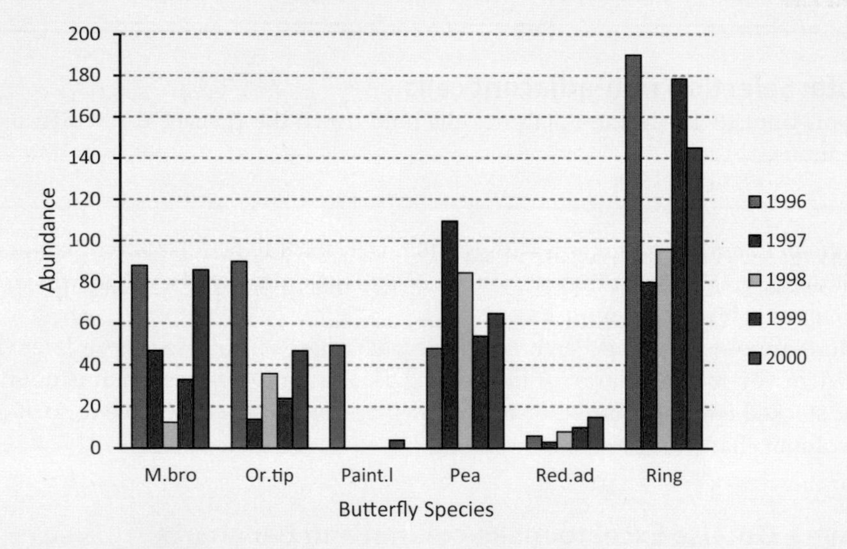

Figure 6.11 Clustered column chart. Abundance of some butterfly species over several years, clustered by species.

9. Now click again on the chart to make sure you have the *Chart Tools* menus activated. Click the *Design > Switch Row/Column* button. This switches the data around so that now the data are clustered by year instead of by species.
10. Click the *x*-axis title and alter it to read `Sample year`. After a bit more editing you should have a chart that resembles Figure 6.12.

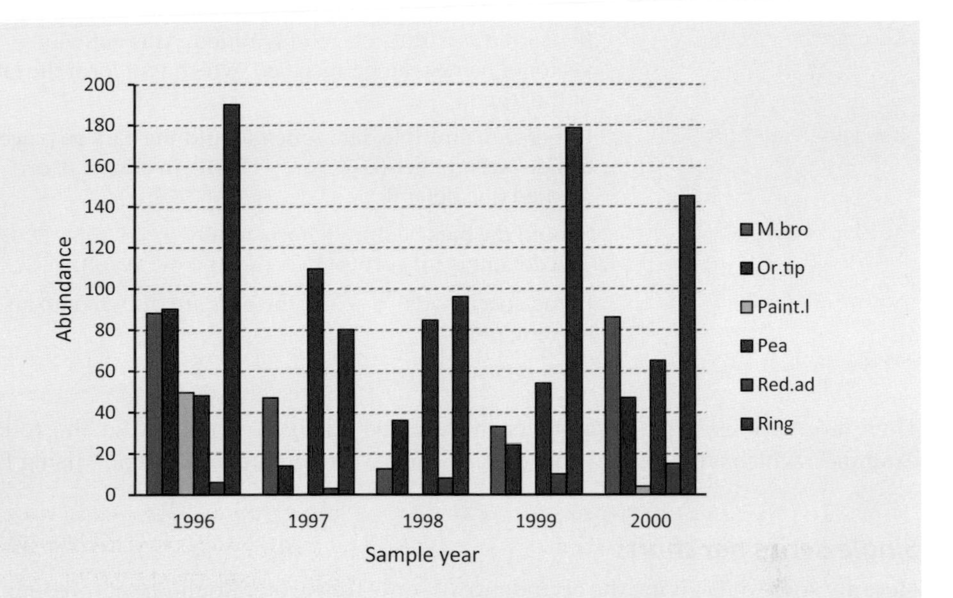

Figure 6.12 Clustered column chart. Abundance of some butterfly species over several years, clustered by year.

Try various editing tools and try and familiarize yourself with what they can do for you.

The charts shown in the preceding exercise (Figure 6.11 and Figure 6.12) do not have error bars because each one is based on a single value (the total abundance).

Bar charts in R

The general command to draw bar charts in R is `barplot()`. The command can accept a range of additional parameters that allow you to alter the appearance of the resulting plot. You can make bar charts of a single data series, or of several with the bars clustered in groups. You can also draw a chart with the bars horizontal. The general form of the command is as follows:

```
barplot(height, names, legend, beside, horiz, add)
```

`height`	The data to be plotted (the height of the bars). Either a single vector of values or a matrix, which can contain multiple rows and columns. If a matrix then multiple data series are plotted.
`names = NULL`	An optional vector of names to assign to the bars. If not specified (the default) then the `names` attribute of the data is used.

`legend = FALSE`	If `legend = TRUE` a legend is added. Alternatively a vector of names can be specified, which will form the labels of the legend.
`beside = FALSE`	If there are multiple data series should the bars be placed beside one another (`beside = TRUE`, in clusters), or stacked (the default, `beside = FALSE`)?
`horiz = FALSE`	Should the bars be drawn horizontally (`horiz = TRUE`)? The default is for vertical bars (`horiz = TRUE`).
`add = FALSE`	If you specify `add = TRUE` the bars are drawn onto an existing plot.

There are various other parameters that can be used, as you'll see in the following examples, which will use the same data as you saw in the previous section using Excel.

Single-series bar charts

Here are some data giving the abundance of some freshwater hoglouse at three sampling sites:

```
> hog3
  Upper Mid Lower
1     3   4    11
2     4   3    12
3     5   7     9
4     9   9    10
5     8  11    11
6    10  NA    NA
7     9  NA    NA
```

Notice that the "missing" data are assigned a value `NA`. R does not like empty cells, so unlike Excel the blanks are given a special assignment (recall this from Section 4.1). The first task is to determine the average for each sample; the median is appropriate.

```
> hlm = apply(hog3, MARGIN = 2, FUN = median, na.rm = TRUE)
> hlm
Upper   Mid Lower
    8     7    11
```

Note the `na.rm = TRUE` part of the command, which essentially tells R to ignore `NA` items in the median calculation. Now you have an item `hlm`, which contains the median values along with the sample names. Use the `barplot()` command to make a basic bar chart (Figure 6.13):

```
> barplot(hlm)
```

The chart is adequate but rather plain and has no axis titles; a few extra parameters will attend to that (Figure 6.14):

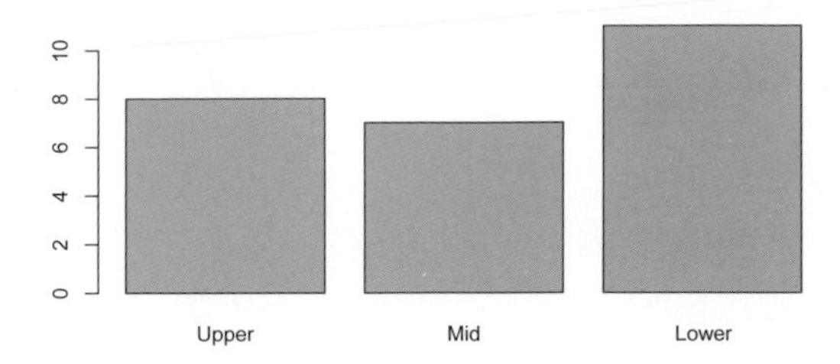

Figure 6.13 A basic bar chart in R. Median abundance for freshwater hoglouse at three sites.

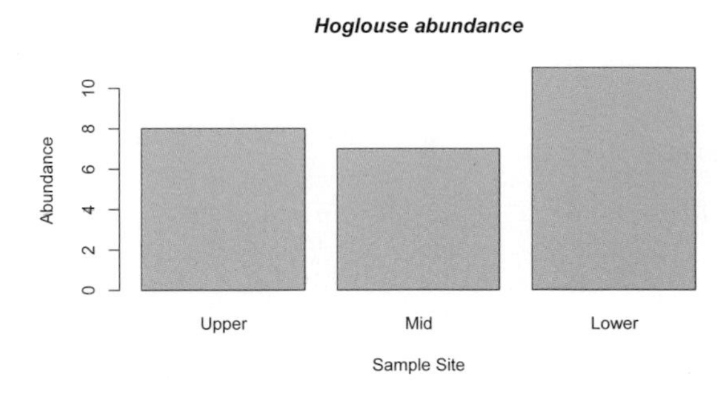

Figure 6.14 A bar chart from R with custom titles and colours.

```
> barplot(hlm, xlab = "Sample Site", ylab = "Abundance",
          col = "lightblue")
> title(main = "Hoglouse abundance", font.main = 4)
```

Note the `font.main` parameter in the `title()` command; a value of 4 gives italic font – try some other values.

Error bars using R

You can add error bars using R but there is no special command to do this (unlike Excel). What you must do is to use the `segments()` or `arrows()` commands to draw the appropriate sections of line onto the plot (see Table 6.6). In the following exercise you can have a go at drawing a bar chart with error bars for yourself.

Have a Go: Make a bar chart and error bars using R

You'll need the *hog3* data for this exercise. The data are in Table 6.7 but you can get them from the support website, where there are also instructions about how to import them into R. Alternatively the *S4E2e.RData* file already contains the data.

Download available from the support website.

1. If you import the data call the result *hog3*. Check the data look like the following:

```
> hog3
  Upper Mid Lower
1     3   4    11
2     4   3    12
3     5   7     9
4     9   9    10
5     8  11    11
6    10  NA    NA
7     9  NA    NA
```

2. Now compute the median values and store the result:

```
> hlm = apply(hog3, MARGIN = 2, FUN = median, na.rm =
TRUE)
> hlm
Upper   Mid Lower
    8     7    11
```

3. Work out the upper quartile (the 0.75 quantile) like so:

```
> up = apply(hog3, MARGIN = 2, quantile, prob = 0.75,
na.rm = TRUE)
> up
Upper   Mid Lower
    9     9    11
```

4. Now do something similar to step 3 but for the lower quartile (0.25 quantile):

```
> dn = apply(hog3, MARGIN = 2, quantile, prob = 0.25,
na.rm = TRUE)
> dn
Upper   Mid Lower
  4.5   4.0  10.0
```

5. The upper error bars will stick up above the bars so you need to make sure you know how tall the *y*-axis needs to be:

```
> max(c(hlm, up))
[1] 11
```

6. Now you can plot the bar chart and use the `ylim` parameter to ensure there is room for the bars. Note that you need to give the plot a name (see step 7 for the reason):

```
> bp = barplot(hlm, ylim = c(0,11), col = "tan")
> title(ylab = "Abundance", xlab = "Sample Site")
```

7. The `segments()` command can add bits of straight line but you would have to add the "hats" separately. So use the `arrows()` command and make

"flat" arrowheads for the "hats". The `arrows()` command requires the *x,y* co-ordinates of the start and end points. You can take the *x*-coordinates from the *bp* object (the barplot). Look at the *bp* result before you make the error bars:

```
> bp
     [,1]
[1,]  0.7
[2,]  1.9
[3,]  3.1
```

8. The *bp* result gives the *x*-coordinates of the centre of each bar. Use these along with the *up* and *dn* values (which will be the *y*-coordinates) to make the error bars (your final plot should resemble Figure 6.15):

```
> arrows(bp,up, bp,dn, length = 0.1, angle = 90,
code = 3)
```

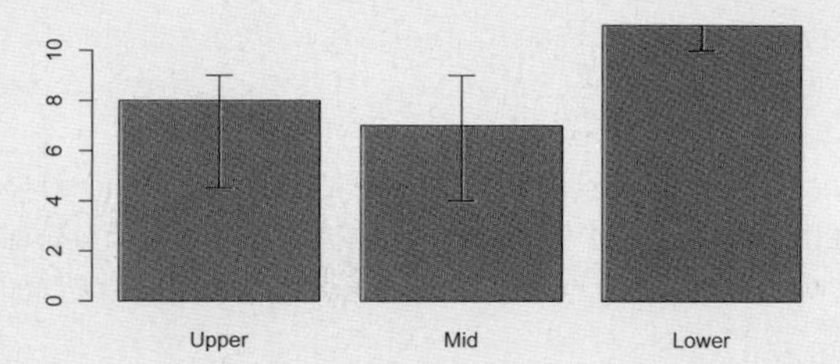

Figure 6.15 A bar chart with error bars. Bars show median values for freshwater hoglouse abundance. Error bars (drawn using `arrows()` command) show IQR.

The `length` parameter specified the size of the arrowheads (the error bar "hats"), the `angle` was set to 90° to make them "flat", and the `code = 3` part draws an arrowhead at both ends.

Tip: Resizing R plot windows

You can use the mouse to resize an R graphic window. The chart elements will resize to fit but the text will remain the same size. If you want a specific size of window then use the `windows()` command (or `quartz()` or `x11()` on Mac and Linux) to make a new plot window with `width` and `height` as you require.

Error bars are a bit fiddly but once you get the hang of it you can add them quite quickly. You could of course keep the commands in a text file and copy/paste them into R as necessary.

Multiple data series bar charts

When you have more than one data series you may well wish to plot them on the same chart. R treats each row as a separate data series and the columns as the categories. However, your data need to be in a special format. Look at the following example that uses the same data as in Table 6.10:

```
> butterfly
         1996 1997 1998 1999 2000
M.bro      88   47   13   33   86
Or.tip     90   14   36   24   47
Paint.l    50    0    0    0    4
Pea        48  110   85   54   65
Red.ad      6    3    8   10   15
Ring      190   80   96  179  145
> barplot(butterfly)
Error in barplot.default(butterfly) :
  'height' must be a vector or a matrix
```

You get an error because the data are stored in R as something called a `data.frame`. The `barplot()` command needs to operate on `matrix` objects (or single rows of data called `vectors`).

When you import data from a CSV file using the `read.csv()` command you get a `data.frame` as a result. You can use the `class()` command to check what kind of object you have and the `as.matrix()` command to convert a `data.frame` to a matrix.

```
> class(butterfly)
[1] "data.frame"
> bf = as.matrix(butterfly)
> bf
         1996 1997 1998 1999 2000
M.bro      88   47   13   33   86
Or.tip     90   14   36   24   47
Paint.l    50    0    0    0    4
Pea        48  110   85   54   65
Red.ad      6    3    8   10   15
Ring      190   80   96  179  145
> class(bf)
[1] "matrix"
```

You can see that the `matrix` looks exactly the same as the `data.frame`! However, there are internal differences and, although not apparent, these are important. So, if your `barplot()` command gives an error you should check the data and convert if necessary.

Once you have your data in the appropriate format you can plot them using the `barplot()` command. To get a clustered bar chart, with bars beside one another in groups, you need to add `beside = TRUE` to the command.

You can specify particular rows and/or columns by appending the row and/or column numbers in square brackets, `[rows, columns]`. Note that rows are always given first and if you leave an item blank R assumes you want all rows (or columns).

Note: Overwriting data

You can replace a data object with a new version of itself simply by assigning the result to the same name. For example:

```
> butterfly = as.matrix(butterfly)
```

This will replace the original (`data.frame`) version of the *butterfly* object with a new (`matrix`) version. There is no warning!

R expects the data series to be the rows but you can rotate (transpose) the data using the `t()` command, which simply swaps rows for columns and vice versa.

Adding a legend can be done with the `legend` parameter or afterwards with the `legend()` command. There are a host of parameters that attend the `legend()` command; type `help(legend)` to explore some of these yourself.

The `col` parameter can be used to specify colours for your bar chart. The default uses a palette of greys. There are built-in colour palettes and you can specify your own `palette()`; look at the `colors()` command for a list of colours that you can use.

Note: Using colour in graphs

On the support website you can find some additional notes about using colour in graphs and charts.

In the following exercise you can have a go at producing some bar charts for yourself, which will give you an opportunity to try out some of the graphical capabilities of R:

Have a Go: Use R for multiple series bar charts

You'll need the data from Table 6.10 for this exercise. The data are available to download on the support website as a file called *butterfly table.csv*. The website includes instructions about importing the data into R.

Download material available from the support website.

1. Import the data to R using the `read.csv()` command and call the result something sensible, *butterfly* will do; note that you'll need to add `check.names = FALSE` as a parameter when you import the data. Then have a look at the data.

```
> butterfly
         1996 1997 1998 1999 2000
M.bro      88   47   13   33   86
Or.tip     90   14   36   24   47
Paint.l    50    0    0    0    4
Pea        48  110   85   54   65
Red.ad      6    3    8   10   15
Ring      190   80   96  179  145
```

2. The data will be in the form of a `data.frame` so check, then convert to a `matrix`:

```
> class(butterfly)
[1] "data.frame"
> butterfly = as.matrix(butterfly)
```

3. Make a bar chart of the first row (the *M.bro* data):

```
> barplot(butterflyf[1,])
> title(xlab = "Sampling Year", ylab = "Abundance")
```

4. Now try a bar chart of the fifth row (*Red.ad*); this time use the name of the row:

```
> barplot(butterfly["Red.ad",])
> title(xlab = "Sampling Year", ylab = "Abundance")
```

5. Now plot all the species but for a single year (the first column):

```
> barplot(butterfly[,1])
> barplot(butterfly[,1], las = 1, ylab = "Abundance",
xlab = "Species")
```

6. Name a year explicitly:

```
> barplot(butterfly[,"1998"], main = "Year 1998",
font.main = 4)
```

7. Plot all the data and have the bars plotted beside one another using default colours; your plot should resemble Figure 6.16:

```
> barplot(butterfly, beside = TRUE)
> title(xlab = "Sample Year", ylab = "Abundance")
```

8. Add a legend and some colour. You will need to extend the *y*-axis to make room for the legend (Figure 6.17):

```
> barplot(butterfly, beside = TRUE, col = terrain.
colors(6),
          ylim = c(0,250), legend = TRUE,
          args.legend = list(bty = "n", x = "top",
ncol = 3))
> title(xlab = "Sample Year", ylab = "Abundance")
```

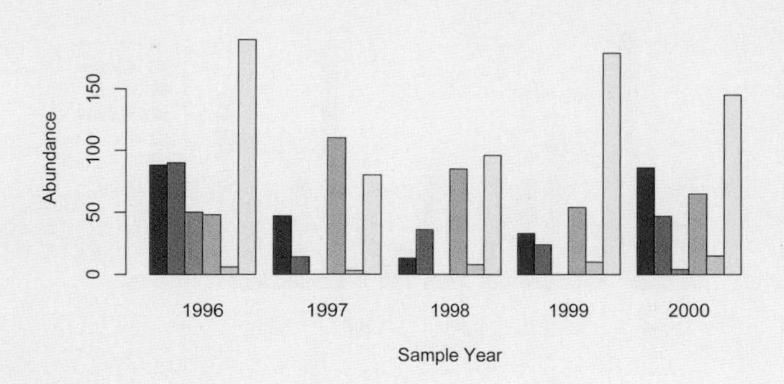

Figure 6.16 A clustered bar chart using defaults of the `barplot()` command. Axis titles added afterwards using the `title()` command.

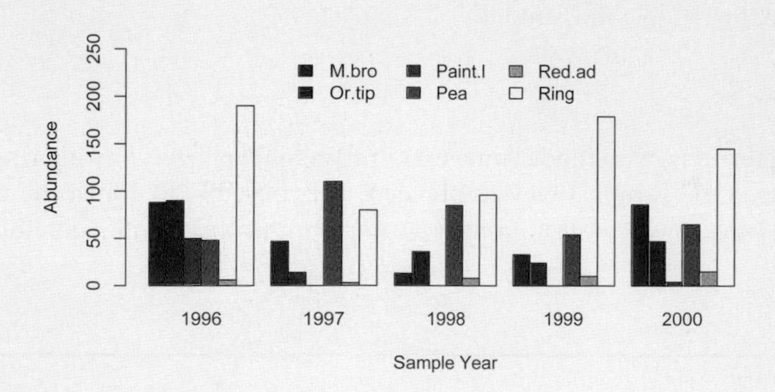

Figure 6.17 A clustered bar chart using customized colours and a legend.

The `terrain.colors()` command produces six colours from a special range of values. The `ylim` parameter allows you to specify the limits of the axis; you always give the start and end values in the form `c(start, end)`. The `args.legend` parameter allows you to pass parameters to the `legend()` command; note that this has to be in the form of a `list()`. The `bty` part specifies no box around the legend. The `x` part sets the location. The `ncol` part forces the legend to be in three columns.

9. Now try making your own list of colours to use. This time use the `xlim` parameter to leave room on the right for the legend. Your plot should resemble Figure 6.18:

```
> mycols = c("tan", "orange1", "magenta", "cyan",
"red", "sandybrown")
> barplot(butterfly, beside = TRUE, col = mycols,
        legend = TRUE, xlim = c(0,45))
```

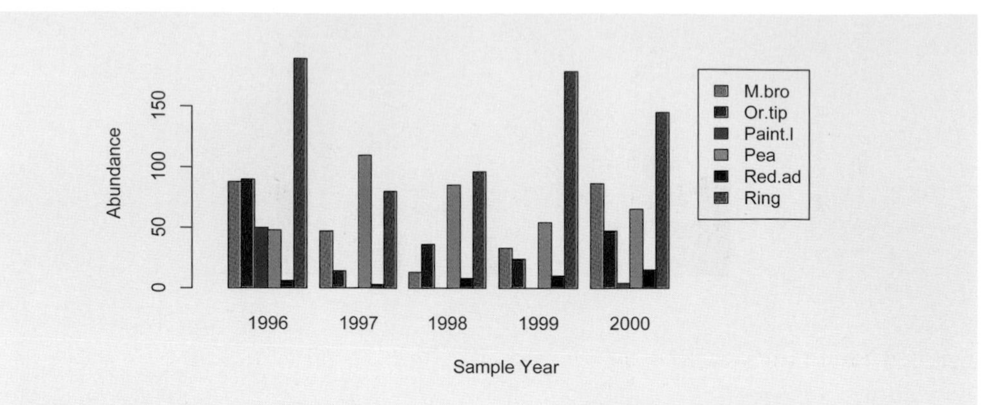

Figure 6.18 A clustered bar chart with custom colours. The *x*-axis has been resized to allow the legend to fit.

The `legend()` command is complicated! There are many options, which you can see by typing `help(legend)` into R.

In the preceding exercise you made some extra room for the legend by altering the scale of the axes. This is the simplest way to proceed. It is possible to adjust the margins of the plot and to force the legend to occupy a margin, but this is not something for the faint-hearted.

Note: Legends in R plots

See the support website for some more details about using legends in R plots.

Horizontal bar charts

Making the bars horizontal is easy: just add `horiz = TRUE` to the `barplot()` command. However, note that for the purposes of adding titles and so on the bottom side is still the *x*-axis and the left side is the *y*-axis!

Gridlines

You can easily add gridlines to your plot, so that the reader can assess the height of the bars more easily. You can use the `abline()` command, which adds straight lines to an existing chart (see Table 6.6). The command can use intercept and slope coefficients to produce a line of best fit but you can also use it to add horizontal or vertical lines. You can alter the style of lines too, using various additional parameters (many are in Table 6.5).

The gridlines will overplot the bars but with a little "trick" you can redraw the bars back over the top (Figure 6.19):

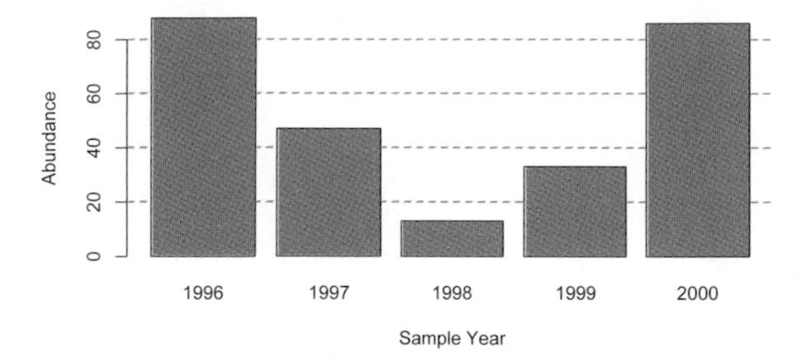

Figure 6.19 Gridlines added to a bar chart using the `abline()` command.

```
> barplot(bf[1,], col = "tan")
> title(xlab = "Sample Year", ylab = "Abundance")
> abline(h = seq(20,80,20), lty = "dashed", col = "gray50")
> barplot(bf[1,], col = "tan", add = TRUE)
```

Notice that in the example the `seq()` command was used to give a sequence of values:

```
seq(from, to, by)
```

You specify the start, end and interval. The line style (dashed) and colour (a mid grey) are specified in the `abline()` command.

The final line of the example is the same as the first except that `add = TRUE` was appended. This redraws the plot over the gridlines (you cannot easily draw the gridlines before the original bar chart).

Note: Gridlines in charts

On the support website you can find some additional notes about gridlines in graphs and charts.

6.3.2 Box–whisker plots to illustrate differences

A box–whisker plot (also called simply a boxplot) is a graph that conveys a lot of information in a small space. In general the boxplot displays the median values for data in sample groups, so enables you to compare samples. The variability is shown in two ways: the box part shows the inter-quartile range and the whiskers show the max–min (Figure 6.20).

In Figure 6.20 you can see the abundance of three samples of a freshwater invertebrate. The box–whisker plot enables you to see how the samples relate to one another as well as an impression of the general shape of the data distribution (Section 4.3).

Although boxplots use non-parametric data they are also used for data that are normally distributed. If a sample is normally distributed the mean and median will coincide and the inter-quartile range (IQR), whilst not directly analogous to standard error, gives a good impression of variability.

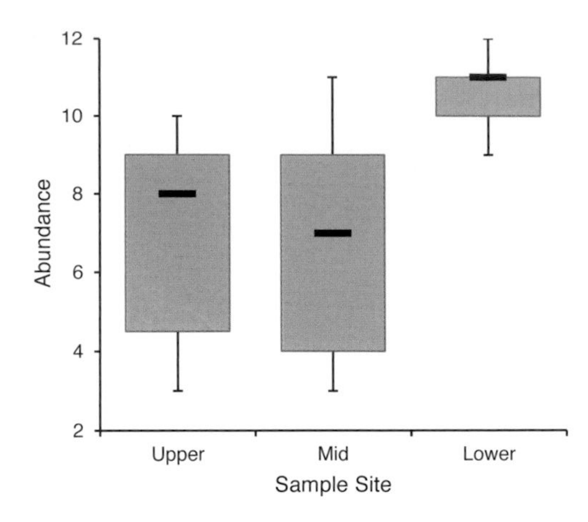

Figure 6.20 A box–whisker plot conveys a lot of information. Abundance of freshwater hoglouse at three sampling sites: points = median, boxes = IQR, whiskers = range.

> **Note: Parametric boxplots**
> It is possible to plot a box–whisker plot to show the actual parametric summary. Make the points the means, the boxes the standard error and the whiskers the standard deviation.

Both Excel and R can produce box–whisker plots, as you will see shortly.

Box–whisker plots in Excel

Excel is able to produce quite effective box–whisker plots (Figure 6.20 was produced using Excel 2013) but you need to use a little "persuasion" to produce the chart you require:

- Data need to be in a particular form and order.
- Use a Stock Chart (*Open-High-Low-Close*) via the *Insert* > *Other Charts* button (Figure 6.21).

Before you can make a box–whisker plot you'll need to prepare the appropriate summary statistics. This means you'll need to arrange your data into sample groups using a pivot table. There is no boxplot button in Excel so you'll have to press into service a different sort of chart, a *Stock Chart*. The type you need is called an *Open-High-Low-Close* chart and requires four values (as the name suggests). The four values correspond to the summary statistics like so:

- Open – forms one edge of the "box", you require one of the inter-quartiles.
- High – forms a point (usually not displayed), you require the median.

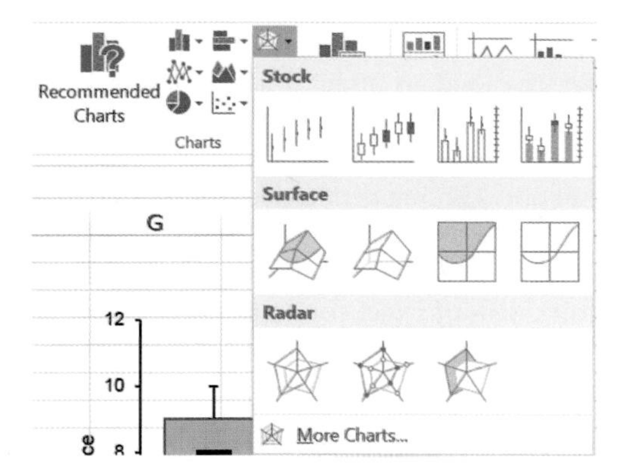

Figure 6.21 You can make a box–whisker plot using a Stock chart, which is in a section along with other miscellaneous chart types.

- Low – forms a point (usually not displayed), you require the median.
- Close – forms one edge of the "box", you require one of the inter-quartiles.

Essentially you use the *Open* and *Close* values to form the box part. You use the *High* and *Low* values to make a point by displaying only one of the two values. The whiskers are formed by using error bars. To work out the size of the error bars you'll have to work out the difference between the median and the max and min values. You did something similar earlier when you built a bar chart with error bars.

The data for the hoglouse boxplot (Figure 6.20) were prepared and laid out in the spreadsheet as shown in Table 6.11.

Table 6.11 Data prepared for assembly as a box–whisker plot need to be in a particular layout. The first four rows are used for the main chart (boxes and median points). The last four rows are used to make the error bars (the whiskers). Each column forms a sample group.

	Upper	Mid	Lower
UQ	9	9	11
Median	8	7	11
Median	8	7	11
LQ	4.5	4	10
Max	10	11	12
Min	3	3	9
Up	2	4	1
Dn	5	4	2

You can see from Table 6.11 that you need to calculate all the quartiles (the median twice). You'll use the inter-quartiles and the median(s) to make the main boxes. The max and min values are needed to construct the error bars (the deflections from the median to the extremes).

In order to build your box–whisker chart you'll need to follow these steps:

1. Start by preparing your data. You'll need four sets of summary statistics – the upper quartile, the median, the median again and the lower quartile. The max and min values are not required directly; you need them solely to calculate the error bars, which are added after the chart is made.
2. Your data are best set out with the different samples in columns, with rows for each of the summary statistics (as in Table 6.11).
3. Highlight the data that form the summary statistics, including their label headings. This is where the quirkiness of Excel comes to the surface. You have to highlight the four rows of summary statistics of course. You also have to have highlighted at least four columns of values (i.e. five columns including the labels). This means you may have to select empty spreadsheet cells!
4. Once you've highlighted the "correct" number of columns and rows you can click the *Insert > Other Charts* button. Then select the *Stock* chart that is labelled *Open-High-Low-Close* – this is the second one along.
5. You now get a chart but there will probably be additional blank data series. You'll need to edit the chart extensively (Figure 6.22).

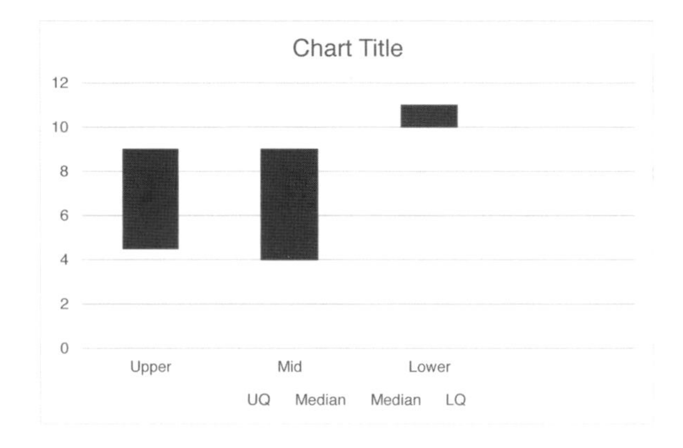

Figure 6.22 The starting point for a box–whisker plot, a stock chart with little formatting, which will need extensive editing.

6. Use the *Select Data* button to open the *Select Data Source* menu window. In the *Chart data range* box you'll see the currently selected data. Delete this and highlight all the data and their labels. Click *OK* and your chart will be updated and the empty data series gone.
7. You should now have a "primitive" chart. Most likely all you'll see are black rectangles, which represent the inter-quartile ranges of the various sample groups. Click on the chart to activate the *Chart Tools* menus. Then go to the *Format* menu. On the left you will be able to select the various chart elements; this is a good deal easier than trying to click on the items directly in the chart.
8. Select the item *Down-Bars 1* then click the *Format Selection* button. You want to set the fill colour of the boxes to something lighter. From the *Fill* section choose *Solid fill* and then pick a medium grey. Click *Close* once you are done and now the boxes will appear in your selected colour.
9. Now select the item *Series "Median"* from the list in the *Layout* menu. There will be two items; select the top one. You want to make a marker appear so click *Format*

Selection then go to the *Marker* section. You'll see that the *None* option is selected. You can use an *Automatic* marker or choose one yourself; click the *Built-in* option then change to a type you prefer (I like a kind of dash). Then change the *Marker Fill*, make it *Solid fill* and choose a dark colour (black is simplest and clearest). Now go to the *Border* section and set that to *Solid line* (or *None*) and change the colour. Your chart now has median markers and inter-quartile boxes.

10. Now you can add the error bars. Click on the chart then go to the *Design* menu. Click the *Add Chart Element* > *Error Bars* button and choose *More Error Bars Options*. If you use the + button as a shortcut to add the error bars you'll have to choose a data series to base the bars on. Select *Median*.
11. Click the *Custom* option then the *Specify Value* button to open the *Custom Error Bars* menu window (Figure 6.23).

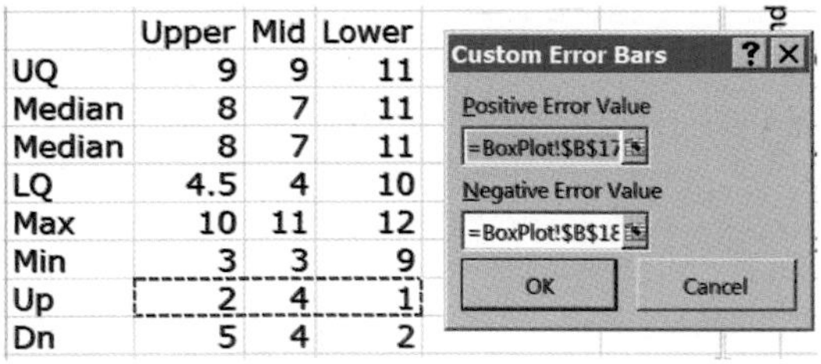

	Upper	Mid	Lower
UQ	9	9	11
Median	8	7	11
Median	8	7	11
LQ	4.5	4	10
Max	10	11	12
Min	3	3	9
Up	2	4	1
Dn	5	4	2

Custom Error Bars
Positive Error Value
=BoxPlot!$B$17
Negative Error Value
=BoxPlot!$B$18
OK
Cancel

Figure 6.23 The *Custom Error Bars* option allows you to choose the values for the error bars explicitly.

12. Use the mouse to select the appropriate values for your error bars. Then click OK to return to the *Format Error Bars* menu window. You can apply other formatting as you like (e.g. make the bars wider).
13. Apply any other formatting you need to make your chart – for example, you can delete the legend and make the gridlines dashed.

The process is slightly involved but once you have the hang of it things are relatively straightforward.

Box–whisker plots in R

The general command to draw box–whisker plots in R is `boxplot()`. R is extremely capable when it comes to boxplots and it is fairly evident that the programmers lean towards this style of graph. The general form of the `boxplot()` command is as follows:

```
boxplot(x, range, col, names, horizontal, add)
```

`x`	The data to be plotted. If `x` is a single sample (a `vector` of values) you will get a single box–whisker. If `x` is a 2-D data object the columns will be treated as separate samples. You can also specify data in the form `y ~ x` where `y` is a response variable and `x` one or more predictors.

`range= 1.5`	The range that the whiskers will cover. The default is to extend 1.5 times the IQR. Any data outside that range are shown as points (outliers). Set `range = 0` to use max–min and to extend the whiskers to their fullest extent.
`Col`	The colour to use in the boxes. The default is for no colour. You can specify colour as a name (in quotes) or as a number, in which case the colour will be taken from the current colour palette.
`Names`	In most cases the names of the boxes are shown on the x-axis. You can specify alternative names using this parameter.
`horizontal = FALSE`	By default the box–whiskers are drawn vertically. If you specify `horizontal = TRUE` they are horizontal. In this case the bottom axis is still the x-axis and the left axis is the y-axis as far as labelling is concerned (e.g. `xlab` labels the bottom axis).
`add = FALSE`	If you specify `add = TRUE` the box–whiskers are drawn onto an existing plot.

There are other graphical parameters you can specify (see Table 6.5); you'll see some examples shortly.

The `boxplot()` command accepts input data in several forms:

- Separate samples.
- Multiple samples arranged by column.
- Multiple samples specified by a response and one or more predictor variables.

In the simplest case you have samples as separate entities; you simply give their names, separated by commas:

```
> sunny
 [1] 17 26 28 27 29 28 25 26 34 32 23 29 24 21 26 31 31 22
[19] 26 19 36 23 21 16 30
> shady
 [1] 18 21 18 51 46 47 50 16 19 15 17 16 52 17 15 16 17 49
[19] 48 16 18 16 17 15 20
> boxplot(sunny, shady, col = "bisque", names = c("Sunny",
"Shady"))
> title(xlab = "Pond site", ylab = "Size in mm")
```

The preceding example produces a plot resembling Figure 6.24. Note that in order to get the samples labelled appropriately the `names` parameter was used. Generally you only have to do this when you have separate samples or you want to use names other than the default (which are taken from the variable names, if present). When you have a multi-sample data object you can plot the data directly.

```
> boxplot(hog3, col = "tan", xlab = "Site", ylab = "Abundance")
```

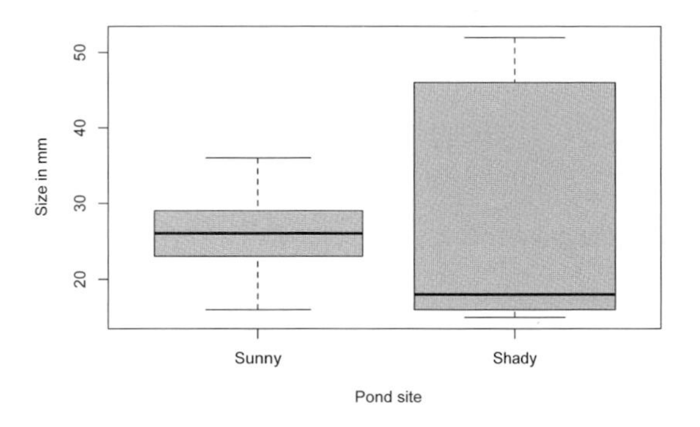

Figure 6.24 A boxplot of two samples of freshwater beetle from different ponds. The stripe shows the median (in mm), boxes show IQR and whiskers the range.

Notice that the names are not specified but are taken from the column headings (you've seen the *hog3* data previously). The titles are also specified as part of the command but could be added afterwards using the `title()` command. The resulting plot looks like Figure 6.25.

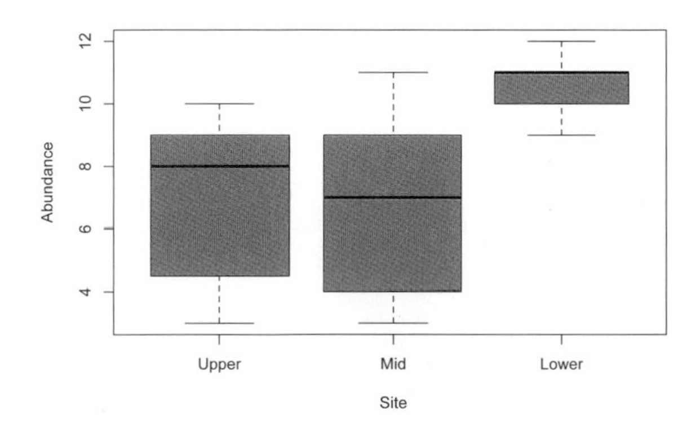

Figure 6.25 A multi-sample data object can be plotted easily using the boxplot() command in R. Abundance of freshwater hoglouse from three sites, data shown: median, IQR, range.

When you have data in a more scientific recording layout, with a column for the response variable and one or more columns for predictor variables, then you can specify the data to plot using a *formula*.

Look again at the *hog2* data:

```
> head(hog2)
  count  site
1     3 Upper
2     4 Upper
3     5 Upper
4     9 Upper
```

```
5      8 Upper
6     10 Upper
```

There is a column for *count* (the response) and a column for *site* (the predictor).

> **Note: Quick data views in R**
> The `head()` command shows the top few rows of a data item (try `tail()` to see the last few rows).

If you simply give the data name (*hog2*) to the `boxplot()` command you'll get a rather strange plot; the command will attempt to plot the two columns as if they were samples, which is plainly incorrect (try it and see).

What you must do is to specify the data as a formula `y ~ x`, where `y` is the response and `x` is the predictor. You can (and generally should) also specify `data = xxx`, where `xxx` is the name of the object that contains the data variable. The following example produces a plot that looks much like Figure 6.26.

```
> boxplot(count ~ site, data = hog2, col = "lightblue")
> title(xlab = "Pond site", ylab = "Size in mm")
```

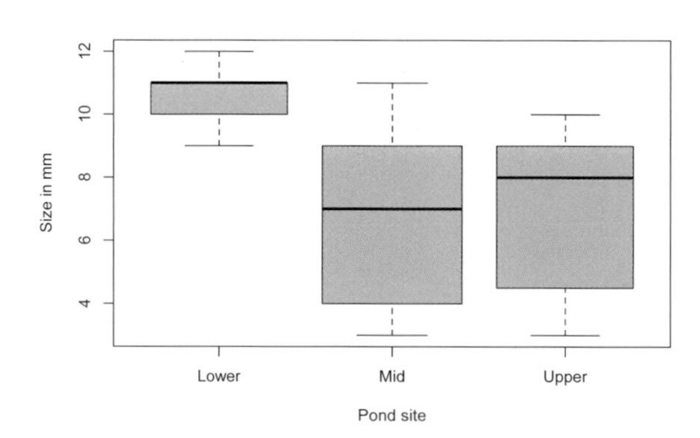

Figure 6.26 Abundance of freshwater hoglouse from three sites, data shown: median, IQR, range. This boxplot was drawn by specifying data in the form of a formula. Note that the samples are in alphabetical order.

Compare Figure 6.26 with Figure 6.25 and you'll see that the samples are in a different order. When the data are in separate columns they are plotted in the order they appear, from left to right. When data are in scientific recording format the data are plotted in alphabetical order. It is possible to alter the order of the plotting but that is a bit more advanced.

> **Note: Ordering data in a boxplot()**
> See the support website for more details about how you can reorder the data when using the R `boxplot()` command.

Horizontal boxplots in R

Sometimes it is useful to have the bars horizontal; this is simply achieved by using `horizontal = TRUE` as part of the command:

```
> bf
        1996 1997 1998 1999 2000
M.bro     88   47   13   33   86
Or.tip    90   14   36   24   47
Paint.l   50    0    0    0    4
Pea       48  110   85   54   65
Red.ad     6    3    8   10   15
Ring     190   80   96  179  145
> boxplot(bf, horizontal = TRUE, las = 1, col = "pink")
> title(xlab = "Total butterfly abundance")
```

In the preceding example the samples are the columns of the butterfly data that you've seen previously. The resulting plot looks like Figure 6.27.

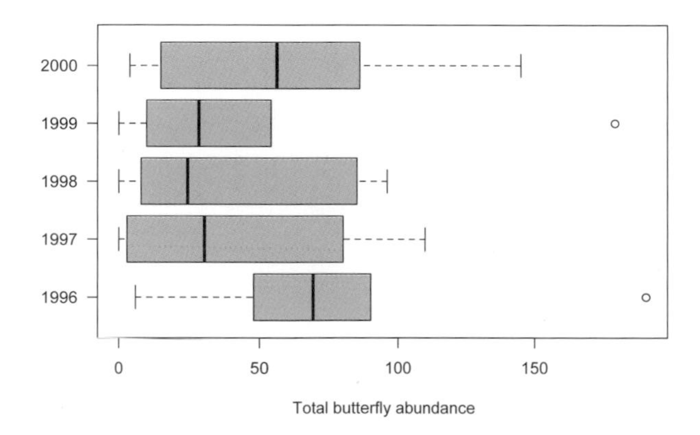

Figure 6.27 Horizontal boxplot; note that the bottom axis is still regarded as the *x*-axis by graphical parameters.

Note that the `xlab` parameter of the `title()` command results in a label to the bottom axis. Technically this is the *y*-axis because it shows the response variable but the parameter does not "know" that.

Notice how the whiskers in Figure 6.27 do not necessarily cover the entire max–min range; there are some points outside the whiskers. This is because the default is for the whiskers to extend to 1.5 times the inter-quartile range. If you want the whiskers to definitely extend to the max–min use `range = 0` in the command, e.g.:

```
> boxplot(bf, horizontal = TRUE, las = 1, col = "pink",
range = 0)
```

It is useful to keep the default, because then you can see data that is a bit "way out". You can easily replot your boxplot by using the up arrow to recall the previous command and append the `range = 0` parameter.

6.4 Graphs to illustrate correlation and regression

A correlation describes the strength and direction of the relationship between two variables. Regression is similar but allows you to describe the mathematical relationship as well as allowing more variables (see Chapter 11). The usual way of showing the link between two variables is with a *scatter plot* (sometimes known as an *x,y* chart). In a scatter plot the *y*-axis is used to display the response variable (the dependent) and the *x*-axis is used for the predictor (independent).

In some cases your predictor (*x*-axis) will be time; in which case you may use something called a *line plot*. The line plot looks at first glance like a scatter plot with the dots joined up. The idea is that the points are joined with a line so you can see how the response variable changes over time. You saw an example of a line plot earlier, when you looked at the running mean (Section 4.7). However, the line plot is more akin to a bar chart than a scatter plot. The *x*-axis is split into equal intervals, like the categories of the bar chart.

It is important to recognize the difference between a true (*x*,*y*) scatter plot and a line chart, as they can look very similar. In fact a scatter plot can have a line with no points, and a line plot can have points with no line! Both Excel and R can draw both types of graph, as you'll see shortly.

6.4.1 Correlation plots using Excel

Excel can produce both scatter plots and line charts; the *Charts* section of the *Insert* menu has a button pertaining to each type (see Figure 4.6). In general the difference between the two chart types will be how your data are arranged.

Scatter plots using Excel

In order to make a scatter plot Excel assumes that your data are in columns, with the *x*-variable first (the predictor) and the *y*-variable second (the response). If your data are in a different order (e.g. Table 6.12) you will not be able to make a chart simply by highlighting the data (as the axes will be transposed).

Table 6.12 The abundance of a freshwater invertebrate and stream speed. A scatter plot would visualize these data but Excel will assume the first column (Abund) is the *x*-variable, which is incorrect.

Abund	Speed
12	18
18	12
17	12
14	7
9	8
7	9
6	6
7	11
5	6
3	2

If you wanted to make a scatter plot of the data in Table 6.12 you have three main options:

- Edit the data so that the columns are in the "correct" order, i.e. *Speed* first (x-variable) then *Abund* (y-variable).
- Highlight the data as they are, which will make the chart with the data the wrong way around. Then edit the chart so that the data are represented correctly.
- Make an empty chart (i.e. one without any data) then select the data afterwards, thus choosing the x and y variables you require.

If you decide to swap the order of the columns then you'll end up with the data in the "correct" order for Excel. Then:

1. Highlight the data, including the column headings.
2. Click the *Insert* menu and select the *Scatter* button; there are several options (Figure 6.28). Usually the first option is best (unmarked scatter) but you can also choose to "join the dots"; it is possible to add connecting lines afterwards (or remove them) so stick to a plain chart to start with.

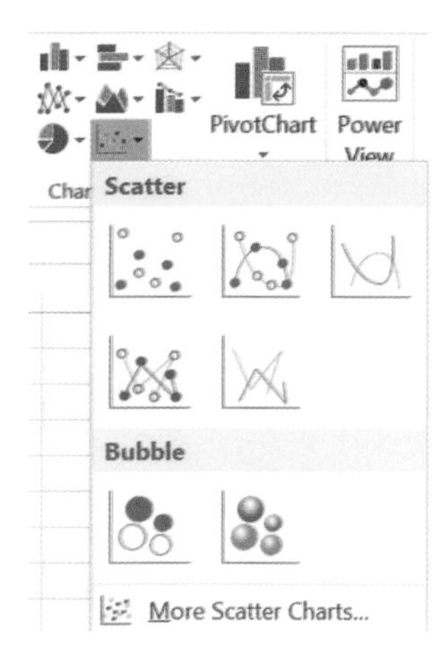

Figure 6.28 The *Insert* > *Scatter* button presents you with several variants of the scatter plot. Generally the unmarked scatter plot is most useful, as connecting lines can be added afterwards.

3. Choose the plain *Scatter* option and your chart will be created immediately.
4. Use the *Chart Tools* menus to edit and alter your chart as required (see Table 6.3).

If you don't rearrange the data, and keep the columns in the "wrong" order, you can change the chart afterwards like so:

1. Highlight the data then click the *Insert* > *Scatter* button and make a plain unmarked scatter plot.
2. Click once on the chart to ensure the *Chart Tools* menus are activated then click the *Select Data* button. This will open the *Select Data Source* dialogue box (Figure 6.29).

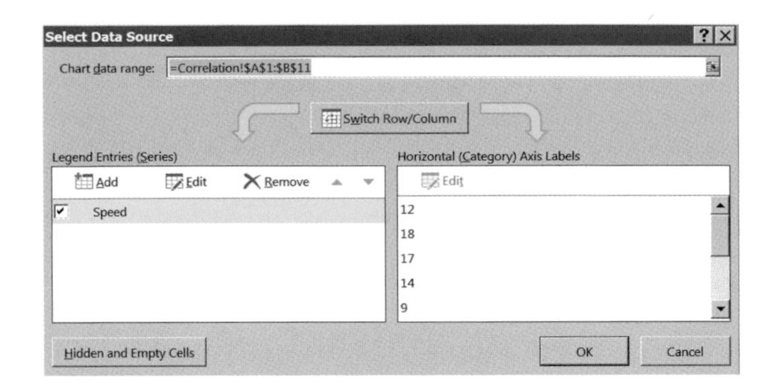

Figure 6.29 The *Chart Tools > Design > Select Data* button allows you to edit or add chart data.

3. Now click the *Edit* button in the *Legend Entries (Series)* section and you can select the correct data for the *x* and *y* axes (Figure 6.30).

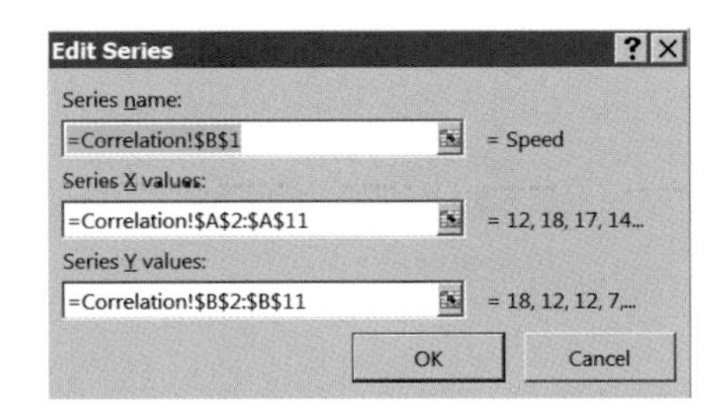

Figure 6.30 The *Edit Series* dialogue box allows you to choose the data to be plotted.

4. Select the appropriate data for each of the boxes in the *Edit Series* dialogue box. Click *OK* to return to the previous dialogue, then *OK* again to finish. Your chart should now be presented with the data on the correct axes.

If you choose to make an empty chart then you proceed along these lines:

1. Click once anywhere in the worksheet that is not in or adjacent to any data. Excel will "search" around the current cell and automatically select data, so by clicking away from any potential data you ensure an empty plot.
2. Click the *Insert > Scatter* button and choose an unmarked *Scatter* plot.
3. Click the *Chart Tools > Design > Select Data* button to open the *Select Data Source* dialogue box (see Figure 6.29).
4. Now click the *Add* button in the *Legend Entries (Series)* section to open the *Edit Series* dialogue box (see Figure 6.30). You can now use the mouse to select the data you want for each axis. Click *OK* to return to the previous dialogue then *OK* again to finish. You should now have a complete plot.

Whichever route you choose you will need to use the *Chart Tools* menus to edit the chart (see Table 6.3). You can add axis titles, change the plotting symbols, alter colours and so on. The data from Table 6.12 are visualized in Figure 6.31.

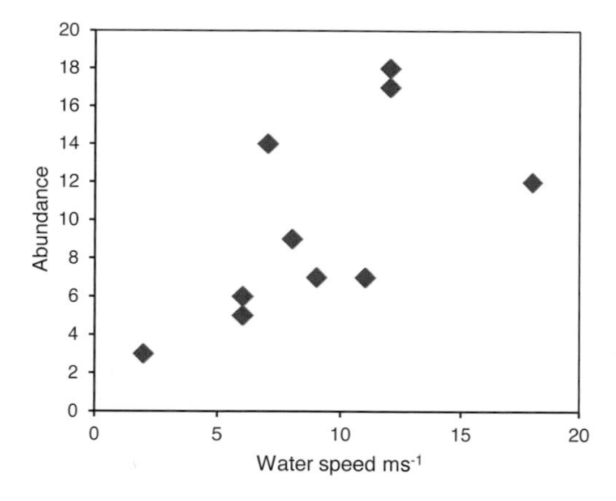

Figure 6.31 The abundance of a freshwater invertebrate and stream speed. This scatter plot was produced using Excel 2013.

Line plots using Excel

Line plots are most useful for data that are collected at discrete time intervals. The *Line Chart* button on the *Insert* menu will make line plots for you (recall Figure 4.23). The layout of the data will affect how the chart appears; by default the rows are assumed to be the categories (usually the time periods) and the columns are taken as the samples (called data *series* in Excel). However, it is easy to switch the chart around using the *Chart Tools > Design > Switch Row/Column* button.

In the following exercise you can have a go at making a line chart for yourself using the butterfly data from Table 6.10.

Have a Go: Use Excel to make a multi-series line chart

You'll need the data from Table 6.10 for this exercise. The data are available to download on the support website as a file called *butterfly table.xlsx*.

Download material available from the support website.

1. Open the spreadsheet *butterfly table.xlsx*; there is a single worksheet that contains the data. Each row gives the abundance for a butterfly species and each column contains data for a single year. Note especially that there is a blank cell in A1.
2. Click once anywhere in the block of data (there is no need to highlight/select any data). Now click the *Insert > Line* button and select a *Line* chart with markers. The chart will be created immediately.
3. Because the rows each contain data for a species the lines on the chart represent a single year. Essentially this chart is the "wrong" way around. Click once on the chart to enable the *Chart Tools* menus. Now click the *Design > Switch Row/ Column* button.

4. You should now have a chart with a line for each species (row). The x-axis will represent the time periods. Use the *Chart Tools* menus to edit the chart and improve its appearance as you like. Your completed chart should resemble Figure 6.32.

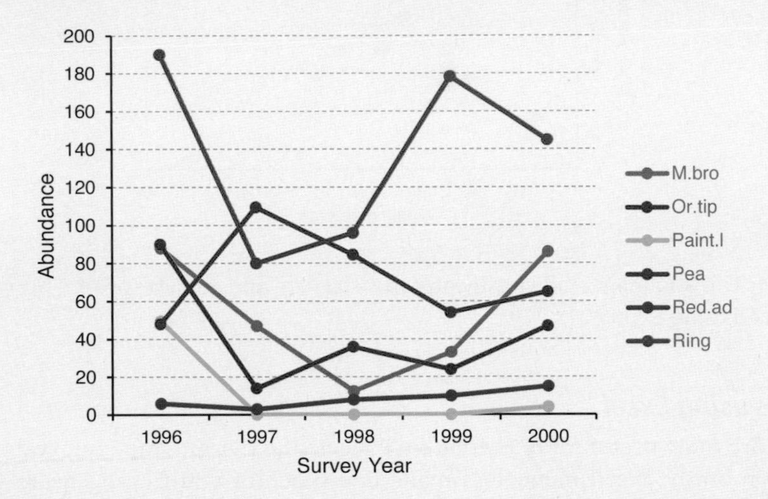

Figure 6.32 This line chart shows the abundance of several butterfly species over several years.

The line plot you created is somewhat congested; with so many lines it can be hard to read. In Excel 2013 you can use the *Select Data* button to open the *Select Data Source* dialogue box. This allows you to untick some of the data series and so reduce clutter. In earlier versions of Excel you can only delete the data series from the chart; with Excel 2013 you can reinstate them by reticking the appropriate box.

The line plot shown in Figure 6.32 has tick-marks displayed on the x-axis; notice how these lie between the labels. This is a good way to see if your chart is a line plot or a scatter plot because in the latter the tick-marks always lie alongside the x-axis labels (see Figure 6.31 for a comparison).

6.4.2 Correlation plots using R

The `plot()` command in R is a very general one and allows you to produce both a scatter plot and a line plot. The general form of the command is:

```
plot(x, y, type = "p", ...)
```

`x, y` The coordinates to plot. You can specify x- and y-coordinates in several ways. The simplest is to give x, y variables separately. You can also give a formula, e.g. `y ~ x`, and optionally specify where these variables are contained (e.g. `data = my.data`).

`type = "p"`	The type of plot to create, the default `type = "p"`, produces points. Use `"l"` for lines only, `"b"` for both (points with line segments between) or `"o"` for over-plotted (points and lines).
`...`	Other graphical parameters, such as those for axis labels, colour and so on (see Table 6.5).

The `plot()` command is very general and it's possible to make special plotting routines to handle all sorts of results. Many of the statistical routines create objects that will produce some special sort of graphical output and by default `plot()` is the one they use. When it comes to making scatter plots or line charts you can specify your data in several ways; most commonly you'll either:

- Specify the *x* and *y* variables separately.
- Give a formula containing the variables, in the form `y ~ x`.

You can add a host of additional graphical parameters to your `plot()` command (see Table 6.5), which can alter the appearance of your plot substantially.

The `plot()` command can produce scatter plots or line plots simply by altering the `type =` parameter, as you'll see shortly.

Scatter plots using R

A scatter plot requires numeric data, with one variable plotted on the *y*-axis and one variable plotted on the *x*-axis. Consider the data from Table 6.12, which show the abundance of a freshwater invertebrate and stream speed. These data are available from the companion website and are called *freshwater correlation.xlsx*. You can easily transfer them into R but the *S4E2e.RData* file also contains the data as an object called *fw*.

Data available from the support website.

```
> fw
   abund speed
1     12    18
2     18    12
3     17    12
4     14     7
5      9     8
6      7     9
7      6     6
8      7    11
9      5     6
10     3     2
```

You can use the `plot()` command in several ways to produce a basic scatter plot.

```
> plot(fw)
```

If you type the name of the data into the command, R will recognize that there are two variables but it takes the first column as the x-variable and the second as the y-variable, which is not correct in this instance.

You need to specify the variables more explicitly:

```
> plot(speed, abund)
Error in plot(speed, abund) : object 'speed' not found
```

This results in an error! Of course the variables are "contained" within the *fw* data item so you need to tell R how to get the variables more explicitly than the previous command. There are several ways you can get your plot; the following all produce more or less the same plot (Figure 6.33).

The first method is to use the `$` to join the variable and data object names:

```
> plot(fw$speed, fw$abund)
```

The next method uses `attach()` to "open" the data object. It is a good idea to use `detach()` when you are done:

```
> attach(fw)
> plot(speed, abund)
> detach(fw)
```

The `with()` command is like `attach()` but is temporary, lasting only for the duration of the current command:

```
> with(fw, plot(speed, abund))
```

Specifying the variables in the formula syntax is probably the best method:

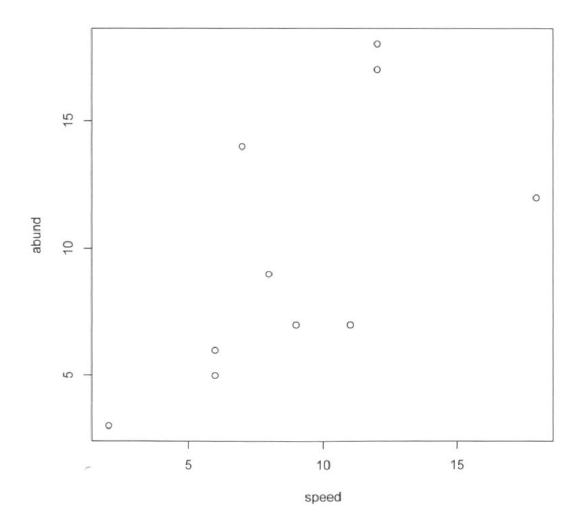

Figure 6.33 Abundance of a freshwater invertebrate and water speed. This scatter plot was drawn in R using the basic `plot()` command.

```
> plot(abund ~ speed, data = fw)
```

The plot is somewhat primitive but you can easily add other parameters to the `plot()` command to alter the basic appearance (see Table 6.5).

You can alter plotting symbols (i.e. plotting characters) using the `pch` parameter. Numerical values from 0 to 25 result in "standard" characters (see Figure 6.34). Values from 32 upwards produce their ASCII equivalent. You can also specify a symbol in quotes. So, if you can type it from the keyboard you can plot it on the chart.

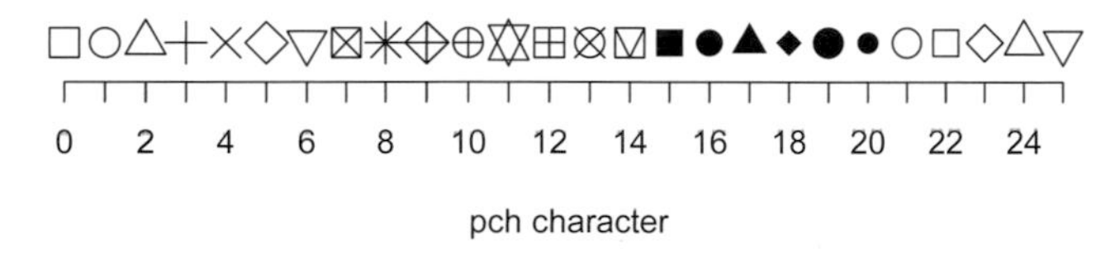

Figure 6.34 R standard plotting characters relating to pch values 0 to 25.

The size of the symbols is controlled using the `cex` parameter. The default is 1; to make the characters larger or smaller use a value >1 or <1, respectively.

The colour of the symbols is controlled using the `col` parameter. Usually this is a named colour (in quotes); type `colors()` into R for a (long) list of available colours.

The axis titles are generally taken from the names of the variables. You can override these using the `xlab` and `ylab` parameters. If you want to specify the titles separately (using the `title()` command), you must use an "empty" title, e.g. `xlab = ""`, in the `plot()` command first.

To create a title with superscripts a special command, `expression()`, is used. In the `expression()` command a ~ acts as a space and anything following a ^ is treated as superscript.

These supplementary parameters can be put together to produce a plot resembling Figure 6.35.

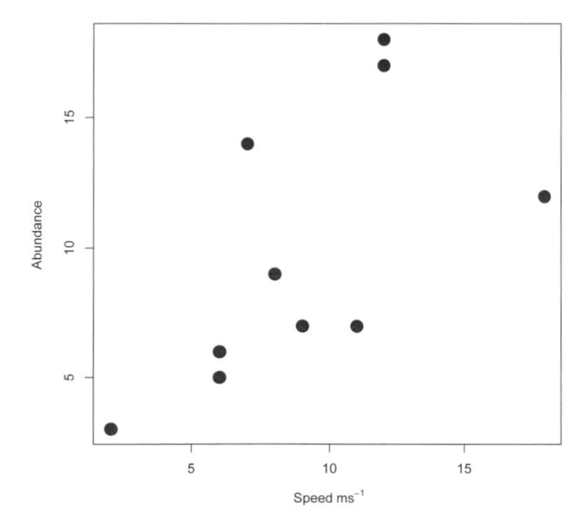

Figure 6.35 Abundance of a freshwater invertebrate and water speed. The `plot()` command has been modified to produce a graph with customized plotting symbols, size, colour and axis labels.

```
> plot(abund ~ speed, data = fw, xlab = "", ylab = "Abundance",
col = "blue", pch = 19, cex = 2)
> title(xlab = expression(Speed~ms^-1))
```

It is possible to add lines of best fit to scatter plots. There are several ways to do this, depending on the sort of data you have. You'll see these methods illustrated in the chapters dealing with regression (Chapters 8 and 11).

> **Note: The expression() command**
>
> The `expression()` command allows complicated text strings to be constructed; go to the support website for more details.

If you need to add other data series to your plot you can use the `points()` command. This operates more or less like `plot()` but only makes the points and does not draw the axes. You'll see `points()` and the closely related `lines()` command shortly.

Line plots using R

Line plots are generally used to show data that was collected across time intervals. They are really more analogous to a bar chart than a scatter plot but their appearance is more commonly confused with a scatter plot. Since the `plot()` command in R can produce both scatter and line plots it seems sensible to explore line plots here.

Usually your data will be in a form where you have a numerical variable and a date-type variable. This date variable could be in the form of a number but often it will be a month name, which can be "problematic" to deal with. The following data show the size of some Sitka spruce trees over time (the data have been modified from some example data from within R itself).

The data are available on the support website as part of the *S4E2e.RData* file, which will load directly into R. The data are named *tree*.

```
> tree
  month     size
1   Mar 5.508333
2   Apr 5.521296
3   May 5.671667
4   Jun 5.905926
5   Jul 6.045370
6   Aug 6.117407
7   Sep 6.127037
8   Oct 6.132037
```

The *size* variable shows the logarithm of the height multiplied by the diameter squared. The *month* variable shows the month of the observation. If you try a `plot()` command on these data you'll end up with something resembling Figure 6.36.

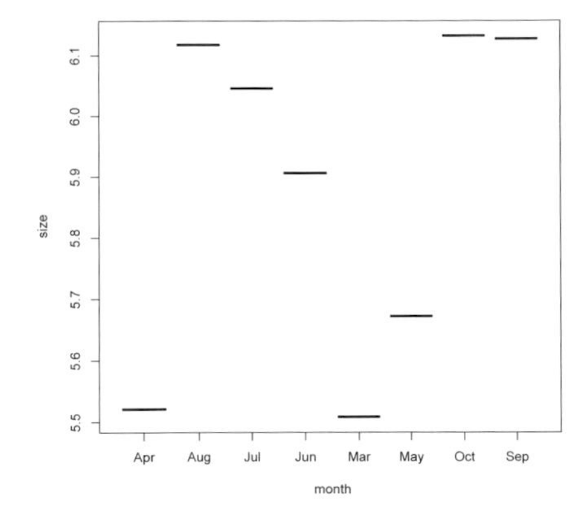

Figure 6.36 Size of Sitka spruce trees log(diameter^2 × height) measured at different months. Because one variable is numeric and the other text (month names) R attempts a box–whisker plot.

```
> plot(tree)
```

Figure 6.36 does not look right at all! The months have been rearranged (alphabetically) and the data points look a little strange. What R has done is to attempt a box–whisker plot (see Section 6.3.2), but because there is only one value per month all you see is the median stripe.

What you must do is to plot the numeric data by itself. If you try it you'll get the points to look okay but the x-axis will simply be an index because nothing was specified (Figure 6.37).

```
> with(tree, plot(size))
```

Now at least you can see that the data are in the right order (look back at the data values). To make this into a line plot you need to specify `type = "l"`, which will produce a line

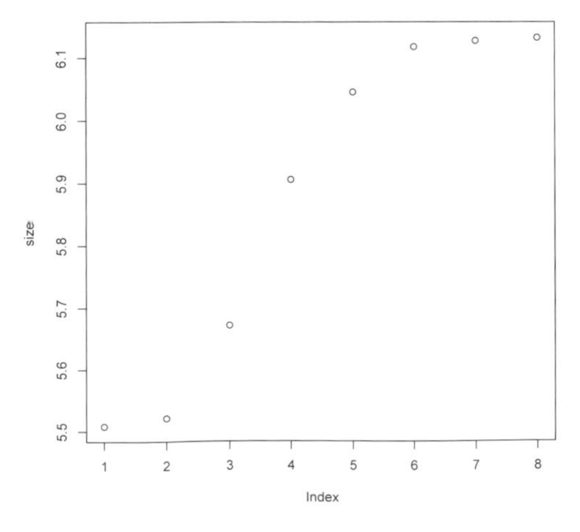

Figure 6.37 Plotting a single numeric variable results in a scatter plot with the x-axis as a simple index.

```
Red.ad     6    3    8   10   15
Ring     190   80   96  179  145
```

2. Start by making a line plot of the first row, which corresponds to the *M.bro* data.

```
> plot(bf[1,], type = "b", xlab = "", ylab = "",
axes = FALSE)
> axis(2)
> axis(1, at = 1:5, labels = colnames(bf))
> title(xlab = "Survey Year", ylab = "Butterfly
Abundance")
```

3. Now add a second data series using the *Or.tip* data row:

```
> points(bf["Or.tip",], type = "b", lty = 2)
```

4. Add a simple legend to the top of the plot to end up with a plot that resembles Figure 6.39:

```
> legend("top", legend = c("Meadow brown", "Orange
tip"), lty = 1:2, box.lty = 0)
```

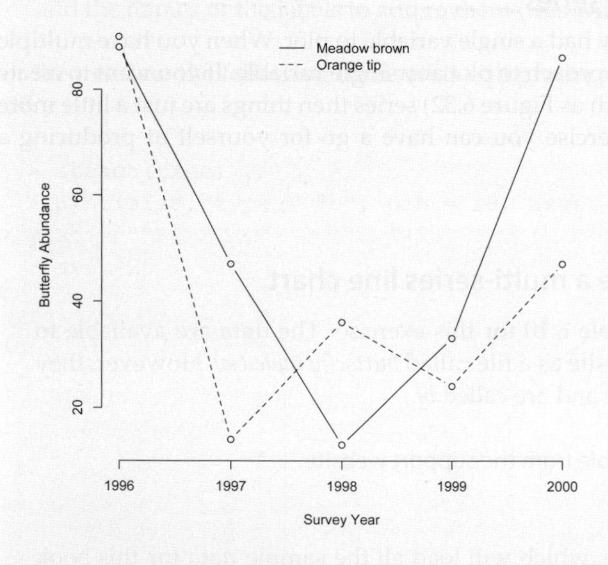

Figure 6.39 A simple multi-series line plot to show abundance of two butterfly species. The second data series was added using the `points()` command.

5. Work out the extent of the *y*-axis across the entire dataset:

```
> range(bf)
[1]   0 190
```

6. Now make a new empty plot using appropriate ranges for the axes:

```
> plot(10,10, type = 'n', xlim = c(1,6),
ylim = c(0,200), axes = FALSE, ylab = "Abundance",
xlab = "Year")
```

Note that you used a range of 0–200 for the *y*-axis so that all the data would fit in the plot. You used a range of 1–6 for the *x*-axis so that there is room for 5 years of data and a bit extra to accommodate the legend.

7. Now add the axes:

```
> axis(side = 1, at = 1:5, labels = colnames(bf))
> axis(side = 2)
```

8. You could add the rows of data one at a time but try a loop:

```
for(i in 1:6) {
        lines(bf[i,], col = i, lwd = 3, pch = 16,
type = "o", cex = 1.5)
        }
```

The *i* in the loop is incremented each time the loop runs, which plots each row in turn. The first part (the `for()` bit) sets the number that *i* will take. The actual loop is between the `{}`.

9. Add the legend to finish the plot, which should resemble Figure 6.40:

```
> legend("right", legend = rownames(bf), fill = 1:6,
bty = "n")
```

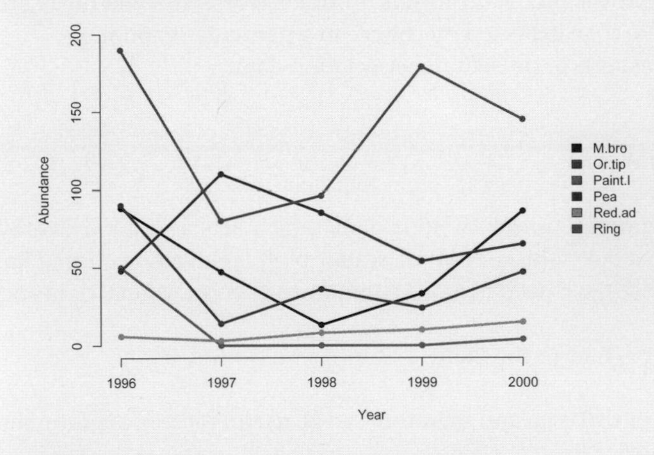

Figure 6.40 Abundance of several butterfly species over 5 years. The multi-series line plot is "built" row by row using a simple loop.

There are many extra options that you could add to the plotting commands to alter line style, plotting character and so on. The `legend()` command must mirror these options so that the colours and character styles and so on match up.

Note: points() and lines() commands

The `points()` command produces points by default because it has an option `type ="p"`. The `lines()` command produces lines by default because it has the option `type ="l"`. This means that you can use either command to produce lines or points by setting the `type` parameter explicitly.

6.5 Graphs to illustrate association

Association analysis is concerned with data that are in categories. Generally you'll have counts of observations (frequencies) that fall into various categorical groupings. The usual analysis looks to see if various categories are associated (positively or negatively), but in *Goodness of fit* tests you look to see if a set of categories "matches up" with another set. The data in Table 6.13 show a typical situation.

Table 6.13 Abundance of common bird species at habitats in Sussex, UK.

	Garden	Hedgerow	Parkland	Pasture	Woodland
Blackbird	47	10	40	2	2
Chaffinch	19	3	5	0	2
Great tit	50	0	10	7	0
House sparrow	46	16	8	4	0
Robin	9	3	0	0	2
Song thrush	4	0	6	0	0

In Table 6.13 you can see that the rows represent categories of bird (species) observed in various habitats, which form the columns. The individual cells represent the number of observations of a species in a particular habitat. As association analysis aims to determine if there are links between the rows and the columns, in other words if some birds are found more often (or less often) than expected by chance in a particular habitat.

There are two important aspects to presenting association data:

- The original data.
- The results of the association test.

Generally you'll produce a summary graph of your data, which acts as a good overall summary of your results (e.g. a box–whisker plot or scatter plot). With association data, however, this is not the case and you'll need to make summaries of your original data and the final results separately.

There are two main sorts of chart to consider:

- Pie charts – show compositional data and therefore can be useful for representing the relative size of your categories.
- Bar charts – can be used to show compositional data in two ways, so are useful to represent your original data as a visual summary. The bar chart can also show the results of association analysis by visualizing the Pearson residuals (see Chapter 9).

You'll see how to undertake the tests of association later (in Chapter 9); this section will focus on the graphical representation of the data and results.

6.5.1 Pie charts

A pie chart shows compositional data; it shows your data as slices of a pie. Each slice of the pie shows the proportion that a single category contributes to the total. So a pie chart is one way to help visualize your data. A pie chart can only sensibly deal with one set of categories at a time, so for the data in Table 6.13, for example, you would produce a pie chart for individual rows or columns. This would mean several charts.

The overall pie covers 360° so the values in your data are converted to a percentage and then each slice takes up that percentage of 360°. The human eye is not really very good at discerning angular data and so the pie chart is not very popular with scientists. The bar char makes a more robust alternative (there are others), but the pie chart remains popular. It is important to know how to make one as you may be required to share your results with others and you might decide that a pie chart is a good choice for your particular audience. You'll see more about presenting results later (Chapter 13).

Using Excel

You can make a pie chart quite easily using Excel. The *Insert* ribbon menu provides a pie chart option (see Figure 4.6 and Figure 6.1). You'll need to have your data arranged in a manner similar to Table 6.13. The general way to proceed is as follows:

1. Open your data and make sure it is arranged appropriately (see e.g. Table 6.13).
2. Highlight the data you want to chart. You will need to select the column (or row) that contains the category names and the column (or row) with the frequency data. If you use the *Ctrl* key you can select non-adjacent columns (or rows).
3. Click the *Insert* ribbon menu and then the *Pie* button. Select a basic 2D pie (usually the first option, Figure 6.41) and a basic chart is created immediately.
4. Use the *Chart Tools* menus to format and alter the chart to fine-tune its appearance.

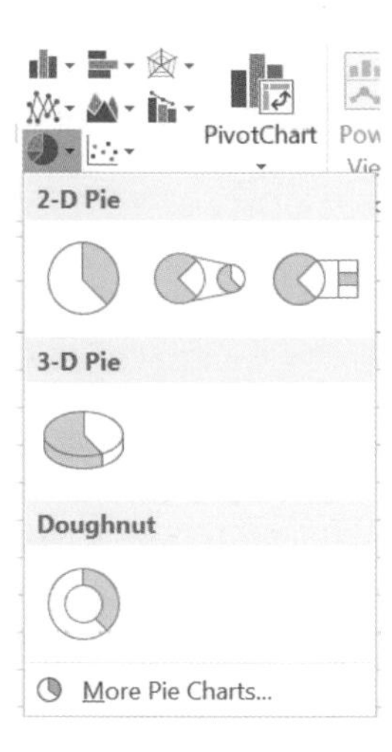

Figure 6.41 Excel can produce various versions of pie chart; a basic 2D chart is generally the most useful.

The basic pie chart can be modified in many ways. The most useful of which is to add data labels to annotate the pie slices. These data labels can be added using the *Data Labels* button; its location depends on the version of Excel you are using:

- Excel 2013: *Chart Tools > Design > Add Chart Element > Data Labels.*
- Excel 2010: *Chart Tools > Layout > Data Labels.*

Once you have the *Data Labels* button "active" you should choose *More Data Labels Options*.

This allows you to choose what labels to present and where they should be placed (Figure 6.42).

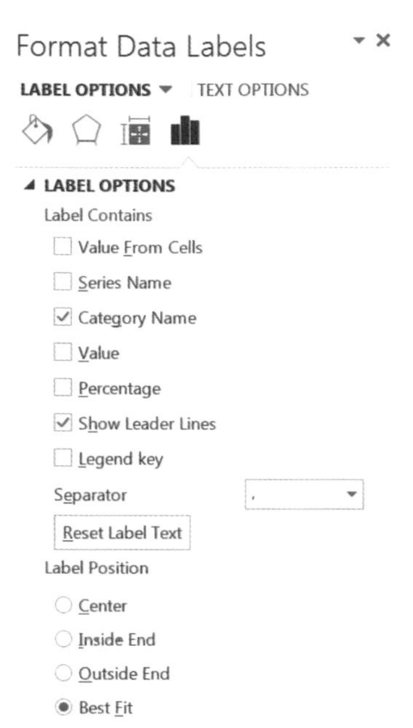

Figure 6.42 Formatting of data labels for pie charts allows fine control over the labels for the pie slices (Excel 2013 version shown).

The most useful option is to produce a label showing the category name. This means you can delete the legend, which can be confusing, especially if the colours are similar. In the following exercise you can have a go at making a pie chart for yourself.

Have a Go: Make a pie chart using Excel

You'll need the data from Table 6.13 for this exercise. The data are available on the support website; the file you want is a CSV file called *bird.csv*. The file should open in Excel and was produced from the "raw" data in the file *birds.xlsx*. See the support website for an exercise in using a Pivot Table to turn data in recording format to a form suitable for analysis and charting.

Download material available from the support website.

1. Open the *bird.csv* file in Excel. You should see the data in a form that resembles Table 6.13.
2. Start with a chart showing the *Garden* habitat. Click in cell A1 and then highlight the data in the first two columns of the bird data, i.e. cells A1:B7.
3. Now click the *Insert* > *Pie* button and choose a basic 2D pie chart; the chart should be created immediately and should resemble Figure 6.43.

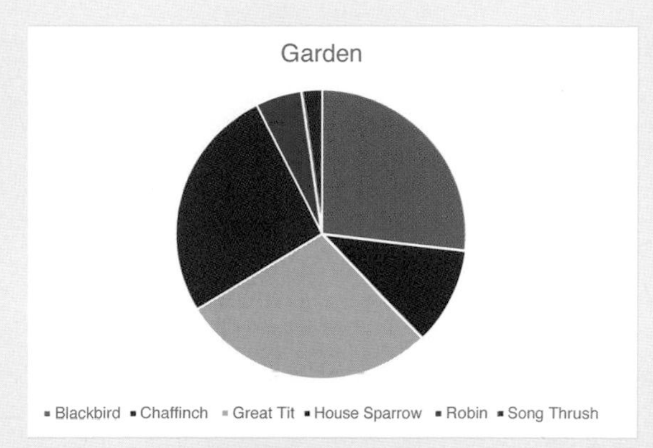

Figure 6.43 The default pie chart in Excel 2013 is adequate but can be improved by using Chart Tools.

4. Delete the chart. Click on it once then use the *Delete* button on the keyboard.
5. Now make a new chart showing data for the *House Sparrow* row. Click in cell A1 then highlight the rest of the row of data. Then hold the *Ctrl* key and use the mouse to highlight the cells A5:F5 (i.e. the row containing the sparrow label and data). You should now have two non-adjacent rows highlighted.
6. Make another basic 2D pie chart with the *Insert > Pie* button.
7. Click once on the chart to ensure the *Chart Tools* menus are active. Now click the *Chart Tools > Design > Add Chart Element* button. Click on the Data Labels option then choose *More Data Labels Options* to open the *Format Data Labels* dialogue box (Figure 6.42).
8. In the *Format Data Labels* dialogue box tick the boxes to enable *Category Name, Show Leader Lines* and *Best Fit*. Ensure the rest are unticked.
9. Close the *Format Data Labels* dialogue box (use the cross at the top right corner). Then click once on the legend in the chart and delete it; use the keyboard delete button or right-click the legend and select *Delete*. Your chart should now resemble Figure 6.44.

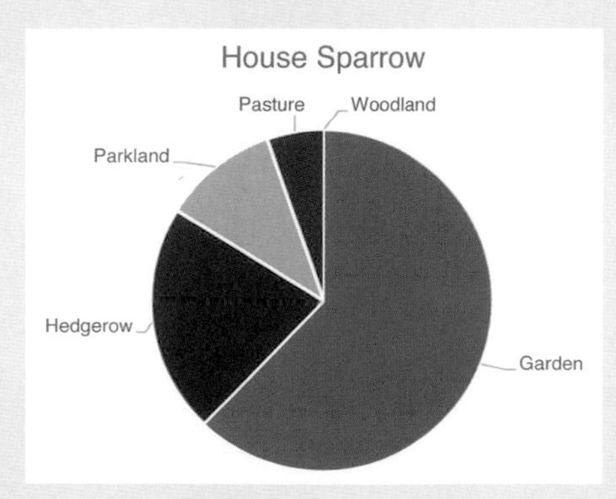

Figure 6.44 A pie chart to show the relative number of observations of house sparrow from different habitats in Sussex. The Excel pie chart has added data labels instead of a legend, which aids readability and interpretation.

You can tweak the labels into slightly different locations using the mouse. Click once on a label to select all labels, and then click again to select a single label. You can then move the label. Try using the *Chart Tools* to alter other features; the *Quick Layout* button can be helpful to apply some standard formatting.

Using R

The `pie()` command will produce pie charts in R. The general form of the command is like so:

```
pie(x, labels, clockwise, init.angle, radius, col, ...
```

`x`	The data to be plotted. This should be a single vector of values. Data in a `data.frame` will not work but you can plot data in a `matrix` object.
`labels`	The labels to use for the slices of pie. The default will take the `names()` attribute of the data.
`clockwise`	By default `clockwise = FALSE`, which results in the slices being plotted in a counter-clockwise direction. If `TRUE`, the slices are plotted clockwise.
`init.angle`	A number that sets the starting point for the plotting. A value of 0 equates to 3 o'clock and 90 to 12 o'clock. The default is 0 unless `clockwise = TRUE` in which case it is 90.
`radius`	The size of the pie in the plot window (to a max of 1). The default is set to 0.8, which allows some "room" for the labels. If you have long labels you can set a smaller value to make room.
`col`	The colour of the slices of pie. The general default uses six pastel shades. If there are more slices than colours the colours are recycled.
`...`	Other graphical commands can be used but these will affect the title and labels only.

The `pie()` command requires that your data are a simple vector but often you'll have the data in a different form, most likely as a `data.frame`. This is what you get when you import data from a CSV file. In order to plot these you need the data to be in the form of a matrix. You encountered this issue previously when looking at multiple category barplots (Section 6.3.1). The easiest way to overcome this problem is to make a copy of the data as a `matrix`. The following example uses the bird data from Table 6.13.

```
> bird
                Garden Hedgerow Parkland Pasture Woodland
Blackbird           47       10       40       2        2
Chaffinch           19        3        5       0        2
Great Tit           50        0       10       7        0
House Sparrow       46       16        8       4        0
Robin                9        3        0       0        2
Song Thrush          4        0        6       0        0

> class(bird)
[1] "data.frame"
> birds = as.matrix(bird)
> class(birds)
[1] "matrix"
```

Now you have a `matrix` data object and can use the `pie()` command to make pie charts of the individual rows or columns.

You use the `data.name[row, column]` syntax to select the data you want to chart. The following commands produce pie charts resembling the plots in Figure 6.45.

```
> pie(birds[1, ], main = "Blackbird")
> pie(birds[,2], main = "Hedgerow")
```

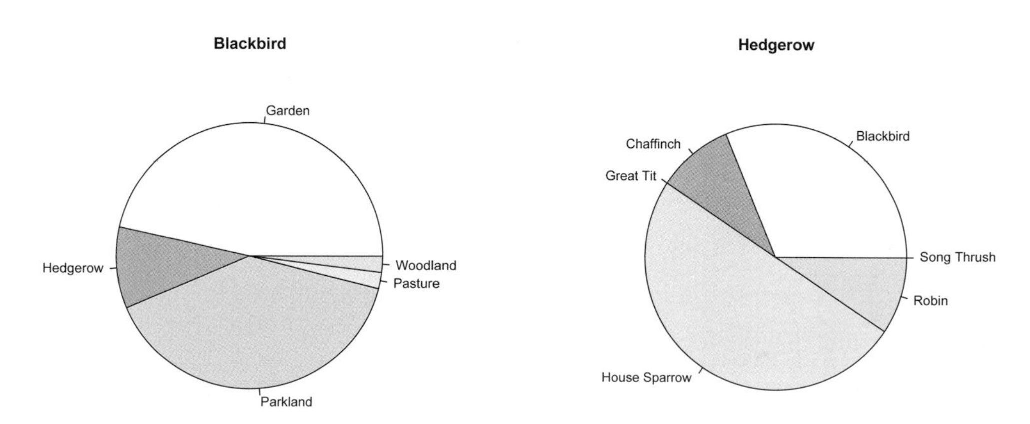

Figure 6.45 Common bird species at a site in Sussex. The `pie()` command produces pie charts. Using `[row, col]` syntax enables individual rows or columns to be plotted. The pie to the left shows the frequency of blackbirds across various habitats. The pie to the right shows the frequency of observation of different bird species in the hedgerow habitat.

It is not trivial a trivial matter to shift the labels in a `pie()` chart. Unlike Excel you cannot drag the labels to a new location. If you have overlapping labels you can try plotting the chart clockwise and/or trying a different `init.angle`. However, this sometimes simply does not do the job. In the following exercise you can have a go at producing a pie chart for yourself with the slices plotted in a different order, to overcome the overlapping labels problem.

Have a Go: Make a pie chart with slices in a custom order

You'll need the bird data for this exercise. The data are on the support website as a file called *bird.csv*.

Download material available from the support website.

1. Open R and import the CSV file from the *bird.csv* data file. Note that you will need to tell R that the first column of the file contains row names (the bird species):

```
> bird = read.csv(file.choose(), row.names = 1)
```

2. Now check the data and see that you have a `data.frame`:

```
> bird
              Garden Hedgerow Parkland Pasture Woodland
Blackbird         47       10       40       2        2
Chaffinch         19        3        5       0        2
Great Tit         50        0       10       7        0
House Sparrow     46       16        8       4        0
Robin              9        3        0       0        2
Song Thrush        4        0        6       0        0
> class(bird)
[1] "data.frame"
```

3. Make a copy of the data as a `matrix` object:

```
> birds = as.matrix(bird)
```

4. Now plot the fifth row (for the *Robin* data):

```
> pie(birds[5,])
```

5. You can see that the labels for *Parkland* and *Pasture* are overlapping. This is because these adjacent data values are both zero and so are very close together. You need to plot the data in a new order so that these items are separated. Make a new pie chart using an explicit column order:

```
> pie(birds[5, c(1,2,3,5,4)])
```

6. Now get the order of the columns in increasing order of overall observation:

```
> order(colSums(birds))
[1] 5 4 2 3 1
```

7. Plot the pie chart again using the order from step 6, your chart should resemble Figure 6.46:

```
> pie(birds[5,order(colSums(birds))])
```

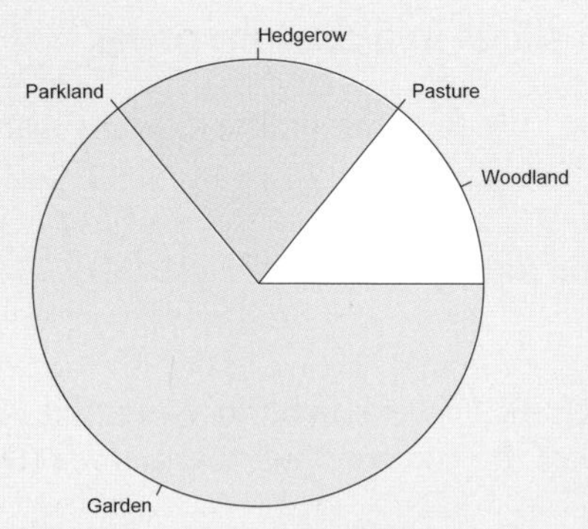

Figure 6.46 A pie chart plotted using custom slice order so that labels do not overlap.

Note: Pie chart labels

See the support website for some additional tips about custom placement of pie chart labels using the `locator()` command in R.

Pie charts are generally considered to be a bit "trivial" by serious scientists! However, they can be useful for presenting findings to non-scientific audiences. If you can make a pie chart then you can present the same data as a bar chart, which is generally a better choice. Your pie chart can present a general summary of the raw data but cannot deal with the results of association analysis. For that you'll need the bar chart, which you'll see next.

6.5.2 Bar charts

The bar chart is a stalwart of scientific presentation and can be pressed into service for many purposes. For association analysis you can use the bar chart in two ways:

- To present compositional data for one or more sets of categories (a pie chart can only represent one set at a time). Essentially a summary of your raw data.
- To present the results of your association analysis by showing the relative strength of the associations.

You've already seen how to produce a multi-sample bar chart when you looked at some butterfly data (Section 6.3.1). When you have association data you can create a summary of the composition of the rows (or columns) of your data in two main ways:

- Bars clustered in groups, each group being a category.
- Bars stacked so that each bar represents a category, which is subdivided. The subdivisions can be presented as regular values or "normalized" so that they add up to 100.

You can present a bar chart with bars that rise up from the main *x*-axis or drop down from it. In other words, you can have positive and negative values and so represent positive and negative associations, thus you can summarize the results of your association analysis.

Using Excel to display compositional data

Excel can produce column charts (i.e. a "regular" bar chart) and bar charts (with bars horizontal) all from the *Charts* section of the *Insert* ribbon menu. The simplest way to proceed is to highlight the data you want to chart then click the appropriate button (see Figure 4.7).

Your basic chart will show the data grouped by column; there are several options:

- Clustered chart: you will have a cluster of bars for each column, each bar being a separate row (see Figure 6.11 and Figure 6.12).

- Stacked chart: you will have a bar for each column; each bar will be split into sections with each section being the value of each row (see Figure 6.48).
- Stacked 100% chart: you will have a bar for each column; each bar will be split into sections with each section being a percentage of the total for that column (see Figure 6.47).

The *Chart Tools > Design* menu gives you some tools to alter your chart:

- *Switch Row/Column*: If you click this button the chart switches from grouping the data by column to row and vice versa.
- *Change Chart Type*: Click this button to alter the chart type without having to reselect the data.

In Excel 2013 you can filter the data, which allows you to show selected parts of your dataset easily. There are two ways to access the filter, both of which are accessible once you've clicked your chart:

- Use the funnel icon beside the chart, which opens a filter dialogue window allowing you to select the data you want to present.
- Click the *Design > Select Data* button, which opens the *Select Data Source* dialogue window. You can use the tick-boxes to select and unselect the items you require.

Note: Selecting non-adjacent columns in Excel

If you want to chart columns that are not next to one another you can use the *Control* key, which allows you to select non-adjacent cells. This is especially useful in older versions of Excel that do not have chart data filtering.

The choice between a regular stacked chart and a 100% chart is a tricky one. In Figure 6.47 you can see a 100% stacked chart of the bird data from Table 6.13.

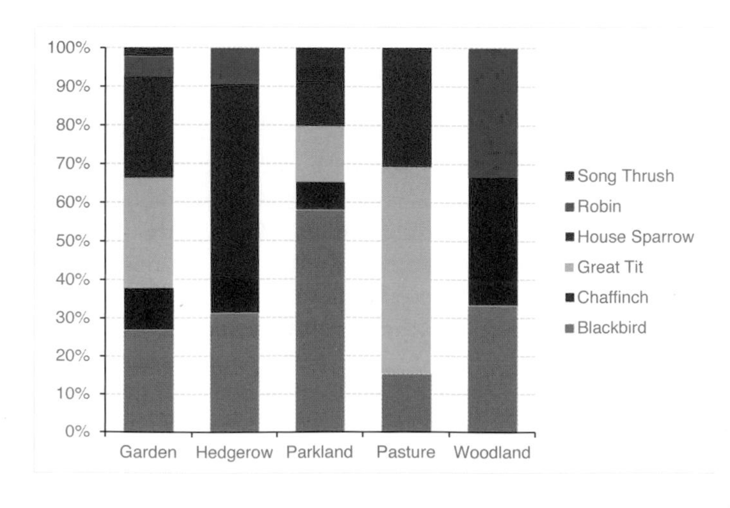

Figure 6.47 Bird species in different habitats in Sussex. Each bar shows the relative abundance of various bird species in a habitat. Because the chart is displaying proportional data, all bars sum to 100%.

The 100% stack allows you to see the composition reasonably clearly but it is not clear that there are big differences in overall abundance. Look at Figure 6.48, which is a regular stacked chart.

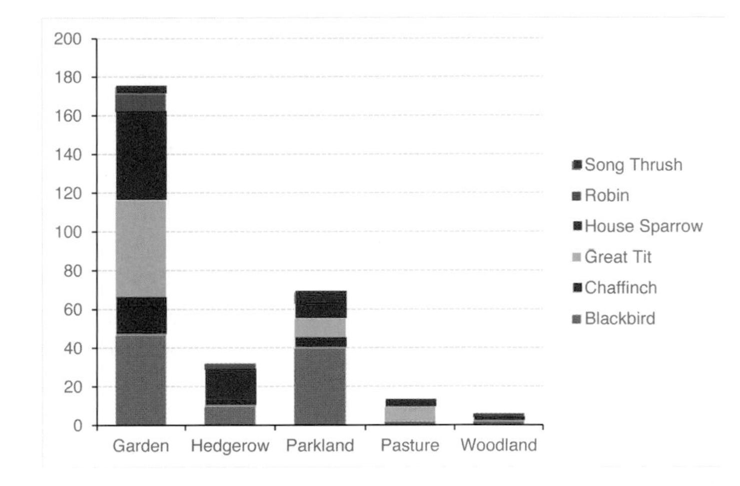

Figure 6.48 Bird species in different habitats in Sussex. Each bar shows the absolute abundance of various bird species in a habitat. It is clear that there are large differences in overall abundance between habitats.

So, you need to be clear about your purpose when producing a chart because different charts accentuate different aspects of the data. In the case of the bird data illustrated here you may decide that a regular grouped chart would be adequate. Try making the charts for yourself; the data are available on the support website as a file *bird.csv*.

Download material available from the support website.

Using Excel to display association analysis results

You'll see how to carry out tests of association in Chapter 9. The best way to visualize the results is to use the Pearson residuals in a bar chart. These residuals give the strength and direction of the association between two items. Thus, positive residuals denote a positive association and negative ones a negative association. Pearson residuals are easily computed (see Chapter 9) and since the critical value is ~2 it is easy to see which residuals are statistically significant.

You can easily construct a column chart in Excel to display the residuals for columns or rows of your results (Figure 6.49). The bar chart in Figure 6.49 was constructed from the data in Table 6.13, which shows the Pearson residuals calculated from the data from Table 6.14.

Once you have your Pearson residuals (see Chapter 9) you can produce a bar chart in the general following manner:

1. Select the data you want to chart. If you need to select non-adjacent columns or rows, use the *Control* key.
2. Click the *Insert > Column Chart* button and select the basic 2D chart. The chart will be created immediately.
3. Use the *Chart Tools* menus to help format the chart as required. Start by making the

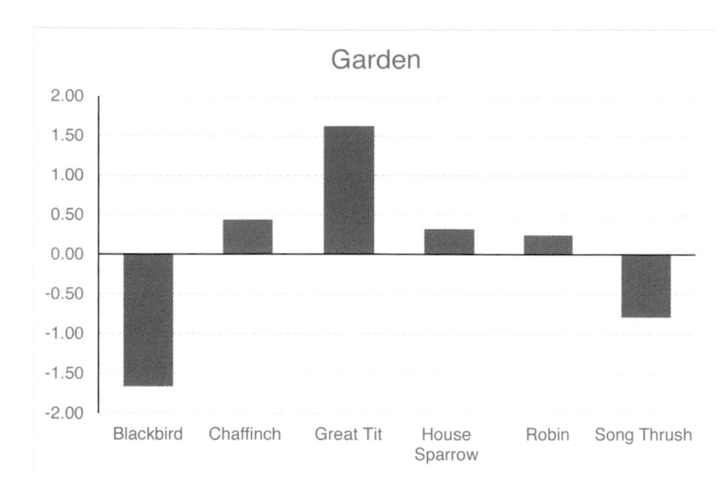

Figure 6.49 Pearson residuals are the simplest and clearest way to present the results of association analysis. Here the bar chart shows the results of association analysis between blackbird and habitat types in Sussex.

Table 6.14 Pearson residuals (a measure of the strength and direction of an association) for some bird species and habitats in Sussex.

Species	Garden	Hedgerow	Parkland	Pasture	Woodland
Blackbird	–1.67	–0.29	3.37	–1.16	–0.04
Chaffinch	0.43	–0.08	–0.68	–1.13	1.84
Great Tit	1.63	–2.70	–1.43	2.36	–1.17
House Sparrow	0.32	2.81	–2.24	0.41	–1.23
Robin	0.24	1.20	–1.81	–0.79	3.21
Song Thrush	–0.79	–1.04	2.39	–0.66	–0.45

x-axis labels appear at the bottom of the chart. Click the *Format* menu and choose the *Horizontal (Category) Axis* option from the *Current Selection* section.

4. Click *Format Selection* to open the *Format Axis* dialogue box.
5. Find the *Labels* section and set the *Label Position* to *Low*.

Once you have set the position of the *x*-axis labels you can use other tools to alter the chart appearance, as you like.

If you want to chart the data by row then highlight the row(s) you want before you use the *Insert > Column Chart* button. If you highlight the entire table of data you can produce a clustered column chart that shows all the residuals, by column. To switch to showing the row residuals you can click the *Design > Switch Row/Column* button.

Using R to display compositional data

You saw the `barplot()` command previously (Section 6.3.1); you can use this command to produce the bar charts you need to display compositional data in association analysis. By default R produces stacked bar charts, and you have to add the parameter `beside = TRUE` in order to produce a grouped bar chart.

If you want to make a 100% stacked bar chart then you'll need to compute the percentages yourself. The `prop.table()` command allows you to compute proportions for rows or columns of a table (you can also compute proportions for the entire table). The command is simple to use; you give the name of the data and then specify `margin = 1` to compute proportions by row or `margin = 2` for the columns.

```
> prop.table(birds, margin = 2)
                  Garden Hedgerow   Parkland   Pasture  Woodland
Blackbird      0.26857143  0.31250 0.57971014 0.1538462 0.3333333
Chaffinch      0.10857143  0.09375 0.07246377 0.0000000 0.3333333
Great Tit      0.28571429  0.00000 0.14492754 0.5384615 0.0000000
House Sparrow  0.26285714  0.50000 0.11594203 0.3076923 0.0000000
Robin          0.05142857  0.09375 0.00000000 0.0000000 0.3333333
Song Thrush    0.02285714  0.00000 0.08695652 0.0000000 0.0000000
```

Your data need to be in `matrix` format to make the plot so you'll need to convert imported data (which will be in `data.frame` format) using the `as.matrix()` command you met previously. If you use the `prop.table()` command you'll get a `matrix` as the result anyhow.

The default colours are shades of grey but you can specify colours explicitly using the `col` parameter (see Table 6.5). You can add a legend directly from the `barplot()` command using the `legend` parameter or add it afterwards using the `legend()` command (recall Section 6.3.1). You'll often have to tweak the axes of the plot to make room for the legend, especially with the 100% stacked plot.

In the following examples you'll see the bird data from the preceding exercise presented in a variety of forms.

Download material available from the support website.

The basic command produces a regular bar chart with stacked groups (Figure 6.50):

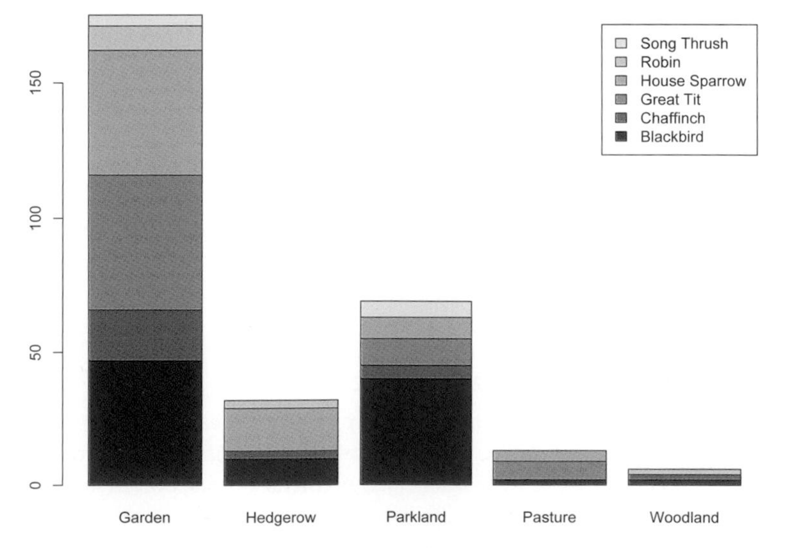

Figure 6.50 The `barplot()` command produces regular stacked charts by default.

```
> barplot(birds, legend = TRUE)
```

You'll need to tweak the *x*-axis to make room for a legend with a 100% stacked chart (Figure 6.51):

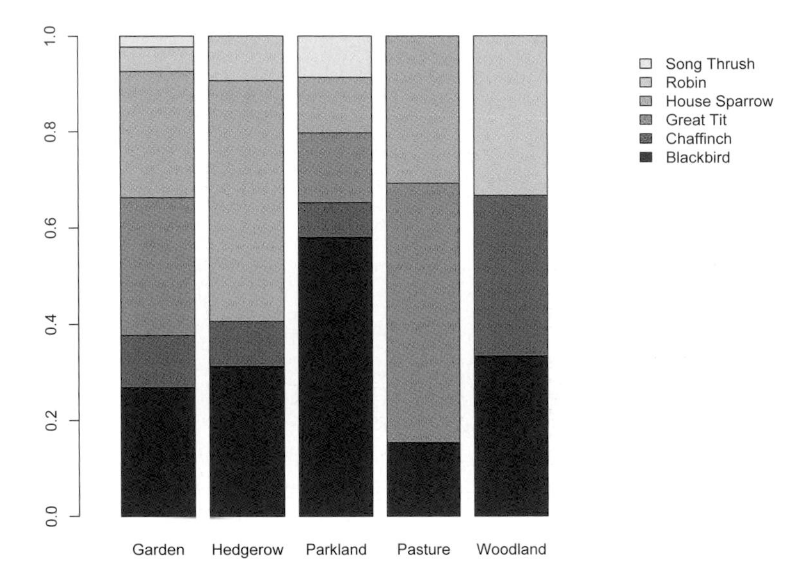

Figure 6.51 The `prop.table()` command is used with parameter `margin = 2` to create the data for `barplot()` to make a 100% stacked bar chart. The *x*-axis needs to be altered to make room for the legend.

```
> barplot(prop.table(birds, margin = 2), xlim = c(0,9),
  legend = TRUE, args.legend = list(bty = "n"))
```

To make a chart of the rows you'll need to determine the proportion for the rows as well as transposing the data (Figure 6.52):

```
> barplot(t(prop.table(birds, margin = 1)), xlim = c(0,9),
  legend = TRUE, args.legend = list(bty = "n"))
```

As with Excel you need to determine which form of plot is best suited to your purpose. You may even decide that the regular grouped chart is best, achieved by adding `beside = TRUE` to your `barplot()` command.

Using R to display association analysis results

You can produce a `barplot()` of the Pearson residuals easily enough. The `chisq.test()` command carries out the calculations of the association, including computing the Pearson residuals. You'll see how to carry out the association analysis in more detail later (Chapter 9).

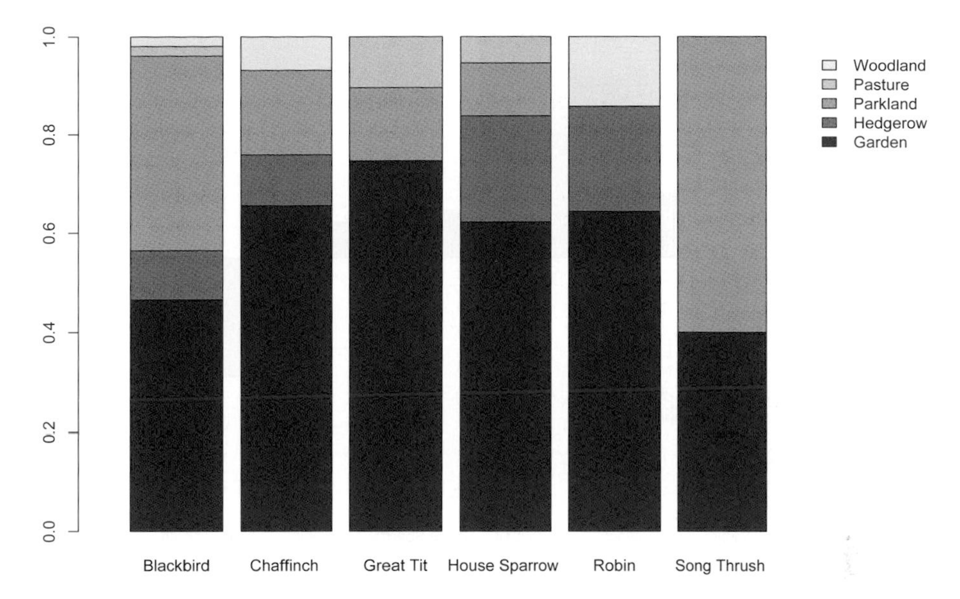

Figure 6.52 Data must be transposed using the `t()` command in order to produce a chart of the row data. If proportions are required the `margin = 1` parameter is added to the `prop.table()` command.

The `chisq.test()` command produces a result with several components, the `$residuals` component contains a table (actually a `matrix` object) of the residuals. You can plot these like any other data.

First of all you need to compute the residuals:

```
> chisq.test(birds)$residuals
                  Garden    Hedgerow   Parkland    Pasture    Woodland
Blackbird     -1.6685309 -0.28880348  3.3693087 -1.1617025 -0.03784188
Chaffinch      0.4331585 -0.08218328 -0.6846225 -1.1304716  1.83614806
Great Tit      1.6265145 -2.69588513 -1.4325935  2.3555061 -1.16735250
House Sparrow 0.3172092  2.81407366 -2.2374283  0.4092210 -1.22681895
Robin          0.2411348  1.20207460 -1.8095790 -0.7854611  3.21439960
Song Thrush   -0.7933135 -1.04151129  2.3938019 -0.6638358 -0.45098762
```

You could save the result separately but it's easy to make a chart like so:

```
> barplot(chisq.test(birds)$residuals[,1])
> abline(h = 0)
```

The preceding commands produce a bar chart of the first column (the *Garden* habitat) that resembles Figure 6.53. Note that the command used the `[row, column]` syntax to select the data to plot. You can plot multiple rows or columns by specifying them in the square brackets. Usually you'll want a clustered bar chart so remember to add `beside = TRUE` to the command. If you want to incorporate a legend you'll need to use `xlim()` in the command to rescale the x-axis to make room.

4. Use `plot()` on the hierarchical cluster result from step 3 to plot the dendrogram.

The following command lines were used to create Figure 6.54:

```
> head(ps)
                       ML1 ML2 MU1 MU2 PL2 PU2 SL1 SL2 SU1 SU2
Achillea millefolium     1   1   1   1   0   1   0   0   0   0
Aegopodium podagraris    0   0   0   0   0   0   0   1   0   0
Agrostis capillaris      0   1   1   1   1   0   1   1   1   0
Agrostis stolonifera     0   0   0   0   0   1   0   1   0   0
Anthriscus sylvestris    0   0   0   0   0   0   0   0   1   1
Arctium minus            0   0   0   0   0   0   0   0   0   1

> d = dist(t(ps), method = "binary")
> h = hclust(d)
> plot(h)
```

The first command merely shows you the first few lines of an object. Here you can see the data are arranged with the species as rows. The `t()` command transposes the data in the next line of command so the rows become the sites. The `dist()` command computes the dissimilarity; there are various algorithms with `"binary"` being equivalent to Jaccard. You'll see more about the methods of making dissimilarity matrix results in Chapter 12.

The `hclust()` command takes a dissimilarity matrix and works out how the various elements join together. The result is an object that can be plotted as a dendrogram, which is what the `plot()` command does at the end.

6.7 Graphs – a summary

There are quite a few different sorts of graph that you can utilize to help visualize your data and make important decisions about the analytical approach (recall Table 6.2). You should also use graphs to illustrate your data, which can make them more comprehensible to readers. When you present graphs you should ensure they are fully labelled and as clear as possible. Even when you use graphs for your own use it is good practice to label and title them fully.

Label axes and include the units.

Do not include too many different elements on a single graph – avoid clutter and if necessary produce two graphs rather than one.

Give a main title explaining what the graph shows. Usually this is done as a caption in a word processor. The caption should enable a reader to understand what the graph shows without having to read the main text. If your graph is in your field notebook then make sure you describe the graph so that someone else can understand it. You'll see more about the presentation of results in Chapter 13.

EXERCISES

Answers to exercises can be found in the Appendix.

1. You have samples of weights of eggs for several species of bird. What would be the most appropriate types of graph to use to visualize these data (there may be more than one)?
   A. Pie chart
   B. Line chart
   C. Bar chart
   D. Scatter plot
   E. Box–whisker chart
2. The box–whisker plot usually displays ____ data, i.e. the ____, ____ and ____.
3. A line plot is used for showing the relationship between two numeric variables. The response variable (dependent) should be plotted on the (vertical) *y*-axis. TRUE or FALSE?
4. You can add a joining line to a scatter plot to make a line chart. This makes it analogous to a bar chart. TRUE or FALSE?
5. When you have compositional data you can use a ____ or a ____ chart to display the results but only a ____ chart allows you to display multiple categories.

Chapter 6: Summary

Topic	Key points
Match type of graph to use	Each type of graph is best suited to a specific purpose. You should choose the most appropriate graph for your needs.
Chart tools in Excel	Charts are made in Excel from the *Insert* menu. Once a chart is made it can be edited and manipulated using the *Chart Tools* menus.
Graphical commands in R	There are many graphical commands in R. Some commands create new graphs and others add elements to existing graphs. There are also many graphical parameters, which can be used to alter the appearance of graphs.
Exploratory graphs	The most common sort of exploratory graph is one that visualizes the distribution, such as a tally plot or histogram. A line plot can be used with a running mean to help determine if sample size is adequate. In Excel the `FREQUENCY` command helps to get the frequency of data in various size classes (bins). You can then use a column chart like a histogram. In R the `hist()` command will draw a histogram directly from the data.

Chapter 6: Summary – *continued*

Topic	Key points
Bar charts	Bar charts (column charts) display the magnitude of items in categories. You can use a bar chart to visualize differences between samples. Multiple categories can be shown in groups or stacked, so they can be used to explore compositional data (e.g. in association projects). A bar chart can be pressed into service as a histogram. In R the `barplot()` command produces bar charts. In Excel use *Insert > Column chart* (or Bar chart). Error bars can be added to show variability in samples (e.g. standard error or confidence interval). When plotting multiple series you need a legend. In Excel use the *Chart Tools* to add a legend. In R use the `legend()` command.
Box-whisker plots	The box–whisker plot (or boxplot) visualizes data in various categories. The boxplot is used to show differences between samples. The boxplot shows three key statistics; median, inter-quartile range and max–min (range). Although it uses non-parametric statistics the boxplot is generally useful, as it displays more information than a bar chart. In R the `boxplot()` command produces box–whisker plots. It is possible to make a boxplot using Excel with one of the Stock charts.
Scatter plots	The scatter plot visualizes the relationship between two numerical variables. Use a scatter plot for correlation and regression. The response variable (dependent) goes along the *y*-axis and the predictor (independent) along the *x*-axis. In Excel use the *Insert > Scatter* button. In R use the `plot()` command. It is possible to join the points but a scatter plot with joined points is not the same as a line chart.
Line charts	A line chart is analogous to a bar chart in that points are used to show the magnitude of samples in discrete categories. These points are joined with a line. The line plot is generally used to display time-series data, where the interval between readings is the same. You can plot multiple series, each having its own symbols, colour and so on, so you can compare directly. In Excel use the *Insert > Line Chart* button. In R use the `plot()` command. Additional data can be added using the `points()` or `lines()` commands.
Pie charts	Pie charts show compositional data, where the proportion of each item is converted to a slice of pie. In Excel use the *Insert > Pie* button. In R use the `pie()` command. A pie chart can only display the composition of one category at a time. Bar charts are usually a better option (you can also display multiple categories).

Chapter 6: Summary – *continued*

Topic	Key points
Dendrogram	The dendrogram is a chart that shows the relationship between multiple samples in terms of their similarity. The dendrogram is used in analysis of communities. In R you need to make a matrix of similarities (e.g. with the `dist()` command) and then a hierarchical cluster result, with `hclust()`. Excel cannot draw a dendrogram automatically (but you could use drawing tools).

7. Tests for differences

In tests for differences you generally have two or more samples that you wish to compare. You usually want to know if they are different, i.e. are their means (or medians) different?

What you will learn in this chapter

» How to conduct the *t*-test for differences
» How to use the *Analysis ToolPak* in Excel for the *t*-test
» How to conduct the *U*-test for differences
» How and when to carry out paired tests
» How to visualize the results of paired tests

In this chapter you will consider three options and in all cases you will compare just two samples. In the first case, where you have two normally distributed samples you will examine the *t*-test, and where you have skewed data you will use the *U*-test. The last section will look at special cases of both tests where you have matched pairs of data. When you have more than two samples the situation becomes a bit more complex and you will see these situations in Chapters 10 and 11.

Of course you should draw a graph to illustrate the situation. When you are looking at differences, you should be thinking in terms of bar charts and box–whisker plots (see Section 6.3). For paired tests you'll see an alternative graphical approach (Section 7.3.5).

7.1 Differences: *t*-test

The *t*-test is a widely used test to determine if two samples are different. It is usually called Student's *t*-test after the nom de plume of the statistician who devised it. The test relies on the properties of the normal distribution (so is called a *parametric* test) and tests for differences in sample means. Figure 7.1 shows the frequency distribution of two samples. You can see they have different means and that there is virtually no overlap between the two samples. In this case, when there is hardly any overlap, you can be pretty certain that there is a real difference between the two samples and it is unlikely that they are really just part of one larger population/sample.

Now look at the next example (Figure 7.2). In Figure 7.2, the means are very close together and there is a great deal of overlap. You might conclude that there is no real difference between the samples. In the two preceding examples, the situation appears fairly clear. In the next example, however, the case is less certain (Figure 7.3).

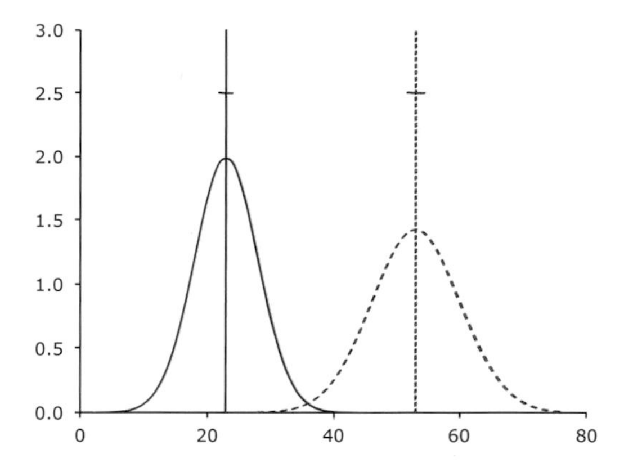

Figure 7.1 Frequency distribution of two samples. You can see very little overlap. Vertical lines represent the mean values and the cross-bars are the standard error.

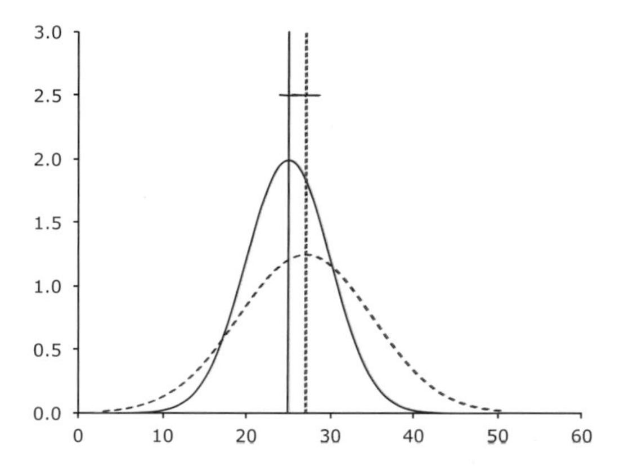

Figure 7.2 Frequency distribution of two samples. This time you see a large overlap, the means are close together and the standard error bars overlap.

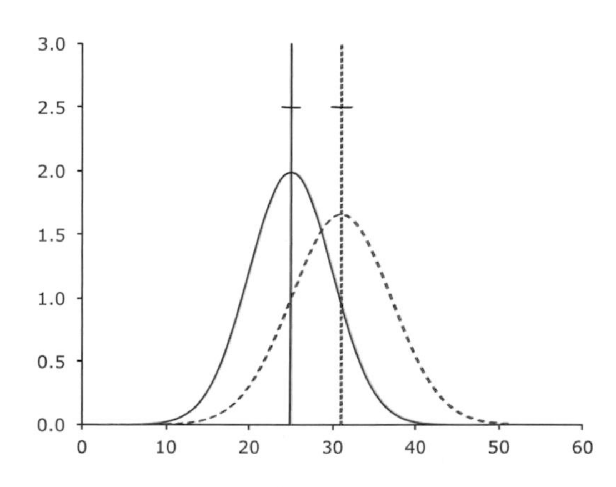

Figure 7.3 Frequency distribution of two samples. There is quite a bit of overlap but the standard error bars do not overlap.

Figure 7.3 shows the frequency distributions of two samples. Their means are moderately far apart and there is a fair amount of overlap. Are these two samples different? This is where a statistical test will help because it will put a value to your uncertainty.

You are looking for differences between two samples and they are both normally distributed (the distribution graphs look symmetrical around the middle). The *t*-test is a suitable test to use to examine these data. This parametric situation (normal distribution) allows you to summarize the data using the mean and standard deviation (for each sample).

The *t*-test uses three pieces of information to determine if the samples are different. The formula for the test is shown in Figure 7.4:

$$t = \frac{\left|\bar{x}_a - \bar{x}_b\right|}{\sqrt{\frac{s_a^2}{n_a} + \frac{s_b^2}{n_b}}}$$

Figure 7.4 The formula for the *t*-test.

On the top you have the means of the two samples. On the bottom you have the variance, which is the standard deviation squared, and the number of items in your sample (the replicates). The vertical braces on the top signify that you ignore the sign of the difference; you are interested only in the magnitude of the difference, i.e. the absolute difference. In Excel and R there is a function that ignores the sign, `ABS` and `abs()`, respectively.

If you look at the top and bottom separately you can see what is happening. If the difference between the means is large, then you have the situation akin to Figure 7.1 where the samples were widely separated. The bigger the difference, the larger your calculated value of *t* will be.

If the standard deviation of the samples is large, then the distributions will be wide and fat and there will likely be more overlap. This situation will be more like Figure 7.2. Since the standard deviations are on the bottom of the equation, the result is that larger *s* will lead to smaller *t*.

So do you want a large *t* or a small one? You want a large one to show that there is a real difference between samples (i.e. with widely separated means like in the Figure 7.1).

It is simple enough to calculate the value of *t* using the formula. Once you have a value, what does this tell you? You need to compare your calculated value with a table of critical values (see Table 7.1).

In the table of critical values you see several columns. The first is labelled degrees of freedom and is related to how many data there are ($df = n_a - 1 + n_b - 1$, i.e. the total number of data items you have minus 2). You look down this column until you reach the one you want and then you read across. The following columns give the critical values of *t* for various levels of significance. You are prepared to accept 5% so if your calculated value is greater than the critical value you can reckon that there is a difference in the sample means. What you are saying is that there is only a 5% chance (or less) that you could have a *t*-value as high as you did by chance. You may get a value greater than the next column; this would then represent only a 2% chance.

Formally you say that you reject your *null hypothesis* (H0) and accept the *alternative hypothesis* (H1). If your calculated value of *t* was lower than the critical value then you

Table 7.1 Critical values for the Student's *t*-test. Reject the null hypothesis if your calculated value is greater than the tabulated value.

Degrees of freedom	Significance level			
	5%	**2%**	**1%**	**0.1%**
1	12.706	31.821	63.657	636.62
2	4.303	6.965	9.925	31.598
3	3.182	4.541	5.841	12.941
4	2.776	3.747	4.604	8.610
5	2.571	3.365	4.032	6.859
6	2.447	3.143	3.707	5.959
7	2.365	2.998	3.499	5.405
8	2.306	2.896	3.355	5.041
9	2.262	2.821	3.250	4.781
10	2.228	2.764	3.169	4.587
11	2.201	2.718	3.106	4.437
12	2.179	2.681	3.055	4.318
13	2.160	2.650	3.012	4.221
14	2.145	2.624	2.977	4.140
15	2.131	2.602	2.947	4.073

know that there was a greater than 5% probability that the result was due to chance so you have to accept your null hypothesis and conclude that there is no significant difference in sample means. You will look at ways of reporting your results in more detail later (Chapter 13).

The *t*-test shown in Figure 7.4 is the "classic" method. There are two main variants of the *t*-test:

- Assumption of equal variance – in this version you calculate a *common variance*, which is used in place of the separate variances.
- Welch (or Satterthwaite) modification – in this version the degrees of freedom are modified (reduced slightly) to make the result a bit more conservative.

The formula for working out the common variance is not especially difficult (Figure 7.5) but it is hardly worth the bother; it is "safer" to assume that the variances of the samples are not equal.

$$S_C^{\ 2} = \frac{S_1^{\ 2}(n_1 - 1) + S_2^{\ 2}(n_2 - 1)}{(n_1 - 1) + (n_2 - 1)}$$

Figure 7.5 Formula to calculate common variance in the variant of the *t*-test that assumes equal variance.

difference between the two samples. In the second example with almost complete overlap (Table 7.4), you can be fairly certain there is no difference between the two samples. What you need is a way to determine how likely there is to be a real difference between two samples when there is some overlap.

The example in Table 7.5 shows you two samples and there is some overlap. Are the two samples different? This is why you need a statistical test to tell you how certain you may be that there is (or is not) a difference. You start by putting the data into numerical order and changing the original value to a rank. This is simplest to illustrate using the first example.

Table 7.5 Non-parametric data. Counts of flour beetles at two sites. Here there is some overlap.

Site *a*	9	12	12	13	14	18		21	21	23	24							
Site *b*							19	21		23		25	28	32	33	34	34	45

In Table 7.6 you can see clearly that all the ranks of the first site are smaller than all the ranks of the second site. When there is a lot of overlap the ranks overlap a lot too.

Table 7.6 Flour beetle count data. The values are first converted to ranks, starting with the smallest. All the ranks of site *a* are smaller than for site *b*.

Ranks R_a	1	2	3	4	5	6	7	8	9	10										
Site *a*	9	12	12	13	14	18	19	21	21	21										
Site *b*											23	23	24	25	28	32	33	34	34	45
Ranks R_b											11	12	13	14	15	16	17	18	19	20

When working out the ranks, you treat all the values as one set and start at the lowest (so the lowest value gets rank 1). The highest value gets the highest rank, so if there were 20 items of data then the highest rank would be 20. Where there are tied values the rank is averaged (from the rank positions the data would occupy). You then step to the next available rank; in Table 7.7, for example, you have two values of 12. These occupy ranks 2 and 3 so you get the average of 2.5. The next available rank is 4 and this is allocated to the 13 (from site *b*).

Table 7.7 Flour beetle count data. After converting to ranks, you see that the ranks also overlap considerably.

Ranks R_a	1	2.5		5		7	9		11.5	13		15		17	18.5	
Site *a*	9	12		14		19	21		23	24		28		33	34	
Site *b*		12	13		18		21	21	23		25		32		34	45
Ranks R_b		2.5	4		6		9	9	11.5		14		16		18.5	20

You carry on adding ranks until they are all done. For the third example (the one with the moderate overlap, Table 7.5) you end up with Table 7.8

Table 7.8 Flour beetle count data. After ranking the values, you see some overlap in the ranks.

Ranks R_a	1	2.5	2.5	4	5	6		9	9	11.5	13							
Site *a*	9	12	12	13	14	18		21	21	23	24							
Site *b*							19	21		23		25	28	32	33	34	34	45
Ranks R_b							7	9		11.5		14	15	16	17	18.5	18.5	20

You can now add up the ranks for each sample (this gives R_a and R_b values). You now calculate your *U*-values; think of this as a measure of overlap. Because you have two samples, you get two *U*-values using the Mann–Whitney formula shown in Figure 7.9:

$$U_1 = n_1 \times n_2 + \frac{n_1 \times (n_1 + 1)}{2} - R_1$$

Figure 7.9 The formula to calculate the *U*-values in the Mann–Whitney *U*-test.

You calculate your first *U*-value then simply switch the subscripts and calculate the second *U*-value. You select the smallest one as your test statistic.

For the example in Table 7.8, you get:

$$-R_a = 63.5,\ R_b = 146.5 \text{ and } U_a = 91.5,\ U_b = 8.5$$

You select 8.5, i.e. the smaller of the two, as your final *U*-value. You now need to compare your test statistic against a table of critical values (Table 7.9).

Table 7.9 Critical values for *U*, the Mann–Whitney *U*-test at the 5% significance level. Reject the null hypothesis if your value of *U* is equal or less than the tabulated value.

n	1	2	3	4	5	6	7	8	9	10
1	–	–	–	–	–	–	–	–	–	–
2	–	–	–	–	–	–	–	0	0	0
3	–	–	–	–	0	1	1	2	2	3
4	–	–	–	0	1	2	3	4	4	5
5	–	–	0	1	2	3	5	6	7	8
6	–	–	1	2	3	5	6	8	10	11
7	–	–	1	3	5	6	8	10	12	14
8	–	0	2	4	6	8	10	13	15	17
9	–	0	2	4	7	10	12	15	17	20
10	–	0	3	5	8	11	14	17	20	23

You look down the first column until you reach the number of data for the first sample, and then look across until you reach the column headed with the number of data items in your second sample (since the table is symmetrical you could read across the header row and then look down). The value in the cell is the *U*-value that you need to equal or be smaller than in order for the difference to be statistically significant. Note that in most statistical tests, you are looking for a large test value to be significant; however, in some (like the *U*-test) you are looking for a low value so it is important to check the requirements.

In this case, you have ten data in both samples so you look in the table and find the critical value to be 23. Since your calculated value is 8.5 you can reject the null hypothesis and accept the alternative hypothesis; there is a statistically significant difference between the two sites. Now look back at the medians and see which site has the most flour beetles (*Tribolium confusum*). Put another way, there is less than a 5% chance that the result you obtained could happen by random chance.

How the U-test works

If you look at the first example, where there was no overlap (Table 7.6), you can see that the ranks add up to $R_a = 55$, $R_b = 155$. This is as good as it gets, with no overlap you will always get a rank sum of 155 (the largest value) when you have $n_a = 10$, $n_b = 10$. If you take the last example, where you had some overlap (Table 7.8), you got a rank sum of $R_b = 146.5$. If you take this from your perfect result you get 8.5, which was the *U*-value you calculated using the equation! So you can summarize by saying that:

$$U = perfect\ rank\ sum - highest\ actual\ rank\ sum$$

If you look at the middle example (Table 7.7) you get rank sums of $R_a = 99.5$, $R_b = 110.5$. This would give you a U of $155 - 110.5 = 44.5$, much higher than the critical value (of 23) which you would expect because there was almost complete overlap.

It is tedious to add up the ranks for the perfect situation every time but it so happens that there is a simple formula that will do it (Figure 7.10).

$$Perfect \sum R = n_a n_b + \frac{n_a(n_a + 1)}{2}$$

Figure 7.10 The perfect rank sum is the basis for the *U*-test.

This is of course the basis of the Mann–Whitney formula; all you do is to subtract the actual rank sum. You do this for both samples, as you cannot always tell which will give you the smallest *U*-value. What you are doing in effect is to measure the overlap based on the ranks of the numbers in your two samples.

Checking the sums

Calculating the ranks and doing the maths is simple enough but when you have lots of data it is easy to make a mistake. There is a quick way to check that you have performed the calculations correctly (usually it is the ranking that is in error):

$$U_a + U_b = n_a \times n_b$$

In other words, the two *U*-values add up to the size of the two samples multiplied together. In your case, *n* is 10 for both samples so $n_a \times n_b = 100$. Your two *U*-values should add up to 100 also.

7.2.1 Using Excel for the *U*-test

There is no built-in function to work out the *U*-test in Excel. What you need to do is to calculate the ranks and then use the rank sums in the *U*-test formula (Figure 7.9). To work out the ranks you can use the `RANK.AVG` function, which takes the following general form:

```
RANK.AVG(number, reference, order)
```

`number`	The cell reference of the number that you wish to get the rank of.
`reference`	The range of cells that contain the data. For the *U*-test you want the data from both samples at the same time.
`order`	The order you want, if 0 or missing the data are sorted in descending order. Usually you want the smallest value to have the smallest rank so you should use a 1 here (actually any value >0 will do), to get an ascending order.

In the following exercise you can have a go at ranking some data and calculating the *U*-values for yourself.

Have a Go: Use Excel to rank data and calculate *U*-values

You'll need the flour beetle data (Table 7.8) for this exercise. The data are on the support website as a file called *flour beetles.xlsx*.

Download material available from the support website.

1. Open the *flour beetles.xlsx* data file. You'll see a single worksheet containing the data (in columns B and C). In cells D1 and E1 type a couple of headings for the appropriate ranks, `Rank W` and `Rank G` will do nicely.
2. In cell D2 type a formula to calculate the rank that the corresponding value (observation 1 for *Woad Fm*) has in the overall dataset: `=RANK.AVG(B2,$B$2:$C$11,1)`
3. Note that the formula in step 2 contains some $, which help to "fix" the cell range when the formula is copied. Start by copying the cell D2 to the clipboard.
4. Now use the mouse to select the cells D2:E11, then paste to place the formula from step 2 into all the cells. You should now have the ranks corresponding to the original data.
5. In cell A12 type a label, *Sum*, for the column totals.

6. In cell B12 type a formula to calculate the column total: `=SUM(B2:B11)`.
7. Copy the formula in B12 across the row to get column totals, including the rank sums.
8. In cell A13 type a label, `n`, for the sample sizes. In cell B13 type a formula to work out the number of observations: `=COUNT(B2:B11)`. Copy the result to cell C13 so you have the number of observations for each of the samples.
9. In cell A14 type a label, `U`, for the *U*-value results.
10. In cell D14 type a formula to calculate *U* (recall Figure 7.9): `=$B13*$C13+B13*(B13+1)/2-D12`.
11. Note that in step 10 the $ sign was used to "fix" the columns that hold the number of observations. Copy the cell D14 to the clipboard and paste it to cell E14. You should now have two *U*-values.
12. The test statistic is the smaller of the two values you calculated.

You can check the results by adding the two *U*-values: 91.5 + 8.5 = 100. This result should equal $n_1 \times n_2$: 10 × 10 = 100. It is a good idea to check this, even if you are using the spreadsheet, just in case of a typo.

Even when you have successfully created the ranks and calculated the *U*-values you cannot automatically get the critical values for the *U*-test. The *Analysis ToolPak* is no help to you either as the routines to carry out the analysis are not built-in. However, you can easily look up the values (see Table 7.9), and the spreadsheet makes it easy to carry out the ranking (which is often where the errors creep in).

7.2.2 Using R for the *U*-test, the wilcox.test() command

When data are not normally distributed you need a *U*-test (Mann–Whitney). This was developed by Wilcoxon originally and R calls the *U*-test the `wilcox.test()`. The basic form of the command is:

```
wilcox.test(formula, data, paired = FALSE, ...)
```

`formula`	The sample data to use. You can specify the data as two separate samples (give their names separated by a comma) or as a formula of the form `y ~ x` where `y` is the response variable and `x` is the predictor.
`data`	If you give the sample data as a formula you can specify the data object that contains the variables.
`paired = FALSE`	If `TRUE` a paired version of the *t*-test is conducted (see Section 7.3).
`...`	There are some other parameters you can specify but the ones given here are the most useful.

The `wilcox.test()` command allows you to enter your samples in several ways, in the same way you saw earlier in the `t.test()` command (Section 7.1.2). The simplest is when you have two separate data samples:

```
> wilcox.test(Woad.Fm, Glebe.Fm)
  Wilcoxon rank sum test with continuity correction
data:  Woad.Fm and Glebe.Fm
W = 8.5, p-value = 0.001887
alternative hypothesis: true location shift is not equal to 0

Warning message:
In wilcox.test.default(Woad.Fm, Glebe.Fm) :
  cannot compute exact p-value with ties
```

You may well get error messages if you have tied ranks (as in this case). This is generally okay unless you are close to $p = 0.05$ when a correction to the p-value should be applied (the correction is usually quite small). The program to do this is not part of the basic package but the routines are available in a separate library (called *exactRankTests*), which can be easily loaded (recall Section 3.1.6 on loading additional packages). You probably don't need to bother with the correction factor; modification of the resulting p-values is usually quite small and mostly makes little practical difference to your conclusions.

7.3 Paired tests

The t-test and the U-test described in the previous sections look for differences between two samples of data. Each sample is *independent* of the other. There are occasions, however, when the two samples may not be totally independent. For example, you may be interested in the difference in lichen abundance on different sides (aspect) of trees. In order to see if there are more lichens on the north compared to the south, you set out with your quadrat and measure abundance on trees. You place your quadrat on the north side of a tree and then move around to the south side. Then you repeat with lots of other trees.

Now you end up with two samples, one from the north and one from the south; however, the two sets of numbers are not completely independent. You can match up the first measurement on the north list with the first measurement of the south list because they come from the same tree. You have a set of matched pair data.

Other examples might include the abundance of a plant species before and after some management treatment. You might set up some quadrats and determine abundance. Then some management is performed. You come back later and as long as your quadrats are in exactly the same spot(s) you could count the data as a matched pair. Each individual quadrat has two values, one for the before and one for the after. You are measuring something over time (but in exactly the same spot).

A further example might be to examine the heart rate of undergraduate students before and after administration of large quantities of caffeine. There will be natural variation in heart rate amongst students and it is logical to pair up measurements and so control variability.

In reality, the occasions where you get matched pair data are fairly limited and if you are not absolutely certain then assume the data are not matched pairs. To be sure, you must be able to match up one measurement in one sample with a specific measurement in the other sample.

There are versions of the t-test and of the U-test that can test for differences in matched pair data. The t-test for matched pairs looks for differences when the data are all normally distributed. The Wilcoxon matched-pairs test looks at differences when the data are not normally distributed (i.e. skewed).

7.3.1 Paired *t*-test for parametric data

When you have normally distributed data you can use a version of the *t*-test to compare matched pair data. Essentially you are looking at differences between each pair of values. If all the values were positive then you could be sure that there was a significant difference. If all the differences were negative then you expect the same result. Often, however, there will be some positive differences and some negative. If the positive differences balance out the negative differences then you can be certain that there is no significant difference.

Table 7.10 shows some paired data. Plastic squares were used to create targets in a greenhouse. Each target was bi-coloured, with a white half and a yellow half. The data show the number of whitefly (a greenhouse pest, actually a hemipteran bug and not a fly at all!) trapped on the respective halves of the targets (which were sticky).

Table 7.10 Matched pair data. Counts of whitefly attracted to coloured targets in a greenhouse. Each target is bi-coloured with white and yellow halves.

White	Yellow	Difference
4	4	0
3	7	–4
4	2	2
1	2	–1
6	7	–1
4	10	–6
6	5	1
4	8	–4

You can see fairly easily that you really have paired data; it makes perfect sense to compare the two halves of each target. If you had a set of white targets and a set of yellow ones then you would not have matched data, simply two samples.

You are interested to know if there is a difference in the number of whitefly attracted to yellow or white targets. The difference column shows the result for white–yellow. You can see that there are some positive values and some negative values as well as a zero difference.

If you look at the original formula for the *t*-test (Figure 7.4), you can see that you have summary statistics for each of the two samples: mean, variance and the number of items in each sample. Because you are looking at differences between pairs of values, the top of the formula can be modified. You can take the difference between each pair of values and calculate the mean of the differences. Similarly you can replace the bottom of the formula with an expression that relates to the variance of the differences. You end up with something that looks like Figure 7.11.

$$t = \frac{\bar{D}}{\sqrt{\frac{s_D^2}{n}}}$$

Figure 7.11 The formula to calculate the matched-pairs *t*-test.

Effectively this is half of the original *t*-test formula (Figure 7.4). If you look at the whitefly data from Table 7.10 you can work out the summary values for all the columns; these results are shown in Table 7.11.

Table 7.11 Mean and variance for samples and differences for whitefly matched-pairs data.

	White	Yellow	Difference
Mean	3.88	5.63	–1.75
Variance	2.98	8.27	8.5

Of course there are eight pairs of data here and therefore eight differences (although one difference is zero). If you substitute these values into the formula you get:

$$t = -1.75 \div \sqrt{(8.5 \div 8)} = 1.698$$

You look this value up in your table of critical values (Table 7.1) and use 7 degrees of freedom, i.e. the number of pairs of observations –1 (8 – 1 = 7). You find that in this case there is no statistical difference between the white and yellow targets.

Your original data were normally distributed; generally this means that the differences between the matched pairs will also be normally distributed but you really should check before you run the test.

7.3.2 Wilcoxon matched-pairs test for skewed data

When your data are not normally distributed, or the differences between the pairs are not normally distributed, the *t*-test is not applicable. You need a non-parametric alternative. Wilcoxon's matched-pairs test is based upon the ranks of the data (like most non-parametric tests). What you do is to look at the differences and give them a rank based on the magnitude of the difference. You can then split them up according to whether they relate to a positive difference or a negative difference.

If most of the ranks are due to positive differences then you will get a fairly large sum of ranks. The other sum will be quite small. Similarly if it is the other way round and most of the differences are negative you will get a large sum of ranks due to the negative differences. Consider the whitefly data again (from Table 7.10). This time the ranks of the differences are also shown. Table 7.12 shows the data and the ranks.

The sum of all the ranks will always come to the same value for a given number of observations. Here you have eight pairs but one of them has a zero difference so you do not count that (this is slightly different to the *t*-test). For seven pairs of observations you will always get a sum of ranks of 28. The bigger the difference between the rank sums

Table 7.12 Matched-pair data. Counts of whitefly attracted to coloured targets in a greenhouse. Each target is bi-coloured with white and yellow halves. Here the non-zero differences have been ranked.

White	Yellow	Difference	All ranks	+ Ranks	- Ranks
4	4	0			
3	7	–5	5.5		5.5
4	2	2	4	4	
1	2	–1	2		2
6	7	–1	2		2
4	10	–6	7		7
6	5	1	2	2	
4	8	–4	5.5		5.5

due to the positive differences and the rank sums due to the negative difference, the more likely it is that you have a significant difference.

This is how the Wilcoxon matched-pairs test works. It compares these rank sums. You can write this formally as:

$$W = \sum R^{+/-}$$

You always end up with two values, one for each rank sum (i.e. one for the positive and one for the negative differences). You take the smaller of the two values as your test statistic. If you add up the ranks for positive and negative differences from Table 7.12 you get 6 and 22 (which adds up to the maximum 28 for seven pairs of differences). You use 6 as your test statistic and compare this to a table of critical values (e.g. Table 7.13).

If your smallest rank sum is less than or equal to the critical value, you can reject the null hypothesis and accept the alternative that there is a significant difference between the two samples. In this case your lowest rank sum was 6 and the critical value is 2. You therefore accept the null hypothesis; there is no difference between the white and yellow targets.

7.3.3 Using Excel for paired tests

In Section 7.1.1 you looked at how Excel could perform a *t*-test. Excel can take paired data too and the formula only has to be modified slightly:

```
TTEST(range1, range2, 2, 1)
```

Here you have your two samples, of course they must be in matched pairs and so each sample will be the same length. You insert a 2 to tell Excel to look at both ends of the normal distribution curve and a 1 to indicate that the data are in matched pairs as opposed to independent samples.

The `TTEST` function only gives you the final *p*-value. You can get the value of *t* without having to do all the calculations the long way by using the `TDIST` function:

```
TDIST(prob, df)
```

Table 7.13 Critical values for Wilcoxon matched pairs. Compare your lowest rank sum to the tabulated value for the appropriate number of non-zero differences. Reject the null hypothesis if your value is equal to or less than the tabulated value.

N_D	Significance level		
	5%	2%	1%
5	–	–	–
6	0	–	–
7	2	0	–
8	3	1	0
9	5	3	1
10	8	5	3
11	10	7	5
12	13	9	7
13	17	12	9
14	21	15	12
15	25	19	15
16	29	23	19
17	34	27	23
18	40	32	27
19	46	37	32
20	52	43	37
25	89	76	68
30	137	120	109

The `prob` part should point to the *p*-value you calculated using `TTEST`, whilst `df` is the number of pairs of data (the degrees of freedom). Of course you can use regular functions to calculate the *t*-value from the mean of the differences and the variance of those differences.

The *Analysis ToolPak* (Section 1.9.1) has a paired *t*-test routine and you may use this to carry out the analysis.

There is no in-built function that will work out the Wilcoxon matched-pairs result for you. However, it is fairly easy to calculate the terms you need:

1. Determine the differences; it does not matter which you subtract from which as long as you do the same for all.
2. Use the `RANK.AVG` function to get the ranks for all the differences. You will need to omit zero differences from the calculations.
3. Once you have all the ranks you can work out which were from positive differences and which from negative.
4. Sum the ranks from the positive differences. Then sum the ranks for the negative differences.
5. The smaller value is the test statistic.

You'll have to look up the critical value for the number of pairs of non-zero differences (Table 7.13).

> **Note: Using Excel for Wilcoxon matched-pairs test**
>
> See the support website for an exercise about conducting the Wilcoxon matched-pairs test using Excel. There are some sample data for you to use and follow along.

7.3.4 Using R for paired tests

The R program has routines for calculating paired tests built in. If you have paired data then both the *t*-test and *U*-test (Wilcoxon matched pairs) may be run in paired form simply by adding `paired = TRUE` to the command, e.g.:

```
> wilcox.test(data1, data2, paired = TRUE)
```

If the two samples are unequal in length then R gives you a helpful error message:

```
Error in wilcox.test.default(data1, data2, paired = T) :
"x" and "y" must have the same length
```

If the samples are the same length but are not really paired, R will carry on regardless; it is a computer program, not a mind reader!

7.3.5 Graphs to illustrate paired tests

You saw in Chapter 6 how to represent the results of regular differences tests (Section 6.3). Usually a bar chart (Section 6.3.1) or box–whisker plot (Section 6.3.2) is what you need. However, when it comes to a matched-pairs test these graphs do not really show how the pairs relate to one another, as they focus on representing the samples independently.

An alternative approach is to use a scatter plot, where you chart one sample against the other. If you add a line with slope 1 and intercept 0, you can see how the points relate to this line, known as an *isocline* (Figure 7.12).

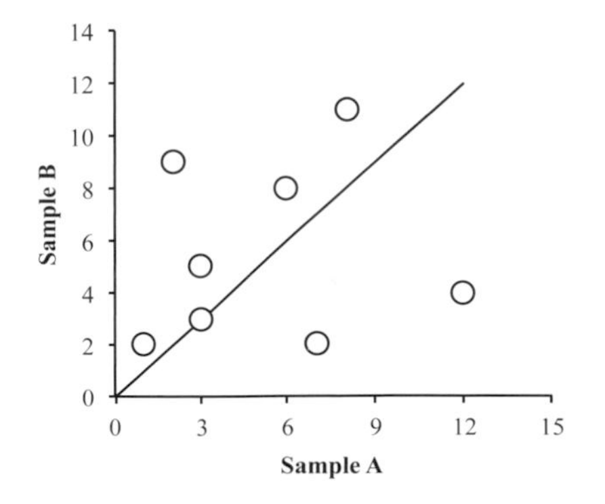

Figure 7.12 A scatter plot used to illustrate a matched-pairs situation. The isocline has slope 1 and intercept 0. If the points are all close to the isocline, or equally distributed around it, the result is likely not significant.

In the example (Figure 7.12) you can see that there are more points to one side of the line than the other, but the points are fairly close to the isocline. The result is not statistically significant ($W = 12$, $W_{crit} = 2$). The scatter plot may not always be the best graph to show when you have a matched pairs situation but it is a potentially useful tool. Look at the comparison between a boxplot and a scatter plot using the same data for example (Figure 7.13).

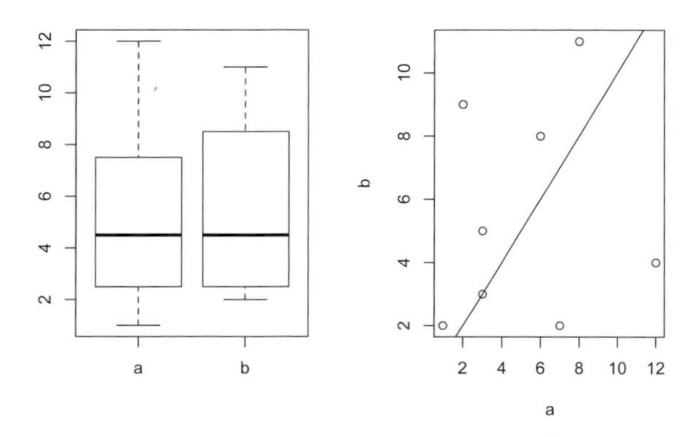

Figure 7.13 Matched pairs data presented in two contrasting ways. The boxplot shows the median, IQR and range for the samples independently. The scatter plot shows one sample plotted against the other. The isocline represents a line of parity (zero difference).

Note: Graphical representation of matched pairs

See the support website for some more information and notes about graphical representation of matched pairs data.

EXERCISES

Answers to exercises can be found in the Appendix.

1. Which of the following items best describes what you need in order to carry out a *U*-test?
   A. Median, range and number of replicates
   B. Mean, variance and number of replicates
   C. Mean, standard deviation and number of replicates
   D. Median and number of replicates
   E. Number of replicates
2. You've counted the number of *Plantago lanceolata* plants in quadrats in two different areas of a field. The first sample is from the main field and the second is from a path through the field. You determine that the data are normally

distributed and the summary statistics are as follows (mean, standard deviation, replicates): *Field* 4.00, 1.41, 8. *Path*: 5.75, 1.67, 8. Calculate a value for *t*. Is the result statistically significant?

3. You've got some data from counts of earthworms in samples of soil. Here are the values: 0 1 0 4 5 0 1 3. Rank these data.
4. If your data are in matched pairs you can use a special version of the *t*-test or *U*-test, depending if the samples are normally distributed or not. TRUE or FALSE?
5. You can rank data in Excel using the ____ function.

Chapter 7: Summary

Topic	Key points
The *t*-test	The *t*-test is used to explore the difference between two samples that are normally distributed. You require the mean and variance of each sample and the number of replicates. There are two main variants of the *t*-test, using equal or unequal sample variance. You can easily calculate the *t*-test result in Excel, various functions can help, e.g. `AVERAGE`, `VAR`, `COUNT`. Other functions include `TTEST`, `TDIST`, `TINV`. The `t.test()` command in R runs the *t*-test.
Using the *Analysis ToolPak* for the *t*-test	The Excel add-in *Analysis ToolPak* can compute the *t*-test statistics for you.
The *U*-test	The *U*-test is used to explore the difference between two samples that are not normally distributed. In order to carry out the test you need to rank the data in size order (smallest first). The *U*-test effectively compares the median values of two samples. You can easily calculate the *U*-test result in Excel. The `RANK.AVG` function allows you to rank the sample data but you cannot automatically determine critical values. The `wilcox.test()` command in R will carry out the *U*-test.
Ranking data	Most non-parametric tests use the ranks of the original data. The smallest original data item gets the lowest rank. Any tied values must get the same rank, so use the mean rank for ties. The next largest value gets the next available rank.
Paired tests	There are versions of the *t*-test and the *U*-test (Wilcoxon matched pairs) for when data are laid out in matched pairs. You can easily calculate the paired tests using Excel. The critical values of *t* are available via `TINV` but there is no way to get critical values for *W* (you need to compare to published tables e.g. Table 7.13). The *Analysis ToolPak* can carry out a paired *t*-test. The `t.test()` and `wilcox.test()` commands in R can carry out matched pair versions.

Chapter 7: Summary – *continued*

Topic	Key points
Visualizing paired test results	A scatter plot can be used to plot one sample against the other when you have matched-pair data. Add an isocline (slope = 1, intercept = 0) to show the line of parity. This is a good alternative to a bar chart or box–whisker plot.

8. Tests for linking data – correlations

When you wish to examine a link between two samples (numeric variables), you are generally looking for a correlation. In this chapter you will look at the simplest situations: finding links (correlations) between two samples.

What you will learn in this chapter

» How to carry out Spearman rank correlation
» How to carry out Pearson's product moment correlation
» How to use Excel for correlations
» How to use R for correlations
» How to calculate curvilinear correlation using polynomial and logarithmic models
» How to add lines of best fit (trendlines) to scatter plots

Generally one of your samples will be a biological variable and the other an environmental variable but this is not always the case. The distribution of the data is an important factor, as this determines which kind of correlation you should go for:

- *Skewed distribution* (non-parametric) – Spearman's rank correlation (Section 8.1).
- *Normal distribution* (parametric) – Pearson's product moment correlation (Section 8.2). This is also sometimes called regression.

You should always show your data graphically – scatter plots are what you need for this (recall Section 6.4). The relationship between the two variables does not have to be a "straight line"; in Section 8.5 you'll see how to deal with relationships that are polynomial or logarithmic. These relationships are described as curvilinear.

When you have one factor and think that there are several other factors that may influence it, you have a more complex situation and this will be dealt with in Chapter 11. If your data are categorical then you will look for an association; this will be the subject of Chapter 9.

8.1 Correlation: Spearman's rank test

Correlation is a way to link together two (numeric) variables. In simple correlation you are simply looking for the strength of the relationship. Here are some examples of where a correlation could be explored:

- Mayfly (Ephemeroptera) abundance in relation to stream flow.
- Density of sheep (grazing pressure) and the abundance of a particular plant.
- The number of sites containing a plant species (e.g. Japanese knotweed) over time.

If you collected some data you would represent it graphically using a scatter plot (see Section 6.4). Your finished graph might look something like the following example (Figure 8.1).

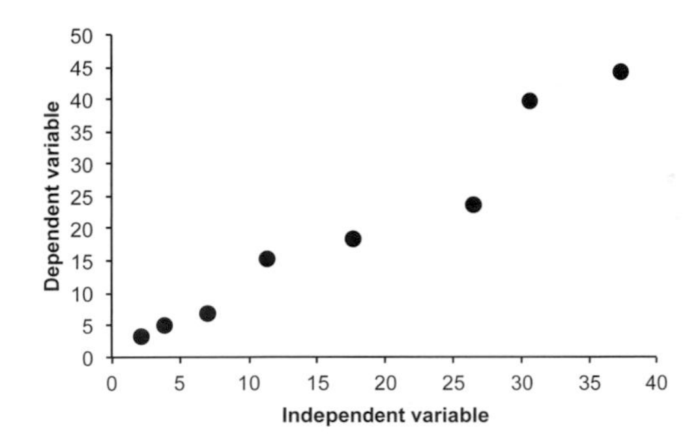

Figure 8.1 Scatter plot. This shows a perfect positive correlation between the independent and dependent variables (the points do not have to be a perfectly straight line).

Here the axes have been labelled as *Independent* variable and *Dependent* variable. They are always this way round, with the *y*-axis showing the dependent variable. Which axis is which? In your examples above, the stream flow, sheep density and year are the independent variables. This independent variable is what you think controls (or explains) the abundance. The dependent variable is the abundance data or number of sites. These variables can also be called the *response* and the *predictor* variables respectively. It is the

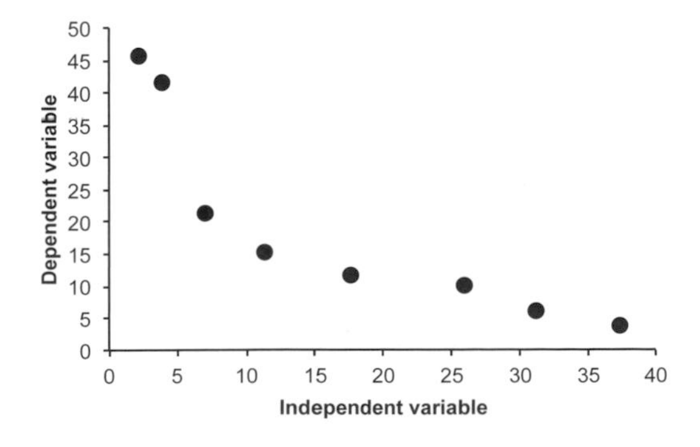

Figure 8.2 Scatter plot showing perfect negative correlation (note that this does not have to be a straight line).

abundance that you are interested in and you want to know if this changes in some predictable manner when the independent variable changes.

There are three possible outcomes. In Figure 8.1 you see an example of a perfect positive correlation. As you move up the x-axis, the next largest value corresponds to the next largest value on the y-axis. As one axis increases so does the other, a positive correlation. You might also get a graph that looks like Figure 8.2. In this case, you see that as you move up the x-axis you go from large y-values to smaller ones. You have a negative correlation. As you move up the x-axis you correspond exactly to steps down the y-axis so you have a perfect negative correlation. There is another possibility:

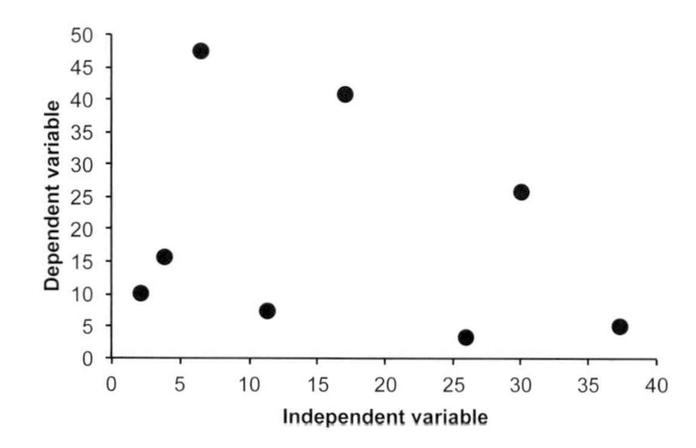

Figure 8.3 Scatter plot showing a perfect mess! There is no evident relationship between the two variables here.

In Figure 8.3 you cannot determine any pattern at all. As you move up the x-axis you do not encounter any regular pattern.

The Spearman's rank correlation test is used when you wish to determine the strength of the link between two sets of values. The link does not have to be in the form of a straight line (look at Figure 8.2 for example) but it should be in the form of a general trend up or down (a U-shape is not good, but see Section 8.5). The test is designed so that you get a value of +1 when you have a perfect positive correlation, –1 with a perfect negative correlation and 0 when you have a perfect mess. As you may guess from the name, the test converts the actual values to ranks; it is a non-parametric test and is used on data that are not normally distributed (i.e. skewed).

Table 8.1 Stream speed (m s^{-1}) and mayfly (*Ecdyonurus dispar*) abundance.

Speed	Abundance
2	6
3	3
5	5
9	23
14	16
24	12
29	48
34	43

Table 8.1 shows data collected about mayflies from a stream and for each sample the stream speed and the number of mayfly were measured. In this case the pattern is not as clear as your perfect illustrations.

If you create a scatter plot of the data in Table 8.1 you see the pattern more clearly (Figure 8.4).

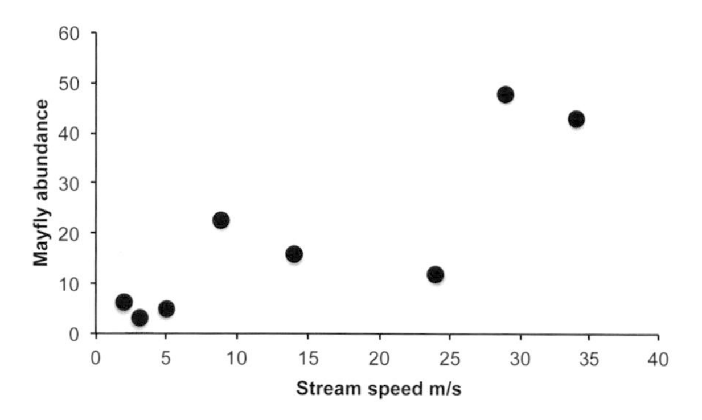

Figure 8.4 Scatter plot of mayfly (*Ecdyonurus dispar*) and stream speed data. The stream speed is the independent axis and the mayfly data the dependent axis.

It looks like there may be a relationship between speed and abundance but there are a number of inconsistencies, it is not a perfect correlation. Is it statistically significant? You use the Spearman's rank test to determine the significance.

You begin with a hypothesis to test. In this case you expect a greater abundance of mayfly as stream speed increases. You would write this formally (your alternative hypothesis H1) as:

"There is a positive correlation between stream speed and mayfly (*Ecdyonurus dispar*) abundance."

You are saying that the relationship is a positive one because of the previous research you have done (reading papers and so on). Your null hypothesis (H0) now becomes:

"There is no correlation between stream speed and mayfly abundance."

Note that H0 is not the opposite of H1; you did not say that the correlation is negative (look back at Sections 1.4 and 5.2).

8.1.1 Calculating the Spearman coefficient

The next thing you must do is to assign ranks to your data. In the *U*-test (looking for differences, Section 7.2) you ranked the whole lot as one set. Here you are looking to compare the ranks of the *y*-data with the ranks of the *x*-data so you rank each set separately. The smallest value gets the smallest rank. Any tied values get the averages ranks (see the examples of ranking in Section 7.2). Table 8.2 shows the data with the ranks added in.

You look down the first column of Table 8.5 and find the number of data (in this case 8) then you read the critical value from the next column (in this case it is 0.738). If your value is greater or equal to this then you can reject your null hypothesis and accept the alternative. In this instance, you can say that there is a statistically significant positive correlation between stream flow and abundance.

You could look over to the next column (headed 2%) but you see that you do not exceed this value. Your correlation is significant at the 5% level so you assume that there is a less than 5% probability that the result was due to random chance. If you get a negative correlation coefficient, then you simply ignore the sign when looking up the table of critical values. You can see that it is much easier to cross the "finishing post" when you have more data. This is something to consider when designing your project.

How the Spearman's rank test works

When you have a perfect positive correlation, the ranks of the x- and y-values will all agree perfectly. Look at the example in Table 8.6.

Table 8.6 Example of perfect positive correlation. The ranks for the independent factor agree perfectly with the ranks of the dependent factor.

Speed	Rank speed	Abundance	Rank abundance	Difference	D^2
2	1	3	1	0	0
3	2	5	2	0	0
5	3	6	3	0	0
9	4	12	4	0	0
14	5	16	5	0	0
24	6	23	6	0	0
29	7	43	7	0	0
34	8	48	8	0	0
Sum:				0	0

In Table 8.6 you see that the ranks agree perfectly and as a result the sum of the rank differences squared ($\sum D^2$) is bound to be zero. Now let's look at an example of perfect negative correlation (Table 8.7).

In Table 8.7 (perfect negative correlation) you see that the ranks all disagree as much as possible so the difference of the ranks squared will be as large as possible. Exactly how large will depend on how many data there are. It so happens that the maximum $\sum D^2$ can be determined using the following equation (Figure 8.6):

$$Max \sum D^2 = \frac{n\left(n^2 - 1\right)}{3}$$

Figure 8.6 Maximum differences of squared ranks form the basis for the Spearman's rank correlation test.

Table 8.7 Example of perfect negative correlation. The ranks for the independent factor disagree perfectly with the ranks of the dependent factor.

Speed	Rank speed	Abundance	Rank abundance	Difference	D^2
2	1	48	8	7	49
3	2	43	7	5	25
5	3	23	6	3	9
9	4	16	5	1	1
14	5	12	4	–1	1
24	6	6	3	–3	9
29	7	5	2	–5	25
34	8	3	1	–7	49
Sum:				0	168

Now you want to get to the point where you have values ranging from +1 to –1. At present you have 0 to max $\sum D^2$. You can now create a new equation using max $\sum D^2$ that will give you the required –1 to +1 range (Figure 8.7).

$$r_s = 1 - \frac{2\sum D^2}{Max\sum D^2}$$

Figure 8.7 The Spearman rank correlation can be expressed in terms of the maximum differences in squared ranks.

If you substitute the equation for max $\sum D^2$ (Figure 8.6) into the equation in Figure 8.7 you end up with the final Spearman's rank formula you saw earlier (Figure 8.5).

Correlation may not mean cause

It is important to keep in mind that a statistically significant correlation does not necessarily mean that one factor *causes* the other. There may be some other factor that is actually responsible that you have not determined. For example the stream velocity itself may not be responsible for the abundance of mayfly but the speed of flow may influence the number of food particles washed along or the oxygenation.

You can use R to carry out the Spearman's rank test quite easily, and you can also use Excel, but before that you'll look at another type of correlation.

8.2 Pearson's product moment

When your data are normally distributed (i.e. parametric), you can use a method of correlation called Pearson's product moment, or simply regression. Unlike the non-parametric Spearman rank correlation, regression makes one more assumption, namely that the relationship between the two variables can be described by a simple mathematical formula. The simplest relationship is a straight line, which is described by the following formula:

$$y = mx + c$$

In this case the y variable is your dependent factor (the response variable). The x is the independent factor (the predictor variable). The c represents the point on the y-axis where the line of best-fit crosses. Finally the m represents the slope of the line. Look at Figure 8.8. You see a scatter plot with a series of points representing some x and y measurements, the dependent and independent variables.

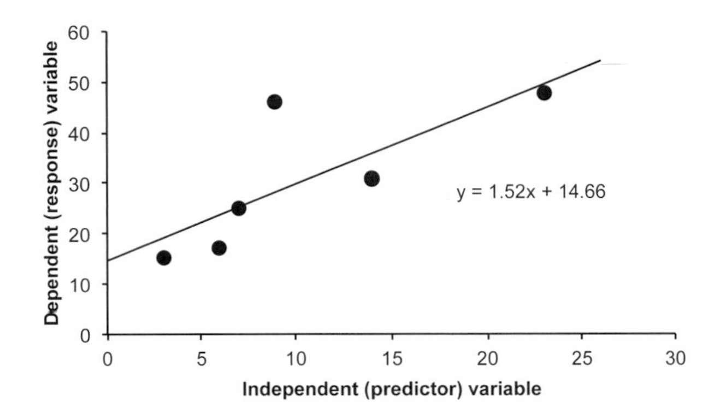

Figure 8.8 A scatter plot to show the elements of a straight-line relationship. The equation shows the relationship in the form $y = mx + c$.

The scatter plot in Figure 8.8 includes a line of best fit (also known as a trendline), you can see that it crosses the y-axis at around 15. You could estimate the slope by reading off the graph (the slope is the change in y-values divided by the change in x-values) but the equation is displayed on the graph to make things easier. You will see how to add lines of best fit later (Section 8.3.1 for Excel, and Section 8.4.2 for R).

Once you know what the equation is, then if you are given a value for x you could calculate the corresponding value of y (and vice versa).

What Pearson's product moment does is to calculate not only the strength of the relationship (like Spearman's rank test) but also the equation that describes the linear fit between the two variables.

To work out the slope, m, you need to evaluate the following formula (Figure 8.9):

$$m = \frac{\sum(x - \bar{x})(y - \bar{y})}{\sum(x - \bar{x})^2}$$

Figure 8.9 The formula to calculate the slope of a straight line.

The formula (Figure 8.9) looks a little scary but actually it is quite simple (just tedious to do). Proceed as follows:

1. Start by evaluating the means for x and y (the dependent and independent factors).
2. Then you take the mean away from each value.
3. On the top line you multiply differences between x and between y (for each pair of data), then add these all together.
4. On the bottom line you examine the x only but square the differences (from the mean) before adding them together.
5. Finish by dividing the top line by the bottom.

The intercept is quite easy to calculate using the formula shown below (Figure 8.10):

$$c = \overline{y} - m\overline{x}$$

Figure 8.10 The formula to calculate the intercept of a straight line, *m* is the slope.

Once you know the slope, you are there, as you already worked out the means of x and y. Now you can reconstruct the formula that represents the line of best fit and could draw it on a graph if you wanted. The line must pass through a point on the graph that is represented by the mean of x and the mean of y. You also know the intercept. If you marked these two points and joined them up, you would make the best-fit line. You could then check the slope to ensure it was correct.

This is only part of the story because you still do not know if your putative linear relationship is statistically significant. In order to achieve this you set up your hypothesis (H1), which will be that there is a correlation between the two factors. Recall from the Spearman's rank test that you also defined the direction of the correlation; you should do the same here. The null hypothesis (H0) is that there is no correlation. The formula that Pearson came up with is in Figure 8.11:

$$r = \frac{\sum(x-\overline{x})(y-\overline{y})}{\sqrt{\sum(x-\overline{x})^2 \sum(y-\overline{y})^2}}$$

Figure 8.11 Formula to calculate the Pearson correlation coefficient.

It is more tedious than really difficult and most often you would use a computer to do it for you. In fact there is a function built into Excel to do this. Because the linear relationship is so commonly used, Excel also has functions that will evaluate the slope and intercept of a set of paired values (see Section 8.3).

Tackle the calculation in steps as follows:

1. Calculate the mean for x and for y.
2. Subtract the mean of x from each x-value to get a series of differences (let's call them dx).
3. Repeat step 2 for the y-values to get dy.

4. Multiply dx and dy for each row of your data, then add the results to get the top of the equation, the numerator.
5. Now take all your dx values and square them. Sum this result.
6. Repeat step 5 for the dy values.
7. Multiply the sum of the dx^2 values and the dy^2 values.
8. Take a square root of the value from step 7 to get the bottom of the equation, the denominator.
9. Divide the numerator (from step 4) by the denominator (step 8) to get the final value of Pearson's product moment, r.

So, you end up with a statistic called r. The final stage is to compare this to a critical value. Table 8.8 gives critical values for the Pearson correlation coefficient. The degrees of freedom are calculated by subtracting 2 from the number of items of data you have. So, if you have three pairs of data, the $df = 1$ (that is 3 y-values and 3 x-values; you cannot have an x-value without a y-value, they are always in pairs).

If your calculated value is greater than or equal to the tabulated value, the relationship is statistically significant. If your r-value is negative just ignore the sign when looking it up in the table. As with the Spearman correlation, a negative value indicates that the correlation is negative, that is as y increases, x decreases.

With the Spearman correlation, you did not make any assumptions regarding the cause and effect. The same is true here; just because there is a linear link between the two variables does not mean necessarily that there is cause and effect.

8.3 Correlation tests using Excel

You can use Excel to carry out tests for parametric correlation (i.e. Pearson product moment); there are several built-in functions as well as the *Analysis ToolPak* (Section 1.9.1). Non-parametric correlation (i.e. Spearman rank) is less well served, although you can calculate the ranks and correlation coefficient easily enough.

8.3.1 Parametric tests of correlation

It is possible to conduct parametric correlation using Excel. There is a function that will work out the Pearson correlation coefficient:

```
CORREL(range1, range2)
```

Here you replace the `range1` and `range2` parts with the cell references that contain your data (it does not matter which is the dependent and which the independent variables). The result given is the coefficient. There is also a `PEARSON` function, which does exactly the same.

You can also determine the slope and intercept:

```
SLOPE(range_y, range_x)
INTERCEPT(range_y, range_x)
```

In these two cases you need to specify which range of data is the dependent (*response*,

Table 8.8 Critical values for the correlation coefficient *r*, Pearson's product moment. Reject the null hypothesis if your value is equal to or greater than the tabulated value.

Degrees of freedom	Significance	
	5%	1%
1	0.997	1
2	0.95	0.99
3	0.878	0.959
4	0.811	0.917
5	0.754	0.874
6	0.707	0.834
7	0.666	0.798
8	0.632	0.765
9	0.602	0.735
10	0.576	0.708
12	0.532	0.661
14	0.497	0.623
16	0.468	0.59
18	0.444	0.561
20	0.423	0.537
22	0.404	0.515
24	0.388	0.496
26	0.374	0.478
28	0.361	0.463
30	0.349	0.449
35	0.325	0.418
40	0.304	0.393
45	0.288	0.372
50	0.273	0.354
60	0.25	0.325
70	0.232	0.302
80	0.217	0.283
90	0.205	0.267
100	0.195	0.254
125	0.174	0.228
150	0.159	0.208
200	0.138	0.181
300	0.113	0.148
400	0.098	0.128
500	0.088	0.115
1000	0.062	0.081

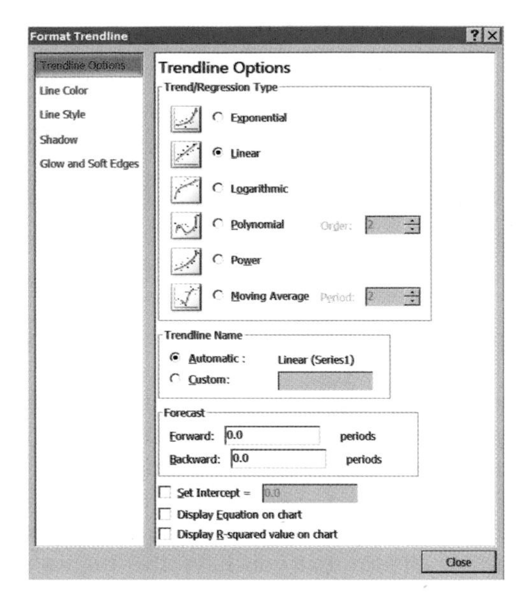

Figure 8.13 The *Format Trendline* dialogue box (here in Excel 2010) gives a range of formatting options for the line of best fit.

2. Subtract the rank of one variable from the rank of the other (it does not matter which way around).
3. Square the rank differences.
4. Sum the squared rank differences.
5. Use the `COUNT` function on one of the variables to see how many items there are (the n term in the Spearman formula).
6. Substitute the values into the Spearman rank formula (Figure 8.7), to obtain r_s, the coefficient.
7. Compare your value of r_s to the critical value (Table 8.5); ignore any minus sign. If your value is equal or greater than the critical value then the result is statistically significant (at $p < 0.05$).

So, there is no shortcut method for Spearman rank correlation in Excel, but the calculations are carried out quite easily. You still have to compare the final result to the table of critical values to determine the significance of the analysis.

Note: The link between Spearman rank and Pearson product moment

The Spearman rank correlation uses ranks of variables to determine a correlation. If you use Pearson's product moment calculations on the ranks, instead of the original data, you obtain the same value (to at least two decimal places). This allows you to use the `CORREL` function, which saves a few mathematical steps. See the support website for an exercise on using Excel for the Spearman rank correlation.

8.3.3 Using the Analysis ToolPak for correlation in Excel

You can carry out correlation using the *Analysis ToolPak* (Section 1.9.1), which allows you to calculate Pearson's product moment (i.e. parametric correlation). You cannot carry out Spearman's rank correlation directly but you can determine the data ranks and use the *Analysis ToolPak* on those.

You start the process by using the *Data > Data Analysis* button, which opens the *Data Analysis* dialogue window. There are a host of routines you can run but for the correlation there are two basic options:

- *Correlation* – gives you the correlation coefficient only.
- *Regression* – calculates the correlation coefficient and the statistical significance, as well as additional results such as slope and intercept.

The *Regression* option is what you need if you are after the statistical significance (parametric correlation is just a simple version of regression, where you only have two variables). In order to proceed, follow these general steps:

1. Click the *Data > Data Analysis* button to open the *Data Analysis* dialogue window.
2. Scroll down and select the *Regression* option, which will open the *Regression* dialogue window (Figure 8.14).

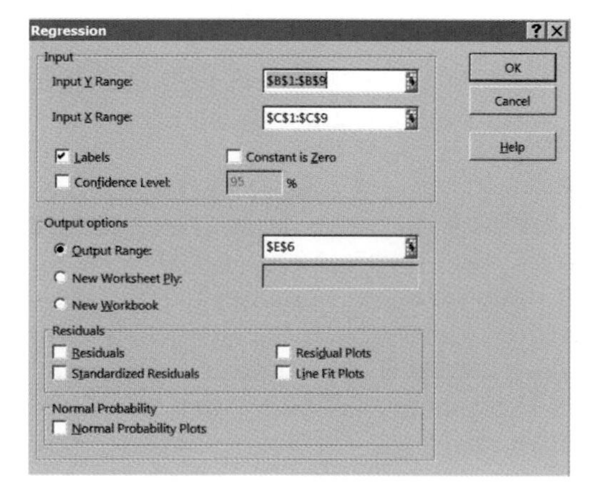

Figure 8.14 The *Regression* dialogue box from the *Analysis ToolPak* allows you to carry out Pearson correlation.

3. Use the mouse to select the x and y data separately. You can include the headings, in which case tick the *Labels* button.
4. In the *Output Options* section choose where you want the results to go.
5. Click OK to finish and place the results where you chose in step 4 (Figure 8.15).

The results are extensive but for basic correlation you are interested in the following:

- Correlation coefficient – this is in the top table, labelled *Multiple R*.
- Number of replicates – this is in the top table, labelled *Observations*.

SUMMARY OUTPUT

Regression Statistics	
Multiple R	0.72372058
R Square	0.52377147
Adjusted R S	0.44440005
Standard Err	10.1613138
Observation:	8

ANOVA

	df	SS	MS	F	Significance F
Regression	1	681.361213	681.361213	6.59899322	0.04240053
Residual	6	619.513787	103.252298		
Total	7	1300.875			

	Coefficients	Standard Erro	t Stat	P-value	Lower 95%	Upper 95%	Lower 95.0%	Upper 95.0%
Intercept	8.25459559	5.8531414	1.41028467	0.2081327	-6.0675255	22.5767166	-6.0675255	22.5767166
flow	0.79136029	0.30806007	2.56885056	0.04240053	0.03756445	1.54515614	0.03756445	1.54515614

Figure 8.15 The results of the *Regression* analysis from the *Analysis ToolPak* are extensive.

- Intercept – this is in the lower table in the *Coefficients* column.
- Slope – this is under the intercept in the *Coefficients* column. The row takes its name from the heading of the x-data (the predictor or independent variable).
- Statistical significance – this is in two places, in the middle table it is labelled *Significance F*, whilst in the bottom table it is labelled *P-value* (make sure to read the correct row, the last one. The other p-value relates to the significance of the intercept). In either event it gives the exact probability of the chance of getting this result by chance so if it is <0.05 the result is statistically significant.

The results give several values you are not really bothered about at this stage. For example, the value of F is a statistic you'll meet later (analysis of variance, Chapter 10). All you need to know now is that $F = t^2$ (recall you calculated a value for t in the correlation earlier).

The lower table also shows you the significance of the regression and you can see that the p-value for the flow is the same as the *Significance* shown in the middle table. This table also has a value for t and if you square this value you get the value for F from the middle table (the F and t statistics are closely related). You also see values for the 95% confidence intervals. When you come to examine multiple regression (see Chapter 11) this bottom table would contain more information (because you have more variables).

If you want to carry out Spearman's rank correlation you could use `RANK.AVG` to get the ranks for each variable then run the *Analysis ToolPak* on the ranks instead of the original values. The p-value for the result will be slightly less conservative than using the critical values from Table 8.5. See the support website for some examples of using Pearson's product moment to calculate Spearman's rank correlation.

8.4 Correlation tests using R

You can carry out correlation in R quite easily using the `cor.test()` command. The command will work with parametric or non-parametric data and its general form is as follows:

`cor.test(x, y, data, method)`

`x, y`	The data to be correlated. You can give the data as two separate variables, or as a formula of the form `~ x + y`.
`data`	If the variables are specified as a formula you can give the name of the `data.frame` that holds the variables.
`method`	The method used for the correlation, one of: `"pearson"`, `"kendall"`, `"spearman"`. You can use an abbreviation. The default is `"pearson"`.

The two variables must be the same length of course (you get an error if they are not). You can specify the variables in the form of a formula but note that this is "one sided", `~ x + y`, with the variables (the order does not matter) both to the right of the tilde.

8.4.1 Non-parametric tests of correlation

The basic `cor.test()` command will work out Spearman's rank or Pearson coefficients. It will also calculate Kendall's coefficient; this is a non-parametric test and broadly similar to Spearman's rank test but less often used in ecology.

To undertake the Spearman rank correlation you specify `method = "spearman"` like so:

```
> Speed
[1]  2  3  5  9 14 24 29 34
> Abund
[1]  6  3  5 23 16 12 48 43
> cor.test(Speed, Abund, method = "spearman")
  Spearman's rank correlation rho
data:  Speed and Abund
S = 16, p-value = 0.02178
alternative hypothesis: true rho is not equal to 0
sample estimates:
      rho
0.8095238
```

The output is sparse but covers all you really need to know. The results give *S*, which is the sum of the differences in ranks squared; *p*-value, which tells you if the correlation is significant; and *rho*, the value of the Spearman's rank correlation coefficient itself.

8.4.2 Parametric tests of correlation

The basic `cor.test()` command will calculate the parametric Pearson's product moment by default:

```
> mayfly
  Speed Abund
1     2     6
2     3     3
```

```
3     5     5
4     9    23
5    14    16
6    24    12
7    29    48
8    34    43
> cor.test(~ Speed + Abund, data = mayfly)
  Pearson's product-moment correlation
data:  Speed and Abund
t = 3.8568, df = 6, p-value = 0.008393
alternative hypothesis: true correlation is not equal to 0
95 percent confidence interval:
 0.3442317 0.9711386
sample estimates:
      cor
0.8441408
```

You can see that the results give you a value for the coefficient (labelled *cor*, at the bottom). You also get a *t*-value, degrees of freedom and the exact *p*-value.

Slope and intercept values

The `cor.test()` command does not compute the slope or intercept; you need to use the `lm()` command to calculate these. The `lm()` command is used for *linear modelling* (i.e. regression), which you will see extensively in Chapter 11.

The `lm()` command will only accept input in the form of a formula: `y ~ x`, where *y* is your response variable (dependent) and *x* is the predictor (independent). However, the variables do not have to be "together" in the same data object:

```
> lm(Abund ~ Speed)
Call:
lm(formula = Abund ~ Speed)
Coefficients:
(Intercept)        Speed
      1.867        1.176
> lm(abund ~speed, data = fw)
Call:
lm(formula = abund ~ speed, data = fw)
Coefficients:
(Intercept)        speed
     3.2338       0.7216
```

In the first example the data are two separate vectors, whilst in the second the variables are in the *fw* data object (the original data are in the *freshwater correlation.xlsx* file on the support website).

You can see that you get two "coefficients", the intercept is labelled as such and the slope is labelled with the name of the predictor variable. If you save the result of the `lm()` command to a named object you can access the slope and intercept later:

```
> mayfly.lm = lm(Abund ~ Speed, data = mayfly)
> coef(mayfly.lm)
(Intercept)       Speed
   1.866728    1.175551
> coef(mayfly.lm)[2]
   Speed
1.175551
```

The `coef()` command accesses the coefficients (intercept and slope) from the regression result. In correlation there are only two coefficients (unlike multiple regression, see Chapter 11) and you can get the intercept only by adding `[1]` to the `coef()` command. In the example `[2]` was used to get the slope.

Adding a trendline in R scatter plots

You should visualize your data using a scatter plot (Section 6.4) and the `plot()` command will do this easily (Section 6.4.2). When you have a linear relationship, which you have described using a formula (e.g. $y = mx + c$), you may elect to show the line of best fit on the graph. The `abline()` command can add a straight line to an existing plot (you met the command earlier when drawing gridlines).

The `abline()` command can draw a line using `a = intercept` and `b = slope` parameters. The command can also "read" the required values (the coefficients) from the result of a `lm()` command. So, to produce a scatter plot with a best-fit line (trendline) you need to proceed in the following manner:

1. Carry out the correlation using the `cor.test()` command.
2. Run the `lm()` command and save the result to a named object.
3. Use the `plot()` command to make the scatter plot.
4. Use `abline()` with the result of step 2 to add the trendline.
5. Optionally you can add the equation using the `text()` command (recall Table 6.6).

For the mayfly data in the current example you'd type the following:

```
> cor.test(~ Speed + Abund, data = mayfly)
> mayfly.lm = lm(Abund ~ Speed, data = mayfly)
> coef(mayfly.lm)
(Intercept)       Speed
   1.866728    1.175551
> plot(Abund ~ Speed, data = mayfly)
> abline(mayfly.lm)
> text(locator(1), "y = 1.175x + 1.867")
```

These commands produce a chart similar to Figure 8.16. The `locator()` command takes the place of a number of x,y-coordinate pairs (in this case 1). Once the command is run R will wait for you to click on the chart: the text is placed (centred) on the spot you click.

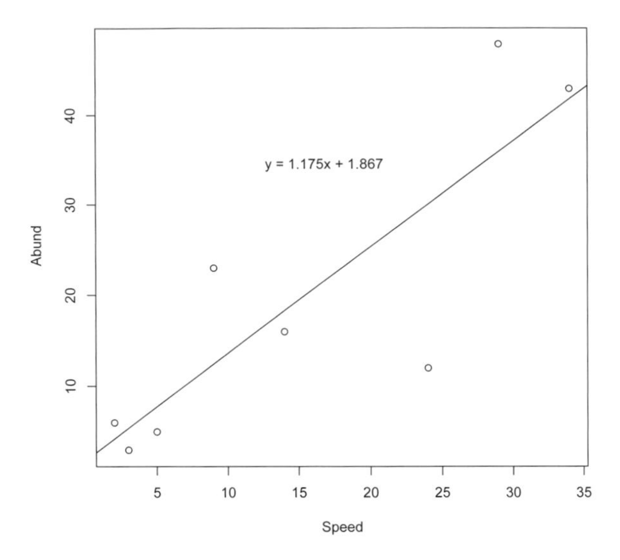

Figure 8.16 Scatter plot and trendline produced using R commands. The `abline()` command adds a line, taking the slope and intercept from the result of a linear model (`lm()` command). The equation is added using the `text()` command.

8.5 Curved linear correlation

When you undertake a correlation you make certain assumptions. If your data are not normally distributed your assumptions are less rigorous than if you have parametric data. In the Spearman rank correlation your only assumption is that the relationship is broadly headed in one direction, the link between the variables does not have to be exactly a straight line (but it should not be a ∪ or ∩ shape).

When the data are normally distributed you can use Pearson's product moment and assume the relationship is in the form $y = mx + c$ (Section 8.2). This assumes that the relationship is in a straight-line form.

However, there are plenty of times when you will get a relationship that is not in the form of a straight line – look at Figure 8.17 for example.

In Figure 8.17 you can see three relationships, highlighted by a line of best fit. The first one (a) is definitely a straight-line type of relationship. The middle plot (b) shows a relationship where the general trend is upwards; however, you can see a definite bend and the line of best fit starts off steep and becomes gradually flatter. The last plot (c) is an inverted U shape.

The standard (parametric) correlations that you have looked at so far would not describe the relationships in Figure 8.17b and 8.17c adequately as they are not in the form of a straight line. You could use Spearman's rank correlation for (a) and (b) but not for (c). In any event there is a better alternative.

In Figure 8.17b the response variable is dependent not on the value of the predictor variable directly but on the logarithm of the variable. Logarithms crop up quite a bit in natural sciences, for example pH, Richter scale and sound (measured in decibels).

In Figure 8.17c you can get this (inverted U) relationship when looking at things that

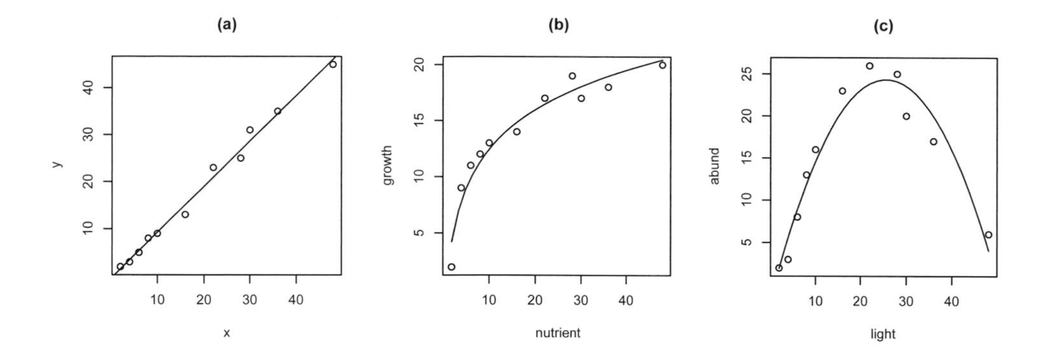

Figure 8.17 Three forms of linear correlation: (a) straight line; (b) logarithmic; (c) polynomial. In all three cases the relationship can be described by a simple linear formula broadly in $y = mx + c$ form.

have an optimum in terms of an environmental factor. For example, a species might be most happy when soil moisture is at a certain level. When the soil becomes too dry or too wet, the abundance of the species declines.

The main point is that rather than $y = mx + c$ you have another mathematical formula that could represent the relationship you observe. It is therefore important to present your data graphically as part of the analysis (see Section 6.4). By drawing a graph you can guess the most correct form of relationship. Two of the most common relationships are:

- Logarithmic (Figure 8.17b): $y = a \text{ Ln}(x) + c$.
- Polynomial (Figure 8.17c): $y = ax + bx^2 + c$.

There are many other potential equations. The way you carry out these correlations is more akin to multiple regression (Chapter 11) than to simple regression between two factors. For this reason you'll see a brief overview here but the details of the analysis will be presented in Chapter 11.

8.5.1 Logarithmic correlation

The equation that describes a logarithmic fit is:

$$y = a \text{ Ln}(x) + c$$

Although you have a different looking formula to the one for a straight line, you can see it has essentially the same elements. You have c, the intercept. You also have a, which is analogous to the slope. The x variable is represented but you use the natural logarithm (any other logarithm would produce a similar result) rather than the plain value. This form of equation produces a curve that starts steeply and then flattens out (Figure 8.17b).

You will look at logarithmic regression again when you examine multiple regression in Chapter 11.

8.5.2 Polynomial correlation

The equation that describes a polynomial is:

$$y = ax + bx^2 + c$$

You have a more complex situation compared to before as you have an extra part to deal with. You can see that x is represented twice. The terms a, b and c are fixed values. The a and b terms are coefficients that are analogous to the slope m in the straight-line equation. The c is the intercept.

You will look at polynomial regression again when you examine multiple regression in Chapter 11.

EXERCISES

Answers to exercises can be found in the Appendix.

1. Correlation describes the ____ and ____ of the ____ between two variables. If your data are normally distributed you can use ____. If data are non-parametric then you use ____ correlation.
2. You have collected data on the number of mayfly nymphs in a stream and the cover of algae on the rocks. What is the Spearman rank correlation coefficient? Is the relationship statistically significant?
   Mayfly: 4 6 12 23 18 35
   Algae: 4 12 23 18 12 35
3. Here are four correlation coefficients (Pearson's r). Each was calculated from samples with 24 observations. Which (if any) of the correlations are statistically significant?
   A. –0.404
   B. 0.765
   C. 0.396
   D. –0.806
4. You have used Excel to calculate the Pearson correlation coefficient, r, between two variables. However, you have not got your table of critical values. How can you assess the statistical significance (without using the Internet)?
5. In order to visualize a correlation you can use a scatter plot with a trendline. TRUE or FALSE?

Chapter 8: Summary

Topic	Key points
Correlation	Correlation explores the strength and direction of the link between two numeric variables. Often one variable is biological and the other environmental, but this does not have to be the case. The distribution of the data is important, as it defines the type of correlation you can use.
Spearman's rank correlation	When data are not normally distributed you use Spearman's rank correlation. This compares the ranks of one variable with the ranks of another variable. A correlation coefficient r_s of +1 indicates a perfect positive correlation, whilst –1 indicates a perfect negative correlation. Values nearer to 0 indicate weak or no correlation. You compare your r_s value to a table of critical values to assess the statistical significance. In Excel you can carry out the calculations, and the `RANK.AVG` function will rank the data. You cannot get the critical values "automatically". In R you can use the `cor.test()` command to carry out the computations and assess significance.
Pearson's product moment	When data are normally distributed you use Pearson's product moment. This uses the properties of the normal distribution to give a result (the correlation coefficient, *r*) between +1 and –1. This type of correlation is also called regression and defines the relationship between variables mathematically: $y = mx + c$, where *m* is the slope and *c* the intercept. In Excel you can get slope, intercept and Pearson's coefficient using `SLOPE`, `INTERCEPT` and `CORREL` functions. You cannot get a critical value directly but can calculate a *t*-value and use `TINV` to get one. In R you can use the `cor.test()` command to carry out all the calculations.
Curvilinear correlation	Relationships between variables do not have to be a straight line (which is $y = mx + c$) but may have some other mathematical form. A logarithmic relationship takes the form: $y = a \text{ Ln}(x) + c$. Here *a* is analogous to the slope and *c* is the intercept. A polynomial relationship takes the form $ax + bx^2 + c$, where *a* and *b* are coefficients analogous to slopes, and *c* is the intercept.
Visualizing correlation	You use a scatter plot to visualize correlation. The response variable (dependent) goes on the (vertical) *y*-axis, and the predictor (independent) goes along the (horizontal) *x*-axis. You can also add a trendline and display the mathematical equation that describes the relationship (if appropriate).
Lines of best fit	If your correlation is described by a mathematical relationship you can add a line of best fit (a trendline) to your scatter plot (curvilinear trendlines are described in Chapter 11). In Excel you can use the *Chart Tools* menu. In R you can use the `abline()` command for straight lines. If you used Spearman's rank correlation you should not use a trendline.

9. Tests for linking data – associations

When you have counts of items in categories, and you want to find links between the categories, you use tests of association.

> **What you will learn in this chapter**
>
> » How to carry out chi-squared association analysis
> » How to visualize results of association analysis using Excel and R
> » How to conduct goodness of fit tests

This chapter deals with two versions of association. The basic chi-squared test examines two sets of categories. When you have a single set of categorical data you can compare these data to a theoretical set. This crops up in genetic studies for example and are called "goodness of fit" tests (see Section 9.2).

9.1 Association: chi-squared test

Imagine the following situation. You are interested to know if certain plant species are found growing in proximity with one another. In other words you wish to know if there is an association between certain species. You could take up a quadrat and go to your field site and look for the two plants. One way to do this would be to note down the presence or absence of the species in each quadrat. When you are done you could generate a table of data similar to Table 9.1, which shows data for two species of heather.

Table 9.1 shows how many of the quadrats contained the first species; this is the column headed *C. vulgaris present*. The number of quadrats that contained both species together goes in the top left cell; the next one down is how many quadrats contained the first species (*C. vulgaris*) by itself. The margins of the table show the totals and you can see that there were 137 quadrats in total.

This sort of table is called a *contingency table*. You can apply this kind of arrangement to other situations. For example, imagine that you are interested in certain invertebrate

Table 9.1 Presence–absence data for two heather species (*Calluna vulgaris*, *Erica cinerea*) in 137 quadrats. From these results you can determine the probability that the two species are associated (positively or negatively).

	C. vulgaris present	*C. vulgaris* absent	Total
***E. cinerea* present**	3	23	26
***E. cinerea* absent**	35	76	111
Total	38	99	137

groups: beetles (Coleoptera), bugs (Hemiptera) and ants (Formicoidea) and want to know if they are associated with a particular habitat. The data look like Table 9.2.

Table 9.2 Habitat selection amongst some invertebrate groups. These data were collected by examining a number of plants; a tally of the number of taxa were recorded against each part of the plant.

	Ant	Bug	Beetle	Total
Upper leaf	15	13	68	96
Lower leaf	12	11	15	38
Stem	65	78	5	148
Bud	3	21	3	27
Total	95	123	91	309

In Table 9.2 you can see that each column contains the number of each invertebrate group that you encountered. Each row represents a habitat and each cell in the table shows how many of that particular invertebrate taxa were found at a particular habitat. The margins show the totals and you can see that the sampling survey found 309 invertebrates in total.

In both of these examples you have *categorical* data. In both examples the categories are represented in the titles of the rows and columns. Because of the way the data were collected (by a simple count or tally) you do not have replicated data. You cannot describe any of the cells in the table in terms of an average or spread using means, medians, standard deviation or range, as you would do with replicated data. Each cell of the table represents a unique combination of two categories; you have a simple *count* or *frequency*. You can represent these data graphically (see Section 6.5), with a couple of options:

- Pie chart – showing the proportion of items in each column or row (Section 6.5.1).
- Bar chart – showing the proportions (Section 6.5.2).

You need to use the *chi-squared test* to analyse these data. You first need to make a hypothesis. For the invertebrate example you might say:

The problem now is that you have no idea how important a particular difference is. A difference of 2, for example, will become 4 when squared. If this difference is based on an expected value of 4 then it is pretty large. If, however, the expected value is 400 then a difference of 4 is actually quite small. You need to take into account the size of the expected value: to do that you can simply divide by the expected value.

The final chi-squared formula is shown in Figure 9.3.

$$X^2 = \frac{(O-E)^2}{E}$$

Figure 9.3 Formula to determine chi-squared in a test for association. *O* are the observed values, *E* the expected values.

If you apply this to your invertebrate data you get something like Table 9.5.

Table 9.5 Final chi-squared values for invertebrate and habitat selection data.

Chi-squared	Ant	Bug	Beetle
Upper leaf	7.1379	16.636	55.827
Lower leaf	0.0086	1.1256	1.2965
Stem	8.3555	6.1842	34.159
Bud	3.3852	9.7801	3.0833

If you add the individual values you get chi-squared = 147 (rounded to be consistent with the original data). This is great but how do you know if this is statistically significant or not? You need to look at a table of critical values (Table 9.6).

Table 9.6 Critical values for the chi-squared test (also used with the Kruskal–Wallis test). Reject the null hypothesis if your value is greater than the tabulated value.

Degrees of freedom	Significance level			
	5%	2%	1%	0.01%
1	3.841	5.412	6.635	10.830
2	5.991	7.824	9.210	13.820
3	7.815	9.837	11.341	16.270
4	9.488	11.668	13.277	18.470
5	11.070	13.388	15.086	20.510
6	12.592	15.033	16.812	22.460
7	14.067	16.622	18.475	24.320
8	15.507	18.168	20.090	26.130
9	16.909	19.679	21.666	27.880
10	18.307	21.161	23.209	29.590

First of all you look down the left column until you find your degrees of freedom. This is related to the amount of data. In the case of chi-squared it is related to the original contingency table (Table 9.2). The margins contain the row and column totals (and you also have the grand total). If you had a blank table and only had these totals, how many cells would you need to be filled in before you could work out the rest (it is a bit like Sudoku)? The *degrees of freedom* is this value.

A simple way to determine degrees of freedom is to calculate it as:

$$\textit{degrees of freedom} = (\#\ \textit{columns} - 1) \times (\#\ \textit{rows} - 1)$$

Here you have three columns (invertebrate categories) and four rows (habitat categories) so $df = (3 - 1) \times (4 - 1) = 6$. You look down to the 6 and read across. The second column is headed 5% and gives you the value you need to exceed. Here you have a value of 12.592 and you comfortably exceed that. You can now reject your null hypothesis and conclude that there is an association between invertebrate type and habitat (you accept the alternative hypothesis). Of course you have a much larger value so you can look across the table row further and find your calculated value is larger than the final column (headed 0.01%). You can now conclude that there is less than 0.01% probability that your result was due to chance.

It is all very well to have a statistically significant association but you are not necessarily saying that all the associations in your contingency table are significant. You must look back at the observed and expected values and see which associations are positive and which are negative. Then you need to look at the individual chi-squared values. Recall that to get your final statistic you added all the values together. You can now look to see which of the individual values makes the greatest contribution to the total. You can see by looking back to Table 9.5 that the first row has fairly large values; the second row has fairly small values. You can thus gain an idea of the relative strengths of the associations by comparing the sizes of these individual chi-squared values. In fact if you look at the table of critical values (Table 9.6) you can see that something of the order of 3.8 would be likely to be significant for a pairwise comparison.

Pearson residuals

You can look at the differences between observed and expected values in a slightly different way. This is called the Pearson residual (after Karl Pearson). In short, instead of using the standard way to calculate chi-squared you use a different formula shown in Figure 9.4.

$$R_p = \frac{O - E}{\sqrt{E}}$$

Figure 9.4 Formula to calculate the Pearson residuals in a test for association.

The formula for the residuals is essentially the square root of the chi-squared formula (Figure 9.3). If you apply this formula to your data on invertebrates and habitat selection you end up with something like Table 9.7.

Table 9.7 Pearson residuals for invertebrate and habitat selection data.

Residuals	Ant	Bug	Beetle
Upper leaf	–2.67	–4.08	7.47
Lower leaf	0.09	–1.06	1.14
Stem	2.89	2.49	–5.84
Bud	–1.84	3.13	–1.76

You can see at a glance which of the associations are positive and which are negative. In the table you can see negative associations as negative numbers. You can also see the relative sizes of these associations and gain some insight into which ones are most important. Any individual residual that is ~2 is likely to be a significant one (Pearson residuals have a distribution similar to the *t*-test with large degrees of freedom, so a value of 1.96 is the generally accepted critical value).

Yates' correction for 2 × 2 contingency tables

When you have small contingency tables (2 × 2) like the heather example in Table 9.1, it is usual to apply a small correction factor to improve the reliability of the result. This is generally known as *Yates' correction*.

During the calculation, each *observed* – *expected* value is reduced by 0.5. This reduced value is then squared and divided by the expected value in the normal manner. You can write the correction down formally as:

$$|O - E| - 0.5$$

The effect is to subtract 0.5 if the value is positive and to add 0.5 if the value is negative. Once you have modified the $O - E$ values the chi-squared test continues as normal.

You can access the heather example from the support website, the file is called *heather.csv*.

9.2 Goodness of fit test

In the two examples earlier, the plant presence/absence data (Table 9.1) and the invertebrate/habitat data (Table 9.2), you collected data (the observed values) and worked out the expected values using the row and column totals. The chi-squared test can also be used in situations where you already have these expected values. The most common example would be in genetic studies. In a simple single locus study you would expect offspring in the ratio of 2:1:1 and you can use these as your expected values in a chi-squared calculation.

Imagine the situation shown in Table 9.8 where you are looking at the genetics of peas. Here you have two categories of colour and two of the coat texture. Your knowledge of Mendelian genetics would lead you to expect ratios of 9:3:3:1 and you can now calculate expected numbers as shown in Table 9.9.

If you add up the ratios, you get 9 + 3 + 3 + 1 = 16 so each expected value is a certain number of 1/16ths of the total observed (200). You can now use the standard chi-squared

Table 9.8 Count of pea seeds for goodness of fit test on pea genetic data.

	Green	Yellow	Total
Smooth	116	31	
Wrinkled	40	13	
Total			200

Table 9.9 Expected numbers in pea genetic study using goodness of fit. The expected values are determined by multiplying the expected ratio by the total number of observations.

Ratio	Category	Calculation	Expected #
9	Green smooth	9/16 × 200	112.5
3	Green wrinkled	3/16 × 200	37.5
3	Yellow smooth	3/16 x 200	37.5
1	Yellow wrinkled	1/16 × 200	12.5

formula: $(O - E)^2/E$ to work out the significance. In Table 9.10 you see the results of the calculations. The first column shows the names of the categories, the second shows the observed numbers in each category. The third column shows the expected values determined from above. The final column shows the individual chi-squared values.

Table 9.10 Goodness of fit test results for pea genetics study.

	Obs	Exp	$(O - E)^2/E$
Green smooth	116	112.5	0.11
Green wrinkled	40	37.5	0.17
Yellow smooth	31	37.5	1.13
Yellow wrinkled	13	12.5	0.02
Total:	200	200	1.42

The total chi-squared value is shown at the bottom. You see a value of 1.42. You need to compare this to a table of critical values (Table 9.6) and see if the result is statistically significant. You have four categories here and so the degrees of freedom are 3 (i.e. # categories – 1). The critical value from the table is 7.815 so you can say that the observed values are not statistically different from the expected values.

The pea genetics data are available on the support website as a file *peas.csv*.

9.3 Using R for chi-squared tests

R will calculate chi-squared tests quite simply but your data need to be in an appropriate form. Since the test involves a contingency table your data also need to be in a table. The simplest way is to make a CSV file and read this into R using the `read.csv()` command:

```
> inv.hab = read.csv(file.choose(), row.names = 1)
```

Here you make a new object (which you will do your chi-squared test on) but you tell R to use the first column as row names. The `file.choose()` part will open a browser-like window so you can select the file.

Note: Chi-squared example data

The data used in the chi-squared example are available on the support website. The file is called *invert habitat.csv*. The data are also as part of the *S4E2e.RData* file, which will open directly in R. The data from Table 9.2 are also available as a file *invert.csv*.

Download material available from the support website.

By default R takes the first row as column names so now you have an object, `inv.hab`, that has both row and column names. You have your contingency table. To see the data you can type its name:

```
> inv.hab
       Upper Lower Stem
Aphid    230   175  321
Bug       34    31   35
Beetle    72    23  101
Spider    11     3    5
Ant       12     9   15
```

Notice that you do not need the row and column totals. In this example you have some more data on invertebrate taxa and habitat selection. Your contingency table shows three columns and five rows. The chi-squared test is run easily using the `chisq.test()` command:

```
> chisq.test(inv.hab)
```

The result is displayed quite simply:

```
    Pearson's Chi-squared test
data:  inv.hab
X-squared = 25.1296, df = 8, p-value = 0.001478
Warning message:
In chisq.test(inv.hab) : Chi-squared approximation may be
incorrect
```

R actually creates more information but in this case it only displays the basic information.

You can get additional information more easily if you make a new object to hold your chi-squared test result.

```
> my.chi = chisq.test(inv.hab)
```

Now you have a new item called `data6.chi` and to see the result try:

```
> names(my.chi)
[1] "statistic" "parameter" "p.value"   "method"    "data.name"
"observed"
[7] "expected"  "residuals" "stdres"
```

This command shows you that there are a number of items you haven't seen. The most useful ones are `observed`, `expected` and `residuals`. To view them you type your object name, add a dollar sign and then the name of the bit you want:

```
> my.chi$expected
```

This shows you a table of expected values:

```
            Upper      Lower       Stem
Aphid  242.000000 162.456825 321.543175
Bug     33.333333  22.376973  44.289694
Beetle  65.333333  43.858867  86.807799
Spider   6.333333   4.251625   8.415042
Ant     12.000000   8.055710  15.944290
```

Now you see the reason for the error message, one of your expected values is <5. Try looking at the Pearson residuals:

```
> my.chi$res
            Upper      Lower        Stem
Aphid  -0.7713892  0.9840984 -0.03029148
Bug     0.1154701  1.8228842 -1.39588632
Beetle  0.8247861 -3.1496480  1.52324712
Spider  1.8543453 -0.6070112 -1.17724779
Ant     0.0000000  0.3327004 -0.23648449
```

You do not have to type the full name at the end (after the $ sign) as long as you type enough for R to work out what you want; typing `$p` for example would not work as there are two parts that begin with "p". If you want to display only the *p*-value you must add `$p.` as an absolute minimum.

By default R will apply a special correction factor (Yates' correction) to all the expected values if the contingency table is 2 × 2. You can turn it off by adding `correct = FALSE` to the command.

values for yourself. There are no functions that will calculate the Pearson residuals or the individual chi-squared values but the calculations are very easy. In the following exercise you can have a go at a chi-squared test of association for yourself.

Have a Go: Use Excel for a chi-squared test of association

You'll use the bird and habitat data from Table 6.12 for this exercise. The data are available in a file *bird.xlsx* on the support website. You used the same data (but in the *bird.csv* file) in an earlier exercise in graphing (Section 6.5).

Download material available from the support website.

1. Open the *bird.xlsx* datafile. There are two worksheets; the one entitled *Completed version* is for you to check your working so try not to cheat! Use the worksheet, *Data*. The data are in the form of a contingency table and show some bird species and some habitats. Each cell shows the frequency of observation of a particular species in a particular habitat over the period of the study.
2. Start by calculating row and column sums. For example, in cell B8 `=SUM(B2:B7)` will calculate the sum of observations for the *Garden* column. Add labels, `Totals`, for the totals in cells A8 and G1.
3. Obtain a grand total in cell G8: `=SUM(B2:F7)`. The result should be 295.
4. Now copy the habitat labels (cells B1:F1) to the clipboard and paste them into cells B10:F10. Similarly copy the species names in A2:A7 to cells A11:A16.
5. In cell B11 type a formula to calculate the expected value for frequency of *Blackbird* in *Garden*: `=$G2*B$8/$G$8`. Note the use of $ to "fix" rows and column references.
6. Copy cell B11 to the clipboard and paste the result down the rest of the *Garden* column to fill cells B11:B16. Paste again to place expected values in the other columns (cells C11:F16). You should now have all the expected values. You can calculate the row and column sums if you like to check that they are the same as for the observed values (and the grand total).
7. Repeat step 4 but place the labels to fill cells B19:F19 and A20:A25. You will use this table for the chi-squared values.
8. In cell B20 type a formula to calculate the chi-squared value for the *Garden/Blackbird* combination: `=(B2-B11)^2/B11`. You should get 2.78.
9. Copy the cell B20 to the clipboard and use paste to fill out the rest of the chi-squared values.
10. In cell F26 type a label, `X-val`, for the chi-squared result. In G26 type a formula to get the total chi-squared value: `=SUM(B20:F25)`. You should get 78.3.
11. Repeat step 4 but place the labels to fill cells B28:F28 and A29:A34. You will use this table for the Pearson residuals.
12. In cell B29 type a formula to calculate the residual for the *Garden/Blackbird* combination: `=(B2-B11)/SQRT(B11)`. You should get –1.67.
13. In I3 make a label for your chi-squared result. In J3 type a formula to get the total chi-squared value. This is a repeat of step 10.

14. In I4 type a label, `df`, for the degrees of freedom. In J4 type a formula to calculate *df*: `=(COUNT(B2:F2)-1)*(COUNT(B2:B7)-1)`. The result should be 20 (# of columns – 1 × # of rows – 1).
15. In I5 type a label, `X-crit`, for the critical value of chi-squared. In J5 type a formula to calculate the value: `=CHIINV(0.05,J4)`. You should get 31.41, which is a lot smaller than the calculated chi-squared value (so you know you have a significant result).
16. In I6 type a label, `p-val`, for the exact *p*-value from the chi-squared test. In J6 type a formula to compute the exact *p*-value: `=CHITEST(B2:F7,B11:F16)`. The result is very small (7.69E–09)!
17. In I7 type another `p-val` label. In J7 compute the exact *p*-value using your calculated chi-squared value and *df*: `=CHIDIST(J3,J4)`. You should get the same result as step 16.
18. In I8 type another `X-val` label. In J8 type a formula to calculate the chi-squared value from the result of the `CHITEST` function and *df* (step 16): `=CHIINV(J6,J4)`.

In this exercise you can see that you have a couple of ways to calculate chi-squared and the exact *p*-value. The *Completed version* worksheet contains the completed version, plus a few additional calculations using the "new" functions.

In the preceding exercise you saw how to calculate the chi-squared and exact *p*-value. This tells you if the overall result is significant but you still won't know which of the associations contributes most to the overall chi-squared value, or which of the associations are "important". You should work out the individual chi-squared values for yourself, or better still calculate the Pearson residuals. Once you have the residuals you can visualize the result with a bar chart (recall Section 6.5.2).

Goodness of fit and Excel

The `CHITEST`, `CHIINV` and `CHIDIST` functions will work just as well for goodness of fit tests. Once you have a set of observed values, you need to create your expected values and then proceed from there.

Because of the nature of goodness of fit tests your contingency table will be in two columns, one for the observed frequencies and one for the theoretical. You'll most likely have your data and calculations arranged in a similar fashion to Table 9.11.

Table 9.11 Setting out data and calculations for goodness of fit in Excel.

Colour	Coat	Obs	Ratio	Exp	Chi-sq	Resid
Green	Smooth	116	9	112.5	0.11	0.33
Green	Wrinkled	40	3	37.5	0.17	0.41
Yellow	Smooth	31	3	37.5	1.13	–1.06
Yellow	Wrinkled	13	1	12.5	0.02	0.14
	Σ	200	16		1.42	–0.18

Note that the Table 9.11 column headed *Ratio* sums to 16. If your theoretical values were proportions and summed to unity there would be no difference in how you calculated the expected values. The degrees of freedom are the number of rows (categories) minus 1.

The data in Table 9.11 are available from the support website (*peas.csv*) and there is also an online exercise to give you some practice at calculating goodness of fit using Excel.

Download material available from the support website.

Graphing chi-squared results using Excel

You saw how to present the data and results of chi-squared tests in Section 6.5. Once you have your chi-squared and Pearson residuals you have everything you need. There are various options including, but not limited to, the following:

- Column chart of observed values, as separate bars or stacked to show composition.
- Pie chart of observed values to show relative proportion of items.
- Column chart of observed vs. expected.
- Column chart of Pearson residuals.

Generally you simply need to select the columns you want to chart, and the labels, and then select the chart you require from the *Insert* menu. Note that you can select multiple label columns to produce axis (category) labels with multiple names (Figure 9.6).

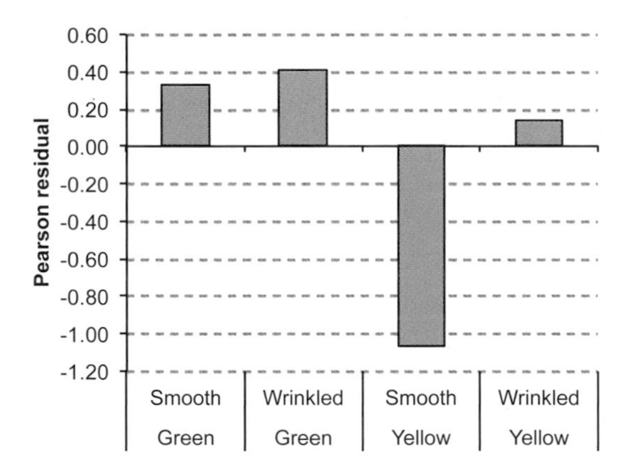

Figure 9.6 Pearson residual plot of goodness of fit data for pea genetics data.

There is an exercise on using Excel for goodness of fit on the support website, which includes producing a chart like that shown in Figure 9.6.

Download material available from the support website.

EXERCISES

Answers to exercises can be found in the Appendix.

1. Association analysis is used to explore ____ between the ____ of items in ____.
2. An old country saying is that you always find dock plants growing near stinging nettles. To test this you have looked for the frequency of occurrence of the two plants. The following table shows a contingency table of the data:

	Nettle.Y	Nettle.N
Dock.Y	96	41
Dock.N	26	57

From the table you can see that 96 quadrats contained both species, whilst 57 contained neither. What is the chi-squared statistic for the association between these species? Is the association a statistically significant one?

3. You have carried out an association analysis between five species of bee and five species of flowering plant. What is the critical value of chi-squared that you will need to exceed for the associations to be statistically significant?
   A. 9.488
   B. 11.070
   C. 15.507
   D. 26.296
   E. 36.415
4. In a study of flower colour in a species you have carried out a goodness of fit test on the assumption of a single gene controlling the colour. You obtain a *p*-value of 0.495. This means that the sample data you obtained is not matched by the genetic theory. TRUE or FALSE?
5. You've carried out a test of association between some species of bee and the flowers they visit. Which kind(s) of graph would be most suitable to present a visual summary of your work?
   A. Pie chart
   B. Bar chart
   C. Scatter plot
   D. Box–whisker plot
   E. Line chart

10.1.1 Post-hoc testing in analysis of variance

When your analysis of variance result shows a significant result you know that the samples are not all from one big group and that there are differences; however, you might want to know if site *a* is different from site *b* and if site *c* is different again or more similar to site *b*. You could run *t*-tests on each pair of samples but this would not be an appropriate approach. The problem is that the more tests you run, the more likely it is that you will get a significant result by chance (e.g. the more dice you roll the more likely you are to get a six. Alternatively the more lottery tickets you buy, the greater chance you have of winning, although it is still a very small chance).

Look at the example shown in Figure 10.6. You see a box–whisker plot of three samples but using the parametric statistics (mean, standard error, *s*) rather than the more usual median, IQR and range. The data are taken from the example used to make the analysis of variance table shown earlier (Table 10.2).

The data used in this example are available on the support website as *Sward height.xlsx*.

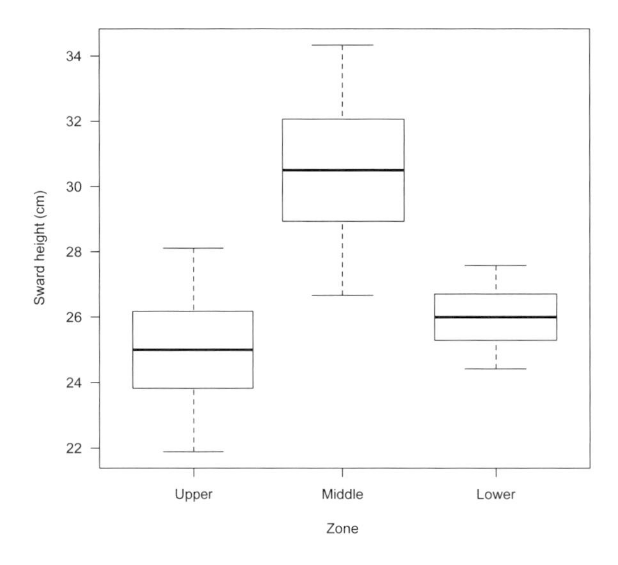

Figure 10.6 Results of ANOVA for mean sward height (cm) at three zones in a meadow in Shropshire. $F_{2,15} = 5.633$, $p < 0.05$. Stripes show means, boxes show standard error and whiskers show standard deviation.

The result of the analysis of variance tells you that there are significant differences between the sward heights at the different zones of this meadow; however, when you look at the graph it would appear that the *Upper* and *Lower* samples are quite close together; with the *Middle* zone looking like it might have a taller sward. The analysis of variance does not tell you anything more than "there is a difference between samples".

The answer is to run a *post-hoc* test, meaning "afterwards" (literally "after this"). What you do is run a special version of the *t*-test but designed to take into account the fact that you are running several tests at once. Many statistical programs will run post-hoc tests (including R of course) and although there are different types they run more or less on the same principle. The Tukey honest significant difference test (Tukey HSD) is a commonly used method (Figure 10.7).

$$t = \frac{|\bar{x}_1 - \bar{x}_2|}{\sqrt{\frac{MS_{within}\left(\frac{1}{n_1} + \frac{1}{n_2}\right)}{2}}}$$

Figure 10.7 The Tukey post-hoc test is a modified version of the *t*-test, and is used to carry out pairwise comparisons after the main ANOVA.

You can see that superficially this looks much like a regular *t*-test. On the top you have the difference between the means of the two samples you wish to compare. On the bottom you would normally have a measure of the variance and sample size of the two samples. In the Tukey test you use the mean square of the within-sample variability (the *error term*) as well as the sizes of the two samples. This version takes into account the variability of all the data as well as differences in sample size.

So for each pair of samples you evaluate this formula; the means and sample sizes vary but the error term remains the same for each comparison. This is one advantage in producing the ANOVA table (like Table 10.2 earlier) as you can readily see the value you need (9.433 in this example, Table 10.2). Once you have your *t*-values you can compare them to a table of critical values for *t* (Table 10.3).

Table 10.3 Table of critical values of *t* used in the Tukey HSD post-hoc test. Reject the null hypothesis if your calculated value is greater than the tabulated value.

5%	2%	1%	0.1%
4.303	6.965	9.925	31.598

There is only one row of values in Table 10.3 since you are always interested in pairs of comparisons (the degrees of freedom = 1). Once you have done your comparisons you need a method of conveying the result to potential readers. You might use a small table showing the various results (*t*-values, critical values, *p*-values), for example Table 10.4.

Table 10.4 Post-hoc *t*-values calculated by the Tukey HSD test for sward heights at three zones in a Shropshire meadow.

	Post-hoc *t*-values		
	Upper	**Middle**	**Lower**
Upper	–	4.552	0.786
Middle		–	3.422

Table 10.4 shows the calculated *t*-values for three samples worked out using the Tukey HSD test. If you compare these to the critical values you can readily see that only one

pair of samples has significantly different means (*Middle* and *Upper*). Table 10.5 shows the exact *p*-values.

Table 10.5 Post-hoc *p*-values calculated by the Tukey HSD test for sward heights at three zones in a Shropshire meadow.

	Post-hoc *p*-values		
	Upper	**Middle**	**Lower**
Upper	–	0.045	0.514
Middle		–	0.076

The *p*-values show you where the significant differences lie. You might have assumed by looking at Figure 10.6 that the *Middle* sample was statistically different from the other two; you would have been incorrect! In Table 10.5 you see the exact *p*-values but you might easily replace the non-significant results with n.s. and the significant one with $p < 0.05$.

Tables are fine but can take up a lot of space so it is common to add the post-hoc information directly onto your graph. The simplest way is to add a symbol to each sample (letters are convenient). Where letters are the same the samples are not significantly different. Where letters differ then the samples are significantly different (as determined by your post-hoc test). Look at Figure 10.8 where the sward height boxplot from Figure 10.6 has been modified to show post-hoc significance.

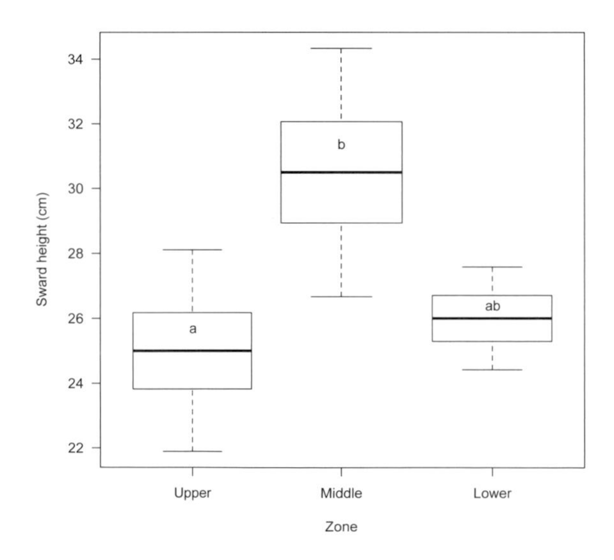

Figure 10.8 Results of ANOVA for mean sward height (cm) at three zones in a meadow in Shropshire. $F_{2,15} = 5.633$, $p < 0.05$. Stripes show means, boxes show standard error, whiskers show standard deviation. Letters show significance as determined by Tukey HSD test. Different letters denote statistical difference at 5%.

Now you have letters above the samples to show post-hoc differences. You can see that you start with sample *Upper* and that gets an a (a good starting point). The next sample

(*Middle*) is significantly different from the first and so you give it a different label, b will do nicely. Now the third sample (*Lower*) is not different from *Upper* so you give it an a, which shows it is not different; however, it is also the same as the *Middle* sample so it needs to have the same letter as that. You add a b and now the third sample has ab as a label, indicating that it is not significantly different from either of the other two samples. When you have a lot of comparisons, the labelling can be tricky to get correct. Start at the left and work your way across and you will hopefully get it right.

You can use R to carry out analysis of variance (Section 10.1.4) and also your spreadsheet (Section 10.1.5). Before that you'll see how analysis of variance is used in more complex situations.

10.1.2 Analysis of variance and several factors

If you are looking for a simple difference in abundance of something over several sites you can run analysis of variance (assuming normally distributed data). What you have is a response variable (e.g. the *abundance*) and a single predictor variable (e.g. the *site*). You would call your analysis a one-way ANOVA to remind you that you have the single predictor variable; however, it may be that there are different management strategies at these sites. You might be interested to see if the *management* also has an effect. Now you have two predictor variables: *site* and *management*. You can use analysis of variance once more but now that you have two predictor variables you call it two-way ANOVA.

You might have even more factors to consider and then you could have three-way or just multi-way ANOVA; however, when you add more factors, things get complicated very quickly!

For example, take the *site* and *management* scenario. You may well find differences in abundance of your species between sites; you may also find that management affects the abundance; however, there is another thing to consider and that is the combined effect. In ANOVA-speak this is called an *interaction* and is often written *site * management* (i.e. the two factors look like they are multiplied by one another). If you had a significant interaction then this would indicate that your management strategy was having a different effect at different sites.

Look at Table 10.6, this shows some data on plant growth under three different water regimes and for two different species.

Now you are looking for differences between treatments and between species but what analysis are you going for? If you examined each site in turn you could apply one-way ANOVA and determine if the water treatment had any effect on growth; however, since you have all the data together you should use an analytical method that uses all the data. This is where two-way ANOVA comes in.

What you do is to split up the data so that you can see how each element contributes to the variability. Table 10.6 is already partitioned like this. You can see that you have two sites. You also have three treatments. In this case you see that you have three observations (replicates) for each of the treatments. This gives a nice balance. More importantly the properties of the variance are such that your calculations become more inexact the greater imbalance you have. You strive therefore to collect data in such a way as to achieve this symmetry.

Usually you would not have your data written out in this format; you should have one column for the response factor and one for each of the predictor factors (*Species* and *Water*); you'll see this a bit later (Table 10.9).

Table 10.9 Growth of two plant species under differing watering treatments. Data are laid out in scientific recording format, i.e. each column contains a single variable.

Height	Species	Water
9	vulgaris	Low
11	vulgaris	Low
6	vulgaris	Low
14	vulgaris	Mid
17	vulgaris	Mid
19	vulgaris	Mid
28	vulgaris	High
31	vulgaris	High
32	vulgaris	High
7	sativa	Low
6	sativa	Low
5	sativa	Low
14	sativa	Mid
17	sativa	Mid
15	sativa	Mid
44	sativa	High
38	sativa	High
37	sativa	High

complicated situations, you may not be able to set out your data in the on-the-ground layout but the scientific recording format will always work for you.

Post-hoc testing in two-way ANOVA

Much like one-way ANOVA, your final result only tells you if there are significant differences between samples. The graph of the means (Figure 10.9) shows you that some of the differences are pretty close and so likely not significant. You can carry out Tukey HSD tests in the same manner as before (Section 10.1.1) but of course now you have more pairwise comparisons to make.

10.1.3 Multi-way ANOVA

It is certainly possible to add more predictor variables to the mix. In this way you can get three-way ANOVA and more. Calculating the results becomes more and more difficult as does the interpretation. If you are going to attempt a multi-way ANOVA, then you are likely to want to use a dedicated computer program to carry out the calculations. Your obvious choice is to use R (Excel cannot handle more than two predictor variables).

10.1.4 Using R for ANOVA

You can use R to carry out ANOVA quite easily; there is a generalized command for running ANOVA (the `aov()` command) and this will run the calculations you need. For post-hoc testing you have another command `TukeyHSD()`.

ANOVA, aov() command

ANOVA allows you to examine the differences between more than two samples at one go. It is not correct to run a series of *t*-tests. R uses the model syntax to drive the routines in the `aov()` command:

```
aov(formula, data)
```

`formula`	A formula of the form `y ~ x` where `y` is a response variable and `x` is one or more predictors. If you have more than one predictor you use + to separate them. The * calculates interactions so `y ~ a * b` would be a two-way ANOVA with interactions.
`data`	The name of the data object (usually a `data.frame`) that contains the variables.

The `aov()` command uses a `formula` to specify what sort of analysis of variance to calculate. There are many options for a formula; you can see a few examples in Table 10.10.

Table 10.10 Formula syntax for `aov()` commands to carry out analysis of variance in R.

Formula	Meaning
`y ~ x`	One-way ANOVA, y is the response and x is the single predictor.
`y ~ x + z`	Two-way ANOVA without interactions, response variables x and z are treated independently.
`y ~ x * z` `y ~ x + z + x:z`	Two-way ANOVA with interactions. The second form of the formula is a long-hand way of showing `x * z`.
`y ~ a * b * c` `y ~ a + b + c + a:b + a:c + b:c + a:b:c`	Three-way ANOVA with interactions. The second form is a longer version of the same thing.

You'll see the formula syntax used again when you look at multiple regression (Chapter 11). The main thing to remember is that your *response* goes to the left of the ~ and *predictors* go to the right. If you have multiple predictors then usually you'll use the * to separate them, indicating that you want to calculate the interactions.

First of all you need a set of data, which must be in scientific recording format (e.g. Table 10.9). Your columns should each represent a separate variable and you'll usually have a column for the response variable (e.g. *Height* in Table 10.9) and one or more columns for the predictors (e.g. *Plant* and *Water*). Each row therefore represents a separate observation (replicate).

Note: Altering data layout

If your data are in the "wrong" layout for `aov()` you need to rearrange them into the proper scientific recording layout. It is often easiest to use Excel for this kind of data manipulation but the `stack()` command can be used to combine multiple samples into the appropriate form. See the support website for an exercise on using the `stack()` command for this purpose.

You'll start by reading in your data, from a CSV file:

```
> sward2 = read.csv(file = file.choose())
```

Use the `names()` command to remind you of the variable names (remember R is case sensitive).

```
> names(sward2)
[1] "Height" "Site"
```

It is best to save the result of your `aov()` to a named object as this allows easier access to the results:

```
> sward2.aov = aov(Height ~ Site, data = sward2)
> summary(sward2.aov)
            Df Sum Sq Mean Sq F value Pr(>F)
Site         2  106.3   53.14   5.633  0.015 *
Residuals   15  141.5    9.43
---
Signif. codes:  0 '***' 0.001 '**' 0.01 '*' 0.05 '.' 0.1 ' ' 1
```

Note that the `summary()` command produces the result as an ANOVA table (recall Table 10.2 and Table 10.8). You can see the degrees of freedom, sums of squares, mean squares and the final F-value and exact p-value for the result. R also adds a code to help you spot significant results (the bottom line is a reminder of the code, * is equivalent to $p < 0.05$).

In the summary ANOVA table the row labelled *Residuals* gives the within-groups results (the error term). The other row takes its label from the predictor name. If you have a more complicated data setup then your results will be also more complicated. For a two-way ANOVA you need to specify two predictor variables when you enter the formula, separated by the * symbol to indicate interactions.

The following example uses the data for plant height in response to watering treatment (Table 10.6):

```
> head(pw)
  height    plant water
1      9 vulgaris    lo
2     11 vulgaris    lo
3      6 vulgaris    lo
4     14 vulgaris   mid
5     17 vulgaris   mid
6     19 vulgaris   mid
> pw.aov = aov(height ~ plant * water, data = pw)
> summary(pw.aov)
            Df Sum Sq Mean Sq F value   Pr(>F)
plant        1   14.2    14.2   2.462  0.14264
water        2 2403.1  1201.6 207.962 4.86e-10 ***
plant:water  2  129.8    64.9  11.231  0.00178 **
Residuals   12   69.3     5.8
---
Signif. codes:  0 '***' 0.001 '**' 0.01 '*' 0.05 '.' 0.1 ' ' 1
```

The `head()` command shows the first few rows of the data, so you get a reminder of the layout. Note that the `plant * water` part of the formula specifies that you want to compute interactions. The ANOVA table shows the two predictor variables, with an additional row for the interaction (recall Table 10.10).

This time you can see that there are results for the two predictor variables (*plant* and *water*) and the interaction (*plant:water*). As before, the *Residuals* row contains the error term. The result for the *plant* row is not significant but the result for *water* is. This means that if you look at the plants without taking watering treatment into account, there is no difference. When you look at watering treatment (without splitting the plant species) you see a significant result. Look back at Figure 10.9 and see how the watering treatment influences growth.

The interaction part is also statistically significant, which indicates that there is a subtly different reaction, with one plant reacting differently to the other. You can see that at the *lo* and *mid* watering levels there is not a lot between the means (and the error bars overlap) but at the *hi* treatment there is a difference between the species.

If the interaction result was not significant then you could re-run your analysis using the predictor variables separately:

```
> aov(height ~ plant + water, data = pw)
```

When you get a significant result for your ANOVA you need to do an additional analysis and run post-hoc tests. These tests delve into the data a little deeper and compare the samples pair by pair, so you can see which are the important differences.

Post-hoc testing in R using TukeyHSD()

You can carry out the Tukey post-hoc test quite simply in R using the `TukeyHSD()` command.

```
TukeyHSD(x, which, ordered = FALSE, conf.level = 0.95)
```

`x`	The result of an `aov()` command.

Note: Changing order of items in a boxplot

See the support website for some notes about rearranging the order of the boxes in a `boxplot()` command.

10.1.5 Using Excel for ANOVA

Excel has some functions related to ANOVA but in the main you'll have to calculate most of the results for yourself (unless you use the *Analysis ToolPak*, Section 1.9.1). The functions that are helpful are as follows:

Function	Description
`FTEST(array1, array2)` `F.TEST(array1, array2)`	Calculates the exact *p*-value for ANOVA between two samples (`array1` and `array2`). The result is identical to the `TTEST` function.
`FINV(F_val, df1, df2)` `F.INV.RT(F_val, df1, df2)`	Calculates an exact *p*-value given a value for the *F*-statistic. You also need the degrees of freedom for within-groups (`df1`) and for the between-groups (error term, `df2`).
`FDIST(p_val, df1, df2)` `F.DIST.RT(p_val, df1, df2)`	Calculates an *F*-value given an exact probability. You also need the degrees of freedom within groups (`df1`) and between groups (error term, `df2`). You can use this to get a critical value for example.
`VAR(data)`	Calculates variance, s^2, of the `data`. Multiply the result by the degrees of freedom to get the sums of squares (SS).
`COUNT(data)`	Counts the number of items. Use this to get degrees of freedom (i.e. $n - 1$).

Later versions of Excel have additional statistical functions; these generally have longer names than their older counterparts. `F.DIST.RT`, for example, calculates an *F*-value for the right-tail of the *F* distribution (`F.DIST` does the left-tail). Since you are almost never interested in the left-tail of the distribution it is easiest to use the shorter named (compatibility) functions!

For most situations you'll have to compute the variance components for yourself; the `VAR` function will do this (multiply variance by degrees of freedom to get sums of squares). It is not too onerous to calculate one-way ANOVA; two-way is a little more involved but possible. Beyond that it is not possible without a great deal of effort (the *Analysis ToolPak* is no help either).

In the following exercise you can have a go at a one-way ANOVA using Excel for yourself.

Have a Go: Use Excel for one-way ANOVA

You'll need the *Sward height.xlsx* data for this exercise. These data show the height of the sward (the general height of the vegetation) at three sites. There are three worksheets; *Sample Format* gives the data in three columns (*Upper, Middle* and *Lower*). The *Recording Format* worksheet gives the data in two columns (*Height* and *Site*), and would be the one to use if you wanted to export the data in CSV format to R. The last worksheet, *Completed anova*, contains the final ANOVA and post-hoc calculations, so you can check your work.

Download material available from the support website.

1. Open the *Sward height.xlsx* spreadsheet and navigate to the *Sample format* worksheet. The three samples are in columns B:D.
2. In cell A9 type a label, `n`, for the number of replicates. In B9 type a formula to determine the number of replicates in the *Upper* sample: `=COUNT(B2:B8)`. You should get 7.
3. Copy and paste the result in B9 to cells C9:D9 so you have the number of observations for each sample.
4. In A10 type a label, `Mean`, for the means. In B10 calculate the mean of the *Upper* sample: `=AVERAGE(B2:B8)`. You should get 25. Copy and paste the result to the other columns so you have all the means (you need them for the post-hoc tests).
5. In A11 type a label, `SS`, for the sums of squares of the individual samples. In B11 type a formula to calculate SS for the *Upper* sample: `=VAR(B2:B8)*(B9-1)`. You should get 58. Copy and paste the result to the other columns to get SS for all samples.
6. In A12 type a label, `Grand Mean`, for the mean of all the data. In C12 type a formula to work out this grand mean: `=AVERAGE(B2:D8)`. You should get 27.1.
7. In A13 type a label, `SS within`, for the total within-sample SS (the error term). In B13 type a formula to calculate the error SS: `=SUM(B11:D11)`. This is the sum of row 11 and you should get 141.5.
8. In A14 type a label, `SS between (indiv)`, for the individual between-groups SS. In B14 type a formula to calculate the SS for the *Upper* sample: `=(B10-$C$12)*(B10-$C$12)*B9`. You should get 31.2. Copy and paste the result to the other cells in the row to get the SS for all three samples.
9. In A15 type a label, `SS between`, for the total between-groups SS. In B15 type a formula to get the total between-groups SS: `=SUM(B14:D14)`. This is the sum of row 14 and you should get 106.28.
10. In A16 type a label, `SS Total`, for the overall SS. In B16 type a formula to calculate overall SS: `=VAR(B2:D8)*(COUNT(B2:D8)-1)`. You should get 247.78.
11. You now have all the variance components. You can now assemble an ANOVA table to display the final results. In A18 type a label, `Source of variation`.

The second option is labelled *Anova: Two-Factor With Replication*, which is two-way ANOVA. To carry out this analysis you need your data arranged in the same manner as Table 10.6, with a column containing one of the predictor variables and other columns containing the samples, split amongst the other predictor (Figure 10.13).

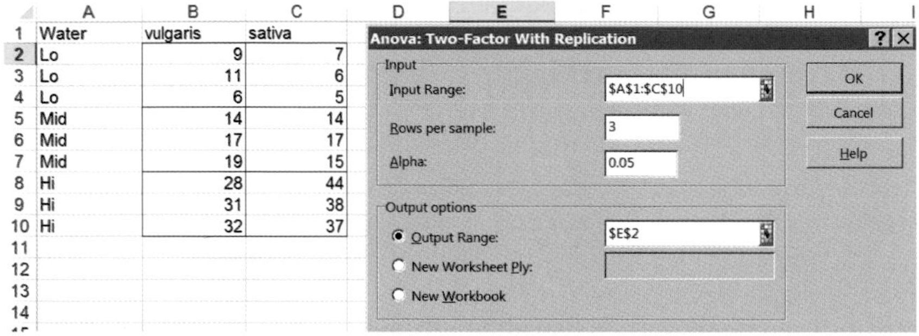

	A	B	C
1	Water	vulgaris	sativa
2	Lo	9	7
3	Lo	11	6
4	Lo	6	5
5	Mid	14	14
6	Mid	17	17
7	Mid	19	15
8	Hi	28	44
9	Hi	31	38
10	Hi	32	37

Anova: Two-Factor With Replication
Input
Input Range: $A$1:$C$10
Rows per sample: 3
Alpha: 0.05
Output options
Output Range: $E$2
New Worksheet Ply:
New Workbook
OK
Cancel
Help

Figure 10.13 Data for two-factor ANOVA with replication must be in a particular layout.

The third option (*Two-Factor Without Replication*) is not very commonly used; this analysis assumes that you only have a single observation per sample/treatment.

In any event your data need to be in a continuous block. You can choose where to place the results (in Figure 10.13 a cell in the current worksheet is selected). The *Alpha* box sets the level of significance for the displayed critical values. Usually you'll simply leave this at the default 0.05, which is the 5% ($p = 0.05$) significance level. For two-way ANOVA with replication you must also tell Excel how many replicates there are for each block (the example shows 3). You must have an equal number of rows for each treatment or the function will simply not run.

The results will appear where you chose and for one-way ANOVA you'll get a standard ANOVA table. For two-way ANOVA you'll also get a table of summary values for each of the blocks (e.g. Table 10.11).

Table 10.11 Example of a summary table as part of the output from *Analysis ToolPak* two-way ANOVA with replication.

ANOVA: two-factor with replication			
Summary	***vulgaris***	***sativa***	**Total**
Lo			
Count	3	3	6
Sum	26	18	44
Average	8.666666667	6	7.333333333
Variance	6.333333333	1	5.066666667

The summary tables (e.g. Table 10.11) show some useful statistics such as mean and variance. The columns match the original data (in this case the plant species). You'll get a summary table for each level of the other predictor variable; in the current example Table

10.11 shows the *Lo* water treatment so there would be other summaries for *Mid* and *Hi* treatments.

The *Analysis ToolPak* cannot carry out post-hoc testing, so you have to use regular Excel functions for yourself.

You cannot undertake more complex analyses than two-way ANOVA and of course your data must be set out in the on-the-ground layout rather than in strict scientific recording layout. The support website contains the plant watering data as a file called *Two way online.xlsx*, which you can use for practice.

Example material available on the support website.

Graphing ANOVA results using Excel

You should always visualize your data/results; a bar chart (Section 6.3.1) or box–whisker plot (Section 6.3.2) would be most sensible for ANOVA. It's easiest to chart the results of one-way ANOVA, as you'll have the mean values easily accessible. You need to calculate some kind of error bar; Figure 10.14 shows a column chart with standard error as the error bars.

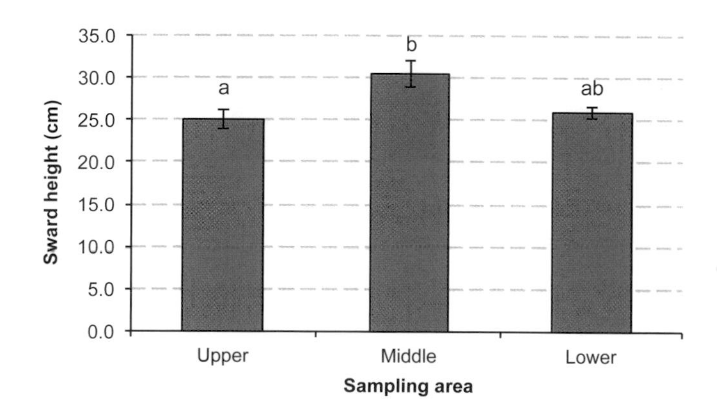

Figure 10.14 A column chart with error bars (here standard error) is easy to produce using Excel. This chart shows mean values for three samples of sward height (a one-way ANOVA). Labels above bars show post-hoc significance (Tukey HSD), with different letters denoting differences at $p = 0.05$.

In Figure 10.14 the error bars were calculated as standard error, which is easy to compute in Excel using `STDEV` and `SQRT` functions. The bar labels are easily added using a text box placed in the chart.

With a two-way ANOVA you have more options about how you present the results. To start with you need to assemble your mean and error bar values. Table 10.12 shows how you might set out the values for the plant watering data you've been looking at.

If you highlight the mean values (and the row/column labels) you'd get a plot resembling Figure 10.15. In Figure 10.15 the bars are grouped by watering treatment, with you can see this especially from the lower sample where the median coincides with the quartile.

The first thing to do is to rank all the data; you lump all the values together for this purpose. In other words you do not rank each sample separately. If there are big differences between samples then the ranks of one sample will tend to be smaller (or bigger) than the others. The more separate the samples, the more separate the ranks. If there are no real differences then the ranks will be all muddled up.

Once you have your data ranked you can start to calculate the Kruskal–Wallis statistic (called H) using the formula given in Figure 10.17.

$$H = \frac{12}{N(N+1)} \sum_{i=1}^{k} \frac{R_i^2}{n_i} - 3(N+1)$$

Figure 10.17 The formula used to determine the Kruskal–Wallis statistic.

The Kruskal–Wallis formula looks pretty horrendous but actually it is not that bad. The numbers 12 and 3 are constants. Uppercase N is the total number of observations (look back at the data in Table 10.13 and see that there are 17). The R refers to the ranks of the observations in each sample and n is the number of observations per sample.

You start by working out the ranks. The lowest value gets the lowest rank. Where there are tied values you determine an average rank; then the next value gets the next available rank. In Table 10.14 you see the data converted to ranks. The two lowest values were 3; these occupy ranks 1 and 2 and so get a tied/average rank of 1.5. The next available rank is therefore 3; however, there are two 4 values and these need to take up ranks 3 and 4, getting an average rank of 3.5.

Table 10.14 Hoglouse at three sites in a stream in Devon. Original data converted to ranks.

Upper	Mid	Lower
1.5	3.5	15
3.5	1.5	17
5	6	9.5
9.5	9.5	12.5
7	15	15
12.5		
9.5		

Next you add up the ranks for each sample. In this case you end up with three rank totals as you have three samples. Now you square each one and divide by the number of observations in each sample. These results are shown in Table 10.15. Now this gives you the R^2/n values from the formula. You add these together (that is the sigma part) giving 1540.29 (the sum of the final row of Table 10.15).

The final step is to combine the rest of the formula. You have the sigma part so now you multiply that by $12/N(N + 1)$ to give $1540.29 \times 12/(17 \times 18) = 1540.29 \times 0.04 = 60.4$. Last

Table 10.15 Hoglouse at three sites in a stream in Devon. Each column is a sample from the dataset. The rows show values calculated from the ranks (*n* is sample size for each).

	Upper	Mid	Lower
Sum	48.5	35.5	69
Sum^2	2352.25	1260.25	4761.00
Sum^2/n	336.04	252.05	952.2

of all you subtract the $3(N + 1)$ at the end (remember that you do all the multiplication and division bits before the adding and subtracting bits). This gives you $60.4 - (3 \times 18) = 60.4 - 54 = 6.404$.

You now have a result, *H*. In order to see if your result is significant you need to compare your calculated value of *H* to a critical value. The Kruskal–Wallis statistic is a funny beast! If you have at least five observations in each sample, then *H* is very close to chi-squared. You can compare your value of *H* to the chi-squared value for appropriate degrees of freedom (Table 9.5). The degrees of freedom are the number of samples – 1. In this example you have three samples so there are two degrees of freedom.

Where the sample sizes are smaller, the *H*-statistic behaves in a rather more peculiar fashion. Fortunately for you, statisticians have worked out exact *H*-values for a range of sample sizes (smaller and larger than 5); therefore you are able to look at tables of values that have been calculated for you. There are two tables, one for equal sample sizes and another for unequal.

The calculated value for the current example is $H = 6.404$, which is greater than the chi-squared critical value (of 5.991) and thus the result is significant at the 5% level.

Note: Critical values for Kruskal–Wallis

You can find critical values for exact sample sizes on the support website. There are two tables, one for equal sample sizes and one for unequal.

If you have tied ranks there is a correction to *H* that you can calculate. The adjustment is usually quite small. See the support website for additional notes related to *H*-value adjustment with tied ranks.

10.2.1 Post-hoc testing in Kruskal–Wallis

Once you have your result, you know if the samples are significantly different or not; however, you do not know if all the samples are different from each other or if some are actually not different from each other (implying that others are). If you look back to the graph (Figure 10.16) you can see that the two sites on the left are overlapping quite a lot and it seems unlikely that they could be significantly different. It looks like the site on the right (labelled *Lower*) might be different from the other two.

As discussed previously (Section 10.1.1) you should not resort to looking at each pair in turn (using a *U*-test for example, since you have non-parametric data) as the very act of performing multiple tests alters the probability of your getting a result. As before, you run a post-hoc test. In the case of Kruskal–Wallis, there is no standard approach and many statistical programs will not even provide a routine to calculate a post-hoc test for the Kruskal–Wallis analysis. You can, however, run a modified version of the *U*-test where the calculation is based loosely on Tukey's approach for his HSD test, and this is suitable for post-hoc testing. What you actually do is run a standard *U*-test but calculate a special critical value for the post-hoc analysis.

Unlike the situation you had with ANOVA (see Section 10.2.1 on post-hoc testing) you don't have an error term. As a result you need a way to take into account the within-groups variability. The formula for the post-hoc test is shown in Figure 10.18.

$$U = \frac{n^2}{2} + Qn\sqrt{\frac{(2n+1)}{24}}$$

Figure 10.18 Post-hoc testing for the Kruskal–Wallis test. *Q* is the Studentized range and *n* the harmonic mean of the sample sizes. The resulting *U*-value is a critical value for the pairwise *U*-test in the post-hoc analysis.

There are only two variables in the formula, *n* and *Q*. The variable *Q* is derived from Student's *t*-statistic and is called the *Studentized range* (values are shown in Table 10.16). The value of *n* is the common sample size and is calculated using the formula shown in Figure 10.19; it is essentially the *harmonic mean* of the sample sizes.

$$n = \frac{2}{\left(\frac{1}{n_1} + \frac{1}{n_2}\right)}$$

Figure 10.19 Calculating the harmonic mean of two samples.

To start with you work out the sample sizes and calculate the common sample size using the formula in Figure 10.19. You then look up the value of *Q*, which depends only upon the number of samples. Even though you are only comparing two samples at a time, you use the value for the total number of samples you have. In this case you have three samples (the three different sites) so your value of $Q = 3.314$ (at the 5% significance level).

Table 10.16 Values of *Q*, the Studentized range.

Number of groups	Significance 5%	1%
2	2.772	3.643
3	3.314	4.120
4	3.633	4.403
5	3.858	4.603
6	4.030	4.757

The value of *U* you have calculated is the critical value (at 5%, although you could easily calculate the 1% value as well), determined for your combination of sample groups and sizes. What you do now is to work out a regular *U*-test for each pair of samples you are interested in. Unlike a regular *U*-test, however, you select the larger of the two *U*-values as your test statistic. Now finally you get to compare your calculated value to the critical value (i.e. the one you worked out previously). If your pairwise *U*-value is equal or greater than the critical value then the post-hoc test is significant (at that level of significance).

If you do this for the hoglouse data (Table 10.13) you find that the only difference is between the *Upper* and *Lower* sites (see Table 10.17). Looking back at Figure 10.16 you might not have guessed this and would probably have assumed that the *Mid* site was also different from the *Lower*. This is of course why you undertook the statistical analysis in the first place!

Table 10.17 Post-hoc test results for Kruskal–Wallis analysis of hoglouse data. Main values shown are calculated *U*-values (upper right of table) and significance (lower left of table).

	Upper	Mid	Lower
Upper	–	18	32.5
Mid	n.s.	–	21.5
Lower	5%	n.s.	–

Table 10.17 shows the main results of the post-hoc test. You can see the calculated *U*-values in the upper-right of the table and the significance in the lower-left.

Post-hoc tests in general are rather conservative and tend to err on the side of caution. The post-hoc procedure outlined here for the Kruskal–Wallis test is a bit long-winded but is really your only option with non-parametric data and small sample sizes. The procedure outlined here probably should be carried out only when the sample sizes are fairly equal. As the sample sizes become more and more unequal, the accuracy of the formula diminishes and you cannot rely on the calculated probabilities being correct.

10.2.2 Using Excel for Kruskal–Wallis

Excel does not have any built-in functions for working out Kruskal–Wallis tests. You can use the `RANK.AVG` function to work out the ranks and then feed the values into the Kruskal–Wallis formula (Figure 10.17). If your samples each have at least five replicates then you can use the `CHIINV` function to determine a critical value, the degrees of freedom being the number of samples you have. In the following exercise you can have a go at a Kruskal–Wallis analysis for yourself.

Have a Go: Use Excel for Kruskal–Wallis testing

You'll need the *Hoglouse.xlsx* data for this exercise. You used the data earlier when you looked at bar charts (Section 6.3.1). The data show the abundance of the freshwater invertebrate at three sampling sites. There are two worksheets; the one labelled *Data* contains the data only. The other worksheet, labelled *Summary Stats and Chart,* contains the summary statistics and an example bar chart.

Example material available on the support website.

1. Open the *Hoglouse.xlsx* spreadsheet. Go to the *Data* worksheet, which shows the data for the three samples in columns B:D.
2. In cell A9 type a label, `n`, for the number of replicates in each sample. In B9 type a formula to calculate the number of replicates for the *Upper* sample: `=COUNT(B2:B8)`. You should get 7.
3. Copy and paste the result in B9 to cells C9:D9 so you get the replicates for all samples.
4. In A10 type a label, `Total Obs`, for the total number of items. In B10 type a formula to work out the total: `=COUNT(B2:D8)`. You should get 17.
5. In A11 type a label, `df`, for the degrees of freedom. In B11 type a formula to work out df, the number of samples minus 1: `=COUNT(B9:D9)-1`. You should get 2.
6. Copy cells B1:D1 and paste to F1:H1 so you have sample labels for the ranks.
7. In F2 type a formula to calculate the rank of the first item in the complete dataset: `=IF(B2="","",RANK.AVG(B2,$B$2:$D$8,1))`. Note you need the $ to "fix" the cell range. The `IF` part ensures that any blank cells are "ignored".
8. Copy the result in F2 to the rest of the *Upper* column and then across the *Mid* and *Lower* columns to fill cells F2:H8. You should now have all the ranks for all the data.
9. In E9 type a label, `Sum R`, for the rank sums. In F9 type a formula to calculate the rank sum for the *Upper* sample: `=SUM(F2:F8)`. You should get 48.5.
10. Copy and paste F9 into cells G9:H9 to get all the rank sums.
11. In E10 type a label, `RSq/n`, for the squares of the rank sums divided by the sample size. In F10 type a formula to work out this value for the *Upper* sample: `=SUM(F2:F8)^2/B9`. You should get 336.04.
12. In E12 type a label, `H`, for the final Kruskal–Wallis statistic. In F12 type a formula to complete the calculation: `=(12/(B10*(B10+1)))*SUM(F10:H10)-(3*(B10+1))`. You should get 6.403.
13. In E13 type a label, `Hcrit`, for the critical value. In F13 type a formula to calculate the critical value from the chi-squared distribution for $p = 0.05$: `=CHIINV(0.05,B11)`. You should get 5.991.

Your calculated value of H (6.403) is greater than the critical value (5.991), so you can say that there is a statistically significant difference between the samples at $p < 0.05$.

In the exercise the data results in some tied ranks. The calculations for the adjustment of H are fairly easy but the execution of them in Excel is difficult to implement. See the support website for additional notes related to H-value adjustment with tied ranks.

Example material available on the support website.

Kruskal–Wallis post-hoc analysis using Excel

There is no short way to carry out post-hoc testing using Excel; you'll need to proceed in the following manner:

- Take the samples pair by pair and run regular *U*-tests (Section 7.2.1).
- Use the larger of the two *U*-values as the test statistic.
- Work out the harmonic mean of the sample sizes (see Figure 10.19).
- Compute a critical value using the *Q*-statistic (Table 10.16) and the post-hoc formula (Figure 10.18).
- If your calculated value is larger or equal to the critical value, then the pairwise comparison is statistically significant.

Graphing Kruskal–Wallis results using Excel

Suitable graphs for visualizing Kruskal–Wallis results are the same as for any "differences" situation (see Section 6.3). Box–whisker or column (bar) charts are most suitable. The preceding exercise used the hoglouse data, which was used earlier to illustrate a column chart with error bars (the results are in the *Hoglouse.xlsx* spreadsheet). Generally column charts are easier to produce in Excel than box–whisker plots (see Section 6.3.2).

10.2.3 Using R for Kruskal–Wallis testing

R has the routines for carrying out Kruskal–Wallis testing via the `kruskal.test()` command. The command allows you to specify the input data in several ways.

- `x` – assumes each column of `x` is a separate sample.
- `x, g` – assumes `x` is a response variable and `g` is a grouping variable (a predictor).
- `formula` – a formula of the form `y ~ x` where `y` is the response and `x` is the predictor.

Here are some examples using the hoglouse data (Table 10.13), which are also in the *S4E2e.RData* datafile.

Download material available on the support website.

The data are in two forms: one is in sample layout, with a column for each sample; the other is in recording layout, with a response variable and a predictor. The first example shows the sample layout.

```
> hog3
  Upper Mid Lower
1     3   4    11
2     4   3    12
3     5   7     9
4     9   9    10
```

```
5      8  11     11
6     10  NA     NA
7      9  NA     NA
> kruskal.test(hog3)
  Kruskal-Wallis rank sum test
data:  hog3
Kruskal-Wallis chi-squared = 6.5396, df = 2, p-value = 0.03801
```

The formula method is best when you have a recording format:

```
> hog2
   count  site
1      3 Upper
2      4 Upper
3      5 Upper
4      9 Upper
5      8 Upper
6     10 Upper
7      9 Upper
8      4   Mid
9      3   Mid
10     7   Mid
11     9   Mid
12    11   Mid
13    11 Lower
14    12 Lower
15     9 Lower
16    10 Lower
17    11 Lower
> kruskal.test(count ~ site, data = hog2)
  Kruskal-Wallis rank sum test
data:  count by site
Kruskal-Wallis chi-squared = 6.5396, df = 2, p-value = 0.03801
```

You get a simple output; R reminds you of the data and gives you the test statistic (the Kruskal–Wallis test results in a distribution similar to chi-squared), the degrees of freedom and the final *p*-value. Here you see that $p = 0.03801$ and since this is <0.05 you can reject the null hypothesis and accept the alternative that the samples are different. In other words, the abundance of the invertebrate is different across the sites.

Post-hoc testing and Kruskal–Wallis using R

There is no built-in command to work out post-hoc comparisons for the Kruskal–Wallis test. What you must do is work out the pairwise *U*-values you are interested in and then use the post-hoc formula you encountered earlier (Figure 10.18).

You would start by running a *U*-test on the pair of samples you wished to examine. Then you need to check which is the largest of the *U*-values. Remember that:

$$U_a + U_b = n_a \times n_b$$

So, if you know the two sample sizes you can work out the other *U*-value once you have your result (the `wilcox.test()` command does not always display the largest *U* value). You can now also determine the harmonic mean of the pair of samples using the formula as before (Figure 10.19).

Once you have your *U*-value you can proceed in one of two ways:

- Calculate a critical *U*-value (Figure 10.18) for your situation, and then compare your calculated value to the critical.
- Rearrange the post-hoc formula to make *Q* the subject (Figure 10.20). Once you have *Q* you can determine the exact *p*-value.

$$Q = \frac{U - \left(\frac{n^2}{2}\right)}{n\sqrt{\frac{(2n+1)}{24}}}$$

Figure 10.20 Using a *U*-value to calculate *Q*, the Studentized range, in a post-hoc Kruskal–Wallis test. Once you know *Q* you can determine an exact *p*-value.

You can work out the value of *Q* that you have to exceed for any given level of significance (a critical value) using the `qtukey()` command. You need to insert the confidence interval (rather than the *p*-value); a value of 0.95 equates to a *p*-value of 0.05.

```
> qtukey(p = 0.95, nmeans = 3, df = Inf)
[1] 3.314493
```

In the example `nmeans` is set to 3, giving a critical value for the 5% significance level in post-hoc test when there were three samples overall (as in the hoglouse dataset). Note that the degrees of freedom (`df`) is set to `Inf` (infinity) for all pairwise comparisons.

You can use the `ptukey()` command to determine the *p*-value if you have solved the post-hoc equation and found *Q*. The result you get is a confidence interval so you need to use 1 – CI to get your *p*-value, therefore:

```
1 - ptukey(Q-value, nmeans, df = Inf)
```

If this sounds a bit complicated, it is! There is no sensible alternative for this kind of analysis. Look at the support website for some additional notes and resources (R code examples) about Kruskal–Wallis post-hoc testing.

Graphing the results of Kruskal–Wallis testing using R

The `boxplot()` command is the easiest way to present a visual summary of your Kruskal–Wallis tests (see Section 6.3.2).

10.2.4 Non-parametric tests for two-way ANOVA

If you have more than two predictor variables and your data are not normally distributed, your options are limited! The best way around this is to design your experiment carefully so that you are likely to get parametric data. If this fails then you may be able to do some cunning maths to alter your data and make it more normal. For example you might take the log of your values or transform your data using some other mathematical process (look back to Section 4.6 on transforming data).

If all else fails you can convert your data to rank values and carry out ANOVA on those.

EXERCISES

Answers to exercises can be found in the Appendix.

1. You have results from an experiment on numbers of insects found in response to several different treatments. There were 6 treatments and insects were counted 12 times for each. The SS for the main predictor variable was calculated to be 2669, whilst the Error SS were 1015. What would the critical value for *F* be in order for the result to be significant (at 5%)? What is the actual *F*-value?
2. In a two-way analysis of variance you have two ____ variables. These can act independently or form ____, which need to be accounted for.
3. When you have non-parametric data you can run the Kruskal–Wallis test as an alternative to ANOVA. TRUE or FALSE?
4. After you have carried out the main analysis you must run ____ tests, which are more in-depth ___ comparisons.
5. You've got some normally distributed data and have carried out a one-way ANOVA. Which of the following graphs would not be a good way to visualize the results?
   A. Bar chart of mean values and letters denoting post-hoc results
   B. Box–whisker plot of median, IQR and range
   C. Bar chart of mean and confidence interval
   D. Bar chart of mean and standard error
   E. Line plot of mean values

Chapter 10: Summary

Topic	Key points
One-way analysis of variance	When you have more than two samples and you want to explore differences (of normally distributed data) you need to use ANOVA. The simplest situation is where you have a response variable and a single predictor variable (with more than two levels). ANOVA uses the variability of samples in various combinations to get a result (an *F*-value). The sum of squares (SS) is a statistic used in the analysis (part of calculating the variance). Excel can carry out the computations but there is no function to "shortcut" the process. Use the `FDIST` function to get critical values for *F*. The *Analysis ToolPak* can carry out one-way ANOVA. Use the `aov()` command in R to carry out ANOVA. The variables are specified as a `formula` of the form `response ~ predictor`.
Two-way ANOVA	When you have more than one predictor variable (and data are normally distributed) you need to extend the ANOVA. In two-way ANOVA you have two predictors. The calculations are more involved than for one-way ANOVA. Predictor variables can act "in concert" with one another so you may need to look at interactions between variables. You can carry out the computations for two-way ANOVA in Excel but there is no shortcut function. The *Analysis ToolPak* can carry out two-way ANOVA but data must be laid out in a particular fashion (on-the-ground layout). You can use the `aov()` command to carry out two-way ANOVA using R.
Post-hoc analysis in ANOVA	The result of ANOVA is an "overall" result and you need to run pair-by-pair comparisons to get additional detail. You need a special process to take into account the fact you are carrying out multiple tests. The most common method is the Tukey honest significant difference (HSD) test. You can use the `TukeyHSD()` command in R to conduct post-hoc analysis. There is no equivalent function in Excel.
Kruskal–Wallis (non-parametric ANOVA)	When you have non-parametric data and a single predictor you can use the Kruskal–Wallis test in lieu of one-way ANOVA. The analysis uses the ranks of the various samples to determine if differences are significant. You can carry out the computations using Excel but there is no way to get critical values. The `RANK.AVG` function can help rank the sample data. You can use the `kruskal.test()` command in R to carry out Kruskal–Wallis analysis.
Post-hoc testing after the Kruskal–Wallis test	The Kruskal–Wallis test gives an "overall" result. After the main test you need to carry out post-hoc analyses, e.g. pair by pair comparisons. There are no universally accepted methods but a modified version of the *U*-test can be used along with the *Q*-statistic (the Studentized range) in a form of Tukey HSD test.

Chapter 10: Summary – *continued*

Topic	Key points
Arranging data	In general the scientific recording layout – i.e. where you have each column as a separate variable and the rows are the observations – is the best way to arrange data. The sample layout is less flexible than the recording layout but is the one required by Excel. In practice you cannot carry out ANOVA with more than two predictor variables in Excel.
Transforming data	If data are non-parametric then transforming data to improve the distribution may be the only option for situations with two (or more) predictor variables. As a last resort you can convert all data to their ranks and then run ANOVA. Any transformation makes interpretation more difficult!
Graphing results of ANOVA	You can use bar charts (with error bars) to visualize differences between samples for all sorts of ANOVA (and Kruskal–Wallis tests). Box–whisker plots are a good alternative, particularly when differences are fairly small. The `boxplot()` command in R makes box–whisker plots easily.

11. Tests for linking several factors

When you are looking for links between several variables you are most likely going to undertake regression.

What you will learn in this chapter

» How to carry out multiple regression
» How to build a regression model
» How to carry out curvilinear regression
» How to carry out logistic regression (binomial regression)
» How to add lines of best fit (straight or curved trendlines) to graphs

You can think of regression as an extended version of correlation, where you determine not only the strength of the relationship(s) but also the mathematical formula that best describes these relationships.

11.1 Multiple regression

In Section 8.1 you looked at correlation, which describes the relationship between factors when you suspect a link (as opposed to looking for differences). Examples might include water speed and stonefly abundance, light levels and abundance of bluebell. You can draw a graph of the relationship and it would naturally be a scatter plot (recall Section 6.4). The relationship might be positive or negative (or there might be no relationship at all).

A correlation coefficient determines the strength and direction of the link between the two variables. In regression, however, you can use the properties of the normal distribution curve to tell you more. The simplest regression is worked out using the Pearson product moment (Section 8.2, often just called r). Using this you can tell the strength of the link (correlation) as well as something about its nature.

For example the equation that describes a straight line on a scatter graph is:

$$y = mx + c$$

y is the *dependent* variable (the vertical axis, also called the *response* factor) and x is the *independent* variable (the horizontal axis, also called the *predictor* factor). In the examples above, y would be stonefly or bluebell abundance whilst x would be speed or light. m is the slope of the line: a steep slope tells you that the abundance changes rapidly with the independent factor (speed or light) whilst a low slope tells you that the abundance changes slowly. This is not related to the correlation coefficient so it is perfectly possible to have a weak link that alters rapidly or a strong link that changes slowly. The correlation coefficient (the link) tells you how close to the best-fit line the data points lie.

When you draw graphs of your correlations you naturally want to emphasise the link between the two factors. For Spearman's rank tests you should never try to show a line of best fit (as the line implies a mathematical relationship between the variables). For the Pearson correlation you may produce a line of best fit if you are looking to determine how quickly one factor changes the other.

Now consider this example: you are looking at butterfly abundance at a number of sites and are trying to work out which environmental factor is most important in determining how many butterflies there are. You might have things like: % cover of food-plant, % cover of trees, max temperature, abundance of crab spiders, number of tourists per hectare (you may be able to think of other factors).

The simplest thing to do is to take each of these potential predictor variables and run a correlation for each one against your response factor (the number of butterflies). You can then simply determine which factor has the highest correlation coefficient. This one would be the most important one in determining butterfly abundance.

As in the example of ANOVA (Section 10.1.2), the factors may interact with one another in unexpected ways. In order to examine this you use multiple regression. In mathematical terms, you can write multiple regression in a similar way to the equation for a straight line so:

$$y = m_1x_1 + m_2x_2 + m_3x_3 + m_nx_n + c$$

Here four factors are shown. In multiple regression, the slopes ($m_1, \ldots, m_n$) are not exactly the same as in the simple example. You can better think of them as expressing how

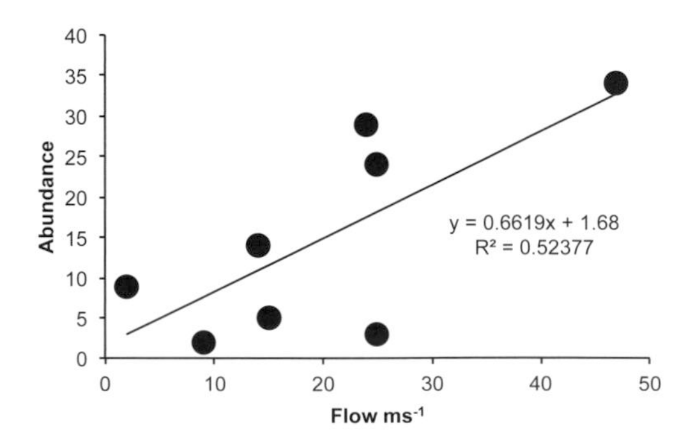

Figure 11.1 The simplest form of linear regression with a single predictor variable. The line of best fit passes through the mean of *x* and the mean of *y* and its slope is adjusted so it minimizes the distance (vertically) to the points. The equation of the linear fit is displayed along with the R^2 value (the regression coefficient squared).

your *predictor* factor (independent variable) is affecting your *response* factor (dependent variable).

It can be hard to visualize what is going on. In the simplest case you have your response variable and a single predictor variable. You can draw a scatter graph and use the $y = mx + c$ formula to help you determine the line of best fit. The graph in Figure 11.1 shows a simple linear situation.

In Figure 11.1 you can see the scatter of points and the line of best fit. This line is fitted by the equation (shown on the figure) and passes through the mean of both variables. The slope is such that the cumulative distance to the points (measured vertically) is minimized (these distances are called *residuals*). The R^2 value is a measure of how close the points lie to the line; it is the square of the correlation coefficient (and varies between 0 and 1). When you add a second predictor variable the situation becomes a little harder. You could draw a 3D graph using the third axis to represent your new variable. Instead of a straight line you now attempt to fit a flat plane (like a sheet of paper or plank of wood) through the 3D cloud of points so that the cumulative distance to those points is minimized. Figure 11.2 shows a plane of best fit in a 3D setting.

The graph in Figure 11.2 was created using the same data as for Figure 11.1 but has a second predictor variable added. This naturally alters the equation as you now have an added factor to take into account (the actual equation is given in the caption). You can just about visualize this 3D situation but if you add more predictor variables you don't have enough planes to work with and it is impossible to draw! Mathematically, however, the situation is just the same as before: trying to fit a plane to all the points to minimise the distance from it to those points.

In simple regression you end up with a correlation coefficient. With multiple regression you end up with several slopes, which are usually called coefficients. You can

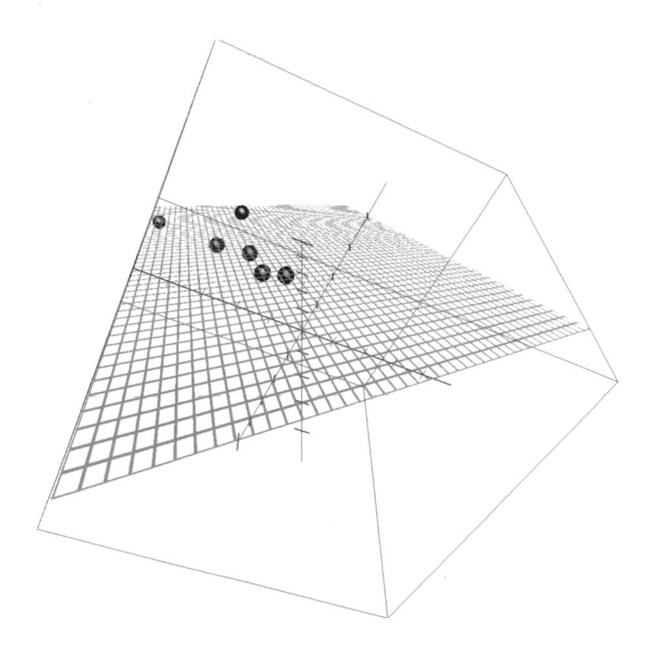

Figure 11.2 A 3D calculated representation of a plane of best fit. The equation for the model fit is $y = 2.39x_1 - 0.83x_2 + 23.04$ calculated by multiple regression.

examine the statistical significance of the various factors and it is usual to remove the non-significant factors and re-run the maths using only those factors that are significant. At the end you finish up with an overall coefficient (usually called R^2 and written as a proportion, e.g. 0.64), which tells you what proportion of the variability is explained by your factors. If you like, it is a way of seeing how good your equation is. If you find that you get several significant terms (another word for the independent factors) but that R^2 is fairly low, then you might think that there are other factors out there that you haven't thought of that might explain the situation better!

You now need to think about how to go about calculating the regression for some data. Table 11.1 shows the data that were used to draw the 3D graph (Figure 11.2). The first column shows the abundance of a butterfly species; this is the dependent (response) variable. The next two columns show the abundance of the larval food plant and the availability of nectar sources (two examples of predictor variables).

Table 11.1 Abundance of butterflies, larval food-plant and nectar plants for a meadow in Shropshire

Count	Food	Nectar
16	2	9
19	5	13
27	7	17
44	9	15
34	12	21
53	23	26

You are going to need to determine the sums of squares for all the columns so you start by determining the mean of each column. Then you'll determine the correlation between all the columns.

The data in Table 11.1 are available on the support website as *bff.csv* and *bff.RData* (which is also part of the *S4E2e.RData*).

Start by evaluating the correlation formula for each pair of columns (recall Section 8.2). The formula is shown again in Figure 11.3.

$$r = \frac{\sum(x-\bar{x})(y-\bar{y})}{\sqrt{\sum(x-\bar{x})^2 \sum(y-\bar{y})^2}}$$

Figure 11.3 Formula to calculate the correlation coefficient.

You need the mean of each column and you then subtract this from each observation. The top line of the formula shows you that you need to multiply each difference by the corresponding difference in the column you are working on (you call one column x and the other column y). The bottom of the formula looks at the sums of squares, which you met when looking at ANOVA in Section 10.2. You multiply the sums of squares for each of the two columns together and take the square root.

In this case you end up with three values: one for the correlation between *Count* and *Food*, one for *Count* and *Nectar*, and one for the correlation between the two predictor variables, *Food* and *Nectar*. Because of the multiple predictor variables, you need to come up with a system for naming the various components. Generally a subscript is used, e.g. r_{y1}, r_{y2}, r_{12}, which gives you the correlation between y and the first predictor variable, between y and the second predictor and finally the correlation between the first and second predictors.

Table 11.2 shows the correlation coefficients and a reminder of which variables they relate to.

Table 11.2 Correlation coefficients for butterflies, larval food-plant and nectar plants for a meadow in Shropshire.

Coefficient label	Coefficient value	Variables
r_{y1}	0.8915	Count vs. food
r_{y2}	0.8148	Count vs. nectar
r_{12}	0.9511	Food vs. nectar

You can see several things already: the coefficients are quite high and all are positive, so as *Food* increases so does *Count* for example. There is a strong link between the two predictor variables. You can also see that *Count* and *Food* have a stronger link than *Count* and *Nectar*. This implies that larval food-plant availability is more important than nectar-plant availability; however, this is getting a little ahead of the game.

You will also need to determine the value of these correlation coefficients squared and the standard deviation of each factor (i.e. column). You will not see all the details of the calculations but you will look at the results in detail, as this will help you interpret the situation. You will also look at how to get Excel to calculate multiple regression for you and also use the R program.

When you have two predictor variables you can work out a *beta coefficient* using the correlations between the columns you already determined. The formula for that is shown in Figure 11.4.

$$b'_{y12} = \frac{\left(r_{1y} - r_{2y} r_{12}\right)}{\left(1 - r_{12}^{2}\right)}$$

Figure 11.4 Calculating a beta coefficient using correlation coefficients.

You can see that you need all three correlation coefficients (the subscripts tell you which variables the coefficients relate to) to determine this one value, which you can think of as being analogous to the slope (for your first predictor variable). There is a similar formula used to determine the other beta coefficients. The beta coefficients (there will be two in this case, one for each predictor variable) are standardized against one another by this formula. Effectively they assume the same units. Now in multiple regression you really want to produce a formula that describes the relationship between your dependent variable and the various predictor variables. You need coefficients that assume the units of the original measurement. You do this by using the standard deviations and the formula shown in Figure 11.5.

$$b_{y12} = b'_{y12} \frac{s_y}{s_{x_1}}$$

Figure 11.5 Calculating regression coefficients from the beta coefficients.

The final coefficient (which you will use to describe the relationship) is determined by taking the beta coefficient and the standard deviation of the two factors you are comparing.

The final list of coefficients for the current example is shown in Table 11.3.

Table 11.3 Regression coefficients for butterflies, larval food-plant and nectar plants for a meadow in Shropshire.

	Coefficients	Beta coefficients
Intercept	23.0468	
Food	2.3889	1.2209
Nectar	–0.8301	–0.3463

You can use the coefficients column to construct the formula that describes the complete relationship between butterfly abundance and the two predictor factors in the following manner:

$$y = 2.38x_1 - 0.830x_2 + 23.05$$

Better still would be to substitute the names of the variables to give:

$$Count = 2.38\ Food - 0.830\ Nectar + 23.05$$

This tells you much about the regression but it is not the final story. The beta coefficients are useful; they are standardized against one another and you can readily see that *Food* is more important than *Nectar*. You have details about the relationship but you do not know if the factors are statistically significant. Nor do you know if the overall relationship is itself significant and how much of the variability (of the dependent factor) is explained by your model.

When you conduct a multiple regression you also determine an overall regression coefficient, called R^2, which is a measure of how good your model is. In this case you get an overall R^2 of 0.806. This means that the two predictor variables explain about 80% of the variability in y, the dependent factor (the butterfly abundance). The R^2 value is calculated by multiplying the *beta* and correlation coefficients for various combinations of variables.

You can also determine the statistical significance of the various elements. You can tell if each individual factor is significant and if your overall model is significant. The methods for computing these statistics "long-hand" are a bit tedious and involve the sums of squares and the R^2 values. You'll see a bit more detail in the sections dealing with calculating regression using Excel (Section 11.1.1) and using R (Section 11.1.2).

Determining the best regression model

When you have several predictor variables you should not simply add them all into your regression. You really want to determine which combination of variables produces the "best" model. That is, you want to find out which variables produce the best fit. Some variables may not make a statistically significant difference, and these variables should be excluded.

A good start is to make a correlation matrix, where you can see which of the pairwise combinations has the highest value. The variable that produces the highest correlation coefficient should be the first one you add to your regression model. Once you have this starting point (your response variable and one predictor) you can add variables to your model and see the effect that they have.

Some computer programs will perform what is called stepwise analysis, i.e. they start with a single factor and keep adding the next best option until the best model is attained. The next best option is usually not the one with the next highest correlation, so it requires some detective work! You cannot easily carry out stepwise analysis using Excel but you'll see later how you can do this in R (Section 11.1.2).

Graphing the results of multiple regression

It is not always easy to represent the results of a regression graphically. When you only have a single predictor variable you can use a scatter plot (Section 6.4). When you have several factors it becomes impossible to represent the raw data in a single graph. What you'll generally do is to present a scatter plot of your "most important" variable and present the regression model as a table of coefficients.

You'll see various examples of how to represent regression in this chapter, especially with regard to curvilinear regression (Section 11.2), where the relationship between variables is not a straight line. You'll see how to add curved trendlines for example.

One way to summarize a regression model is to plot some diagnostic features, the most common being the residual (think of residuals as being how "bad" the fit between the trendline and the points is). The support website has some additional notes and examples of how to draw these diagnostic plots.

Go to the website for additional notes.

11.1.1 Using Excel for multiple regression

It is not surprising to learn that multiple regression is not easy to calculate on your own, as many of the calculation steps are long, involved and tedious (although not in themselves mathematically difficult). However, you can undertake regression fairly easily using Excel and the `LINEST` function. The general form of the function is like so:

```
LINEST(y_values, x_values, intercept, stats)
```

`y_values` the response variable.

`x_values`	the predictor variable(s). These must be in a continuous block, in other words adjacent columns.
`intercept`	how to compute the intercept; a value of `1` indicates a regular intercept. A value of `0` forces the intercept through the origin (you almost never need to do this).
`stats`	do you want additional statistics? A value of `1` produces additional statistics (which generally you want), whilst a value of `0` results in slope and intercept values only.

The `LINEST` function is an *array* function, meaning that it produces a result spread over several cells. It always produces five rows of results and the number of columns matches the number of variables. To run the `LINEST` function you need to proceed as follows:

1. Make sure your predictor variables are in a single block (i.e. adjacent columns). If necessary you will have to copy/paste data.
2. Use the mouse to select where you want the results to go. You need to highlight a block of cells 5 rows deep and a number of columns to match the variables. So, if you had a response variable and 3 predictors you would highlight a block 5 rows deep and 4 columns wide.
3. Type `=LINEST(` to start the function.
4. Now highlight the cells containing the response variable (data only, not the heading).
5. Type a comma and then highlight the predictor values, which have to be in a single block (you cannot use the Ctrl key to select non-adjacent cells).
6. Type another comma and then enter a `1`.
7. Repeat step 6. Then type the closing parenthesis `)`.
8. Press *Ctrl+Shift+Enter* (in Windows) instead of just *Enter* (on a Mac *Cmd+Enter*). The results should now fill the block you highlighted.

You'll see all the results (there will also be some `#N/A` values, which you can ignore) but there are no labels! In Table 11.4 you can see the results of a `LINEST` function applied to the data from Table 11.1. The table has been annotated to demonstrate which results are which.

The first row shows the coefficients and these are shown in reverse order. The first

Table 11.4 Example of output from the `LINEST` function. The function does not produce labels (which are added here for information). See text for more details.

–0.830 Coefficient of last predictor	2.389 Coefficient of first predictor	23.047 The intercept
1.971 Standard error of coefficient	1.609 Standard error of coefficient	19.293 Standard error of intercept
0.806 Total R^2 value	8.190 Standard error of R^2 value	#N/A
6.244 Overall F statistic	3 Degrees of freedom	#N/A
837.617 SS of predictors	201.217 SS of error term	#N/A

coefficient is the last of the predictor factors. The last one is always the intercept. The next row shows the standard errors of the coefficients and they are in the same order as the coefficients. The third row shows you the overall fit of the model, i.e. the R^2 value and its standard error. You then see the *F*-value and the degrees of freedom. The last row shows the sums of squares.

The `LINEST` function gives you a variety of statistics but it does not compute the statistical significance of the overall regression or of the individual factors. You must do these calculations for yourself.

Determine overall model significance by calculating an exact *p*-value with the `FDIST` function:

```
FSIDT(F-value, df_numerator, df_error)
```

The required values are:

- `F-value` – this is in the `LINEST` results (4th row, 1st column).
- `df_numerator` – this is the total number of variable columns (response and predictors) minus 1. Alternatively, think of it as the number of predictor variables you have.
- `df_error` – this is in the `LINEST` results (4th row, 2nd column). The error term is also called *residuals*.

Determine the significance of the individual model terms (the variables) by calculating a *t*-value first:

$$t\text{-value} = \text{coefficient} \div \text{standard error}$$

The coefficients are in the first row of the `LINEST` results. The standard errors are in the second row.

Once you have *t*-values you can use the `TDIST` function to compute exact *p*-values:

```
TIDST(ABS(t-value), df, 2)
```

The degrees of freedom (`df`) are in the `LINEST` results (4th row, 2nd column). Note that the `ABS` function is used because you may get negative *t*-values (the `ABS` function ignores the sign). The final `2` denotes that you require the two-tailed result (both ends of the normal distribution).

The `LINEST` function does not calculate the *beta coefficients* for you, which is a shame but simple to do for yourself. First of all you need to determine the standard deviation for each column. You can use the built-in formula `STDEV` for this. Now you use the standard deviations and the value of the regression coefficients to determine the beta coefficients using the formula shown in Figure 11.6 (which is essentially a rearrangement of the formula in Figure 11.5). In the formula (Figure 11.6) *b* is the regular coefficient. The two standard deviations (s_x and s_y) relate to the *x* and *y* values, i.e. the predictor and response variables.

$$b_{y12} = b'_{y12} \frac{s_y}{s_{x_1}}$$

Figure 11.6 Calculating a beta coefficient using a regression coefficient and the standard deviation of the variables.

Model building using Excel

You can't carry out any kind of automatic model building using Excel. The best you can do is to keep rearranging data columns and re-running your analyses on various combinations of variables. The *Analysis ToolPak* is a great help here, as it can construct a correlation matrix as well as running the multiple regression.

Using the Analysis ToolPak for multiple regression

You can use the *Analysis ToolPak* to carry out multiple regression. If you have this add-in installed (see Section 1.9.1), you can run it by selecting the *Data Analysis* button on the *Data* menu on the ribbon.

You select the *Regression* option from the *Data Analysis* selection window and can then select your data and select other options from the *Regression* dialogue box (Figure 11.7). Note that your predictor variables must be in a single block (you cannot use the *Control* key to select non-adjacent cells).

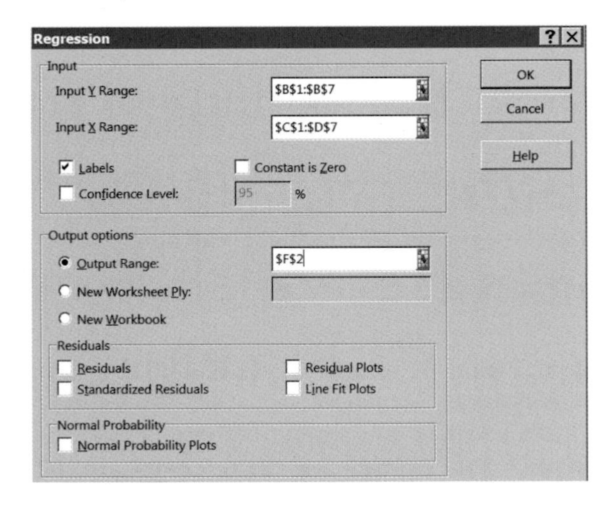

Figure 11.7 The *Regression* dialogue box from the *Analysis ToolPak* allows you to carry out regression analysis.

Once you have selected the data and the options you require, you carry out the analysis by clicking the OK button. The results appear where you selected the output range to be and are comprised of three main tables (Table 11.5 shows an approximation of the screen display).

The top table shows you some overall statistics, including the R^2 value and the number of observations. The middle table shows you the overall significance of your

Table 11.5 The multiple regression results from the *Analysis ToolPak* are quite comprehensive.

SUMMARY OUTPUT

Regression Statistics	
Multiple R	0.898
R Square	0.806
Adjusted R Square	0.677
Standard Error	8.190
Observations	6

ANOVA

	df	*SS*	*MS*	*F*	*Signif. F*
Regression	2	837.617	418.808	6.244	0.085
Residual	3	201.217	67.072		
Total	5	1038.83			

	Coeff	*Standard Error*	*t Stat*	*P-value*	*Lower 95%*	*Upper 95%*	*Lower 95.0%*	*Upper 95.0%*
Intercept	23.047	19.293	1.195	0.318	–38.353	84.446	–38.353	84.446
Food	2.389	1.609	1.485	0.234	–2.731	7.509	–2.731	7.509
Nectar	–0.830	1.971	–0.421	0.702	–7.102	5.442	–7.102	5.442

regression model. The final table shows you the coefficients for the various variables as well as their significance. You see for example that neither of the two predictor variables are statistically significant. The lower table also shows the confidence intervals.

In the lower part of the data selection window (Figure 11.7), you can see that there are some additional options. You can produce some plots and tables of residuals. You can think of the residuals as being the difference between a particular point of data and the idealized regression, which you get when you work out the line of best fit. Ideally you would like your residuals to be normally distributed and you can use the tables of results to check for this. You can also check the *Normal Probability Plot* option to produce a graph that presents the data plotted against the ideal probabilities; essentially this allows you to see if your data are normally distributed. If the points lie more or less on a straight line then you have normality (Figure 11.8).

The plot (Figure 11.8) does not have the trendline added automatically; adding one (Section 8.3.1) makes it easier to see if the points are in a line. The support website gives more information about plotting the results of regression.

Go to the support website for more information.

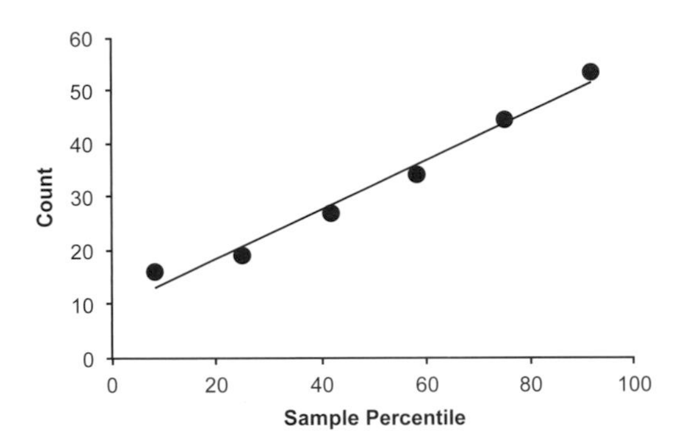

Figure 11.8 Normal probability plot from the *Analysis ToolPak, Regression* routine. If the points lie approximately in a straight line the model is normally distributed.

Note: Correlation matrices using the Analysis ToolPak

You can make a table displaying the correlation coefficients for all combinations of variables. Click the *Data > Data Analysis* button and select *Correlation*. You can then select your data (and their heading labels) and place the results.

11.1.2 Using R to perform multiple regression

R will perform multiple regression quite easily; the basic command to do this is `lm()`. In addition, there are several "helper" commands that allow you to extract results and to carry out model building.

Multiple regression and the lm() command

The `lm()` command uses a model syntax similar to that you met earlier when looking at ANOVA (Section 10.1.4).

```
lm(formula, data)
```

`formula`	A formula of the form `y ~ x` where *y* is a response variable and *x* is one or more predictors. If you have more than one predictor you use + to separate them.
`data`	The name of the data object (usually a `data.frame`) that contains the variables.

The `formula` you specify determines what kind of regression analysis is performed. You place your *response* factor on one side of the equation and list the *predictor* variables on the other. Most often you'll separate the variables with + signs to indicate that the variables are to be treated independently (see Table 11.10). So a formula of the form `y ~ x + z` would be a multiple regression with two predictor variables (`x` and `z`).

Note: Regression formulae in R

The formula syntax allows you to specify complicated regression situations. You can, for example, include interactions between variables, e.g. `y ~ x * z`. However, it is a lot harder to interpret the results of regressions with interactions!

The general way to proceed with a regression is as follows:

1. Have your data ready, usually you'll have used the `read.csv()` command to read the data from a file. You want your data arranged so that each column is a variable. Ideally the first column should be the response but this is not actually essential.
2. Run the regression via the `lm()` command and save the result to a named object so you can extract the component results.
3. Use the `summary()` command to get an overview of the regression result. Other helper commands can be used to extract certain components.

The process is illustrated in the following example, using the butterfly data that you met earlier:

```
> bff
  count food nectar
1    16    2      9
2    19    5     13
3    27    7     17
4    44    9     15
5    34   12     21
6    53   23     26
> bff.lm = lm(count ~ food + nectar, data = bff)
> summary(bff.lm)
Call:
lm(formula = count ~ food + nectar, data = bff)
Residuals:
      1       2       3       4       5       6
-4.3537 -5.2002  1.3423 11.9041 -0.2821 -3.4104
Coefficients:
            Estimate Std. Error t value Pr(>|t|)
(Intercept)  23.0468    19.2932   1.195    0.318
food          2.3890     1.6089   1.485    0.234
nectar       -0.8301     1.9708  -0.421    0.702
Residual standard error: 8.19 on 3 degrees of freedom
Multiple R-squared:  0.8063,   Adjusted R-squared:  0.6772
F-statistic: 6.244 on 2 and 3 DF,  p-value: 0.08525
```

The `summary()` command shows the main results. The rows relate to the intercept and the predictor variables. The column headed `Estimate` shows the coefficients (the intercept and slope values). You can also see the overall significance of the model and the final R^2 value.

Beta coefficients

R does not work out the beta coefficients for you and if you wish to determine what they are, you need to calculate them yourself. In order to do that you need to work out the standard deviations for each of the columns of data and use the formula you met previously (Figure 11.6). In words, this is the regression coefficient multiplied by the standard deviation of the predictor variable divided by the standard deviation of the response variable.

First of all you need a way to get at the individual coefficients from the linear model you created in R. The coefficients are stored inside the main result (called `bff.lm` in the example). The simplest way to get these coefficients is to use the `coef()` helper function:

```
> coef(bff.lm)
(Intercept)        food      nectar
 23.0468094   2.3889877  -0.8301202
```

To get a single coefficient you can add another bit at the end in square brackets:

```
> coef(bff.lm)[2]
    food
2.388988
```

Here you've extracted the second coefficient, which happens to correspond to the one for the food variable. You can also type the name directly, but you must put it in quotes:

```
> coef(bff.lm)["food"]
```

You can determine the standard deviation using the `sd()` command. The final calculation would look like the following:

```
> coef(bff.lm)[2] * sd(bff$food) / sd(bff$count)
    food
1.220935
```

Other beta coefficients can be determined in a similar manner:

```
> coef(bff.lm)["nectar"] * sd(bff$nectar) / sd(bff$count)
    nectar
-0.3463437
```

Note: Calculating beta coefficients with R

Visit the support website for some more notes about beta coefficients. You can download custom R commands that allow you to calculate the coefficients more easily.

Download material and information available on the support website.

Stepwise regression

When you have only a few predictor variables, you can explore the situation by typing in various combinations of factors. The "best" regression model would explain much of the variability in the response variable (i.e. would have a large R^2 value), but would contain only those predictor variables that were statistically significant. You can keep piling in variables to a regression and sooner or later you'll explain most of the variability (this is sometimes called over-fitting). However, you should aim to summarize the situation as succinctly as possible, by producing a *minimum effective model* (or *minimum adequate model*).

Once you get to the best situation you can stop and report the most effective model for your data. As you get more and more variables, it becomes a lot more time consuming to try out every combination of factors. It would be better if you could find a way to determine the best starting point and then add the next best factors one at a time. This approach is called stepwise analysis, and you'll see this process used on some example data that you can download from the support website.

Go to the website for support material. The example data are in CSV form as a file called *mayfly regression.csv*. These data are also pre-loaded into the R datafile *S4E2e. RData* as a `data.frame` called *mf*.

A good starting point is to select the single predictor variable that has the best correlation with the response factor you are considering. Here is a preview of some data:

```
> head(mf)
  Length Speed Algae  NO3 BOD
1     20    12    40 2.25 200
2     21    14    45 2.15 180
3     22    12    45 1.75 135
4     23    16    80 1.95 120
5     21    20    75 1.95 110
6     20    21    65 2.75 120
```

There is a single response factor: the length of a freshwater invertebrate (a species of mayfly). There are four predictor variables: the water speed, percentage cover of algae, nitrate concentration and biological oxygen demand. You could also use the `names()` command to get a preview of the variables:

```
> names(mf)
[1] "Length" "Speed"   "Algae"   "NO3"     "BOD"
```

In your data files it is a good idea to place the response factor first and the predictor factors in subsequent columns. You need to determine the best starting point for your multiple regression. One way is to use the `cor()` command. You encountered this previously (Section 8.4.1) when looking at the correlation between two variables. If your data comprises several columns of data, R will attempt to correlate all the variables with one another:

```
> cor(mf)
           Length       Speed      Algae         NO3        BOD
Length  1.0000000 -0.34322968  0.7650757  0.45476093 -0.8055507
Speed  -0.3432297  1.00000000 -0.1134416  0.02257931  0.1983412
Algae   0.7650757 -0.11344163  1.0000000  0.37706463 -0.8365705
NO3     0.4547609  0.02257931  0.3770646  1.00000000 -0.3751308
BOD    -0.8055507  0.19834122 -0.8365705 -0.37513077  1.0000000
```

The command produces a rectangular matrix, with each pair of variables being represented twice. In this case you have `Length` as the first variable and this is your response factor. You can look down the first column (or along the first row) and see which of the four predictor variables produces the highest correlation coefficient. You can see that `BOD` has the highest coefficient (–0.806) with `Algae` (0.765) coming next.

Tip: Targeted correlation matrix

If you have a lot of variables you may not want to see all the relationships. You can see the correlation between the response variable and the predictors using the following syntax:

```
cor(response, data)
```

You specify the response variable explicitly and the data object containing all the variables. R is "smart" enough to realize what you want:

```
> cor(mf$Length, mf)
     Length      Speed     Algae       NO3        BOD
[1,]      1 -0.3432297 0.7650757 0.4547609 -0.8055507
```

This means that when you start your multiple regression, `BOD` should be the first of the predictor factors included in your model. The `Algae` factor has the second highest correlation coefficient but this may not be the next best factor to add after the `BOD` variable.

Note the high correlation between `BOD` and `Algae`. Adding this variable may not give you much additional information.

What you need to do is to try out each variable and see which has the most influence on the regression result. R has commands that enable you to "try out" various variables and give you a report so you can pick the next best variable to add. The main command that "drives" the process is `add1()`.

```
add1(object, scope, test)
```

`object`	A regression model result.
`scope`	The variables to test in the regression. Usually this is the same as the original data.
`test`	A significance test. Use `test = "F"` for regular linear regression to carry out an *F*-test.

The `add1()` command gives you two particularly important pieces of information:

- *AIC* value – this tells you how "bad" the fit of your model is. You want to pick the variable with the lowest AIC value.
- *F* value – this tells you if the variable is statistically significant when added to the current regression model.

The best factor to add to your regression model is the one with the lowest *AIC*. This is the *Akaike information criterion* (named after a Japanese statistician) and is a measure of how "bad" your model is with this factor in place. You do not need to know much about *AIC* as it is determined from the residual sums of squares (*RSS*) as follows:

$$AIC = 2k + n[\text{Ln}(RSS)]$$

The residuals are worked out from a line (or plane) of best fit. Essentially they are a measure of how far the points lie from the line of best fit. In the *AIC* equation, k is the number of terms in the regression model and n is the number of replicates (observations). From your point of view you only need to know that you are looking for a low *AIC* and this will also go with a low *RSS* and a high sum of squares. The term that has these characteristics will be the best to add to your regression model. However, you should only add the variable if it is statistically significant, which is where the *F*-value comes in.

In general this stepwise regression proceeds along the following lines:

1. Prepare your dataset; this will usually be a `data.frame` that has a column for the response variable and several columns, each containing a predictor variable.
2. Rather than create a linear model with the best predictor factor, you create a blank model, i.e. one with nothing except an intercept:

    ```
    mod = lm(response ~ 1, data)
    ```

3. Use the `add1()` command to determine the effect of adding each term individually to your regression model:

```
add1(mod, data, test = "F")
```

4. From the result of the `add1()` command you will be able to see the "best" variable to add to your regression (significant *F*-value and lowest *AIC*) . Edit the model by replacing the `1` with the variable you identified from step 3.
5. Run step 3 again. The `add1()` command looks through the remaining variables and adds them to your regression model one at a time. The result shows you the effect each variable has on the existing regression model.
6. If you get a variable that adds significantly to your regression model you can edit the model to include it. Then you can run step 3 again. If the `add1()` command shows that none of the variables would add significantly to the model you stop.

In the following exercise you can have a go at running a stepwise regression for yourself.

Have a Go: Use R for stepwise regression

You'll need the *mf* data for this exercise. The data are part of the *S4E2e.RData* file, which is available on the support website. You can also use the *mayfly regression.csv* file if you want to practise importing data to R.

Download material available from the support website.

1. Have a look at the *mf* data object; you can see that there is one response variable *Length* and four predictors:

```
> names(mf)
[1] "Length" "Speed"  "Algae"  "NO3"    "BOD"
```

2. Make a "blank" regression model, i.e. one with the response and intercept only:

```
> mod = lm(Length ~ 1, data = mf)
```

3. Use the `add1()` command to help select the first variable to add to the model:

```
> add1(mod, scope = mf, test = "F")
Single term additions
Model:
Length ~ 1
       Df Sum of Sq     RSS    AIC F value    Pr(>F)
<none>              227.760 57.235
Speed   1    26.832 200.928 56.102  3.0714   0.09300 .
Algae   1   133.317  94.443 37.228 32.4672 8.404e-06 ***
NO3     1    47.102 180.658 53.443  5.9967   0.02237 *
BOD     1   147.796  79.964 33.067 42.5106 1.185e-06 ***
```

4. You can see from step 3 that the *BOD* and *Algae* variables are both significant. However, the *BOD* variable has the lower *AIC* value. Replace the intercept

in the model with the *BOD* variable (you can use the up arrow to recall the previous command, which you can edit):

```
> mod = lm(Length ~ BOD, data = mf)
> summary(mod)
Call:
lm(formula = Length ~ BOD, data = mf)
Residuals:
   Min      1Q Median      3Q     Max
-3.453 -1.073  0.307  1.105  3.343
Coefficients:
              Estimate Std. Error t value Pr(>|t|)
(Intercept) 27.697314   1.290822   21.46  < 2e-16 ***
BOD         -0.055202   0.008467   -6.52 1.18e-06 ***
Residual standard error: 1.865 on 23 degrees of freedom
Multiple R-squared:  0.6489, Adjusted R-squared:  0.6336
F-statistic: 42.51 on 1 and 23 DF,  p-value: 1.185e-06
```

5. Now run the `add1()` command again:

```
> add1(mod, scope = mf, test = "F")
Single term additions
Model:
Length ~ BOD
       Df Sum of Sq    RSS    AIC F value Pr(>F)
<none>              79.964 33.067
Speed   1    7.9794 71.984 32.439  2.4387 0.1326
Algae   1    6.3081 73.656 33.013  1.8841 0.1837
NO3     1    6.1703 73.794 33.060  1.8395 0.1888
```

You can see from step 5 that none of the remaining variables add significantly to the model. This is the point at which to stop!

When you are building a regression model you only retain terms that are themselves statistically significant. In the preceding exercise you might have thought that the `Algae` variable would have a place in your model, as it did have a high correlation. If you look back to the correlation matrix you created earlier, you can see that `BOD` and `Algae` are highly correlated with one another (–0.837). So, adding the `Algae` term to the `BOD` already present is not giving you new information.

Note: Regression models

Sometimes you are not just interested in the "best" model but are interested in certain factors and how they affect your response variable. It is okay to present alternative models but beware of drawing conclusions from variables that are not statistically significant.

So, now you have a way to step along from a bare model to an ever more complex regression. You keep only those terms that are significant and stop when you start to get non-significant terms.

Visit the support website for more information about presenting the results of regression models.

11.2 Curved-linear regression

You first encountered the possibility that a correlation could be based on a trend that was not more or less straight in Section 8.5, where you looked at *polynomial* and *logarithmic* equations. Any mathematical equation that could be employed can be pressed into service and used to create a form of regression. Here you'll see how to carry out polynomial and logarithmic regression (using R and Excel) as the methods involved can be applied easily to other equations.

11.2.1 Polynomial regression

As you saw previously, the equation that describes a polynomial is:

$$y = ax + bx^2 + c$$

This tends to produce an inverted U-shape when plotted as a regular scatter graph. You can see how the equation is analogous to the multiple linear regression formula that you used earlier:

$$y = m_1x_1 + m_2x^2 + c$$

In Table 11.6 you see some data that follow a polynomial trend. You have measured

Table 11.6 Data on abundance of bluebell in response to light levels. The data form a polynomial relationship that can be explored using regression.

Light levels (arbitrary units)	Abundance (plants per m^2)
2	2
4	3
6	8
8	13
10	16
16	23
22	26
28	25
30	20
36	17
48	6

the abundance of a plant (bluebell, *Hyacinthoides non-scripta*) in woodlands in southern Britain. The light levels were also measured. Rather than an exact unit you have a relative unit for light (a digital camera can work out the shutter speed and aperture and this can be used to create a unit for light).

What you see is that at low light levels you have few plants. As the light increases you see more plants. At higher levels of light the abundance decreases once more. The abundance data is your response variable and the light level is the predictor variable; however, you need to express the light in terms of light^2 in addition to light.

Download material available on the support website. The data described in the text are available as an Excel file *Bluebell polynomial.xlsx* and as part of the *S4E2e.RData* R file (the data are called *bbel*).

Polynomial regression using R

You can run a polynomial regression analysis in much the same manner to regular linear regression using the `lm()` command, which is a powerful and flexible tool. Start by plotting the variables (Section 6.4.2), which will show you that the relationship is possibly a polynomial one. You need to use a `formula` in the `lm()` command that describes the variables:

```
> names(bbel)
[1] "abund" "light"
> bbel.lm = lm(abund ~ light + I(light^2), data = bbel)
```

Since the data are called `bbel` it seems logical to call your linear model `bbel.lm` but anything would do. Note that in the `lm()` command the `abund` part is your response variable and `light` is your predictor. You use the `I()` part to tell R that you want to include light^2 (`light^2`) as a separate variable. You have to use this because otherwise R would try to evaluate `light + light^2` as "real maths". The `I()` part *insulates* the bit in the brackets and whatever is inside is evaluated as a regular mathematical expression and a separate variable.

You can look at the main results with a simple `summary()` command.

```
> summary(bbel.lm)
Call:
lm(formula = abund ~ light + I(light^2), data = bbel)
Residuals:
   Min     1Q Median     3Q    Max
-3.538 -1.748  0.909  1.690  2.357
Coefficients:
             Estimate Std. Error t value Pr(>|t|)
(Intercept) -2.004846   1.735268  -1.155    0.281
light        2.060100   0.187506  10.987 4.19e-06 ***
I(light^2)  -0.040290   0.003893 -10.348 6.57e-06 ***
---
Signif. codes:  0 '***' 0.001 '**' 0.01 '*' 0.05 '.' 0.1 ' ' 1
Residual standard error: 2.422 on 8 degrees of freedom
```

```
Multiple R-squared:  0.9382,    Adjusted R-squared:  0.9227
F-statistic: 60.68 on 2 and 8 DF,  p-value: 1.463e-05
```

You interpret these in exactly the same way that you did previously. The first column shows the coefficients (although it is labelled *Estimate*) and you can see the three values that you need to create your polynomial equation. In this case you get:

$$abund = 2.060\ light - 0.040\ light2 - 2.005$$

You also see that both *light* and *light*2 are statistically significant. The bottom part shows details for the overall model, including, for example, the R^2 value, around 0.93, which shows that this model explains quite a lot of the variability. You also see the overall *p*-value, which is highly significant.

Plotting a polynomial relationship using R

You should have made a basic plot of the relationship before you started, in order to see what kind of mathematical regression was most suitable:

```
> plot(abund ~ light, data = bbel)
```

What you could do with is a line of best fit, a trendline, as this would make the relationship clearer. You saw how to add a straight line using the `abline()` command (Section 8.4.2) but that is clearly not going to work here!

What you need to do is to work out the coordinates of the "ideal" polynomial relationship based on the equation you just calculated. R provides a helper command, `fitted()`, which calculates these values from the results of a `lm()` command.

```
> fitted(bbel.lm)
        1         2         3         4         5
 1.954197  5.590923  8.905333 11.897427 14.567205
        6         7         8         9        10        11
20.642642 23.817233 24.090979  23.537596 19.943548  4.052917
```

If you use these "ideal" values instead of the original *y*-values you'll be able to draw the trendline. However, this will simply "join the dots" and you will have a "chunky" line. What you need to do is smooth the joins using the `spline()` command.

To add the line to the plot you'll need the `lines()` command, which adds lines to existing plots (recall Table 6.6). The command needs *x* and *y* coordinates, so you use the original *x*-values and the fitted values (fitted to the mathematical equation) you calculated using `fitted()`, as the *y*-values:

```
> plot(abund ~ light, data = bbel, ylab = "Abundance",
xlab = "Light levels")
> lines(spline(bbel$light, fitted(bbel.lm)), lwd = 2)
```

In the example the `lwd = 2` part makes the trendline a bit "fatter" (recall Table 6.5). The resulting plot resembles Figure 11.9.

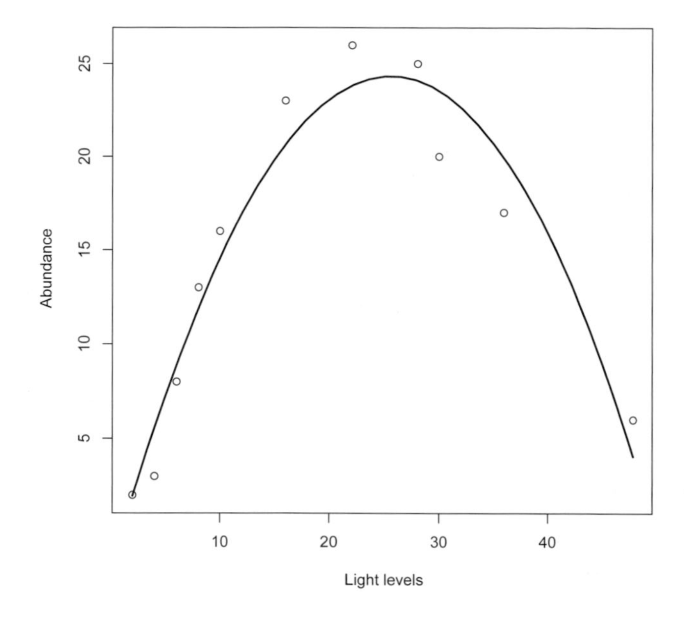

Figure 11.9 The relationship between abundance of a plant (bluebell) and light levels in an English woodland. Polynomial graph with line of best fit. The trendline coordinates are calculated using the `fitted()` command, which reads the results of a regression.

The plot (Figure 11.9) gives you a pretty good impression of the situation and helps you to visualize the data in a way that is simply not possible from a plain text summary.

Polynomial regression using Excel

You can use Excel to determine your regression fairly easily using the `LINEST` function described earlier (Section 11.1.1). First of all you need to add a new column for the x^2 data, then you simply carry out the regression as you did before.

The general way to proceed is as follows:

1. Arrange your data so that the response variable is in the first column, with the remaining predictor variables in subsequent columns.
2. In the third column square the response variable to produce the x^2 values.
3. Highlight a block of cells to hold the `LINEST` result. You need to highlight a block 3 cells wide and 5 rows deep. Leave room to add annotations around the results.
4. Type the `LINEST` formula, e.g. `LINEST(y_val, x_vals, 1,1)`. Remember to press Control+Shift+Enter as `LINEST` is an array function.
5. Annotate the results to help you remember what the components are (see Table 11.7 and Table 11.1).
6. Calculate *t*-values from the coefficients and their standard errors (coefficient ÷ SE, recall Section 11.1.1).
7. From the *t*-values you can determine the significance of the variables and intercept; use the `TDIST` function, e.g. `TIDST(ABS(t-value), df, 2)`.

8. Use the `FDIST` function to calculate an overall significance using the F-value and the degrees of freedom, e.g. `FSIDT(F-value, df_numerator, df_error)`.

In Table 11.7 you can see the data and layout of results using the bluebell and light example data.

Table 11.7 Using `LINEST` in Excel to carry out a polynomial regression. The data are shown on the left and the results on the right. Cells have been annotated (`LINEST` does not add labels).

Abund	Light	Light^2		Light^2	Light	Intercept	
2	2	4	coef	–0.040	2.060	–2.005	
3	4	16	SE	0.004	0.188	1.735	
8	6	36	R-sq	0.938	2.422	#N/A	SE
13	8	64	F-value	60.681	8.000	#N/A	df
16	10	100	SS	711.806	46.921	#N/A	
23	16	256					
26	22	484	t-value	–10.348	10.987	–1.155	
25	28	784	p-value	6.57E–06	4.19E-06	0.281	
20	30	900					
17	36	1296	P-value	1.46E–05			
6	48	2304					

Remember that Excel does not add labels to the `LINEST` results. The result columns are shown backwards! The first one is the second of your response variables (light squared), followed by the first response variable. The last column shows the intercept. Notes regarding the calculations of the t-values, p-values and so on were given above (Section 11.1.1).

Plotting a polynomial relationship using Excel

You'll want to plot the data to show the polynomial nature of the relationship. This will involve adding a trendline (recall Section 8.3.1). Fortunately Excel allows you to fit polynomial trendlines directly from the data. You will need to create a scatter plot using the x and y variables (Section 6.4.1). This might cause you a slight problem as Excel expects the data to be set out with the x-column first and the y-column second. You had them the other way around to make it easier to run the regression calculations. You could rearrange the data but you can also proceed as follows:

1. Click once in your spreadsheet in a blank cell that is not adjacent to any other values. Excel "searches" around the cursor and picks up any data it finds (it is being "helpful"). By having a blank cell you will make a blank chart and can add the data you want afterwards.

2. Click the *Insert > Scatter* button to create a blank scatter plot. Make sure you choose a points-only option (i.e. no joining lines). Your blank chart will be created immediately.
3. Click once on the chart to make sure the *Chart Tools* menus are activated. Then click the *Design > Select Data* button.
4. Click the *Add* button in the *Legend Entries (Series)* section.
5. Use the mouse to select the x and y values as appropriate (you can also select a label if you wish). Click OK to return to the previous dialogue window and OK again to return to the chart. Your data should now be plotted.
6. Click the *Add Chart Element* button (Excel 2013, in 2010 use *Layout > Trendline*). Select *More Trendline Options*. You can also right-click a data point.
7. Click the radio button beside the *Polynomial* option (Figure 8.13). The box beside the option, labelled *Order*, should be set to 2 (a third-order polynomial has another variable and produces a line that switches direction twice).
8. You may elect to display the equation and/or R^2 value on the chart too.

Once you have your basic chart you can use the *Chart Tools* menus, to add axis titles and so on. With a bit of tinkering you can produce a chart that resembles Figure 11.10.

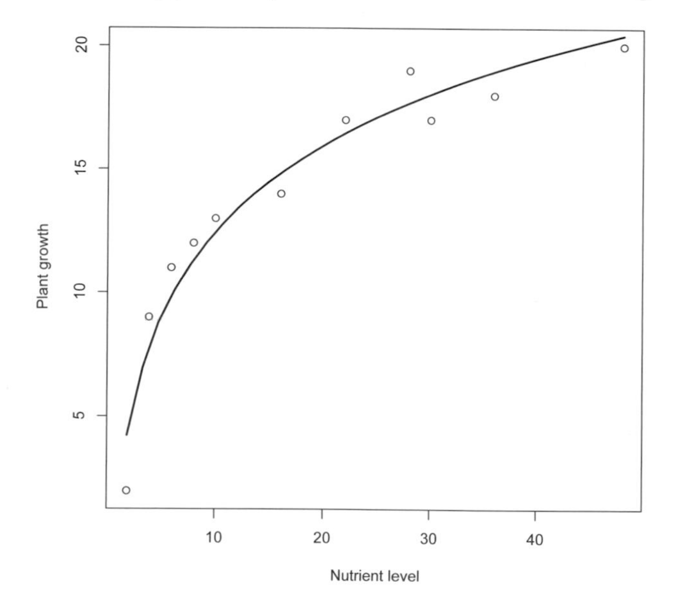

Figure 11.10 Polynomial relationship charted using Excel. The trendline options allow easy addition of a polynomial line as well as the R^2 and equation.

Using the trendline with the equation and R^2 value is a useful way to explore your data before running the `LINEST` function.

Visit the support website for more information about presenting the results of regression models.

Using the Analysis ToolPak for polynomial regression

You can use the *Analysis ToolPak* (Section 1.9.1) to carry out polynomial regression. You'll need to make the x^2 values from the predictor variable as you did before with `LINEST`. You also need the predictor variables to be adjacent to one another.

Once you have your data arranged you can start the process:

1. Click the *Data* > *Data Analysis* button.
2. Select the *Regression* option.
3. Use the mouse to select the y-values (and the labels).
4. Use the mouse to select the x-values (and the labels).
5. Check the box headed *Labels*.
6. Choose where to place the results.
7. Click OK to finish up and complete the regression.

The results are laid out in a similar manner to before (e.g. Table 11.5), with the summary in the top table including the R^2 value, then the overall model significance as an ANOVA table, and finally the regression table and the coefficients and their significance.

11.2.2 11.2.2 Logarithmic regression

As you saw previously, the equation that produces a logarithmic correlation is:

$$y = a \,\text{Log}(x) + c$$

This tends to form a curve that begins quite steeply and then flattens off. In Table 11.8 you see some data that follow this form. You have the growth rate of a plant subjected to different levels of nutrients. Initially the more nutrients you add, the faster the growth rate. At higher concentrations, however, there is a flattening off of the increase (in other words you get diminishing returns).

Table 11.8 Example of plant growth rate in response to nutrient concentration. The data show a logarithmic relationship, which can be explored using regression.

Nutrient concentration	Growth rate
2	2
4	9
6	11
8	12
10	13
16	14
22	17
28	19
30	17
36	18
48	20

Download material available on the support website. The data described in the text are available as an Excel file *Growth logarithmic.xlsx* and as part of the *S4E2e.RData* R file (the data are called *pg*).

As always, you should draw a basic scatter graph of this before you do anything else, as this will help you to visualize what is happening (recall Section 6.4). If you suspect a logarithmic relationship you can explore the regression easily using R or Excel.

Visit the support website for more information about presenting the results of regression models.

11.2.3 Logarithmic regression using R

As before, you use the `lm()` command to carry out regression. In a logarithmic relationship you modify the `formula` to specify the nature of the relationship. There is no need to modify the original data, as you can do this via the `formula`.

```
> names(pg)
[1] "nutrient" "growth"
> pg.lm = lm(growth ~ log(nutrient), data = pg)
```

Note that you do not need to use `I()` like you did with the polynomial. The `summary()` command gives you access to the results:

```
> summary(pg.lm)
Call:
lm(formula = growth ~ log(nutrient), data = pg)
Residuals:
    Min      1Q  Median      3Q     Max
-2.2274 -0.9039  0.5400  0.9344  1.3097
Coefficients:
              Estimate Std. Error t value Pr(>|t|)
(Intercept)     0.6914     1.0596   0.652     0.53
log(nutrient)   5.1014     0.3858  13.223 3.36e-07 ***
---
Signif. codes:  0 '***' 0.001 '**' 0.01 '*' 0.05 '.' 0.1 ' ' 1
Residual standard error: 1.229 on 9 degrees of freedom
Multiple R-squared:  0.951,    Adjusted R-squared:  0.9456
F-statistic: 174.8 on 1 and 9 DF,  p-value: 3.356e-07
```

Here you can see that the relationship can be summarized by:

$$growth = 5.10\ \mathrm{Ln}(nutrient) + 0.69$$

Remember that in R all logarithms are the natural log unless you say otherwise (e.g. `log(x, base = 10)`); there is no separate symbol such as Ln. In the summary formula above you write Ln to make it clear and avoid ambiguity; it would also be possible to write Log_e.

Have a Go: Use R for logistic regression

You'll need the *Newt HSI.csv* file for this exercise. Alternatively the data are contained in the *S4E2e.RData* file as an object called *gcn*.

Example data available from the support website.

1. If you are using the CSV file then read this into R and call the result *gcn*:

```
> gcn = read.csv(file.choose())
```

2. The data contain various predictor variables; you are going to look at the cover of macrophytes, *macro*. Look at the available variables:

```
> names(gcn)
  [1] "presence"       "area"           "dry"            "water"
"shade"
  [6] "bird"           "fish"           "other.ponds" "land"
"macro"
[11] "HSI"
```

3. Make a logistic regression model for the presence/absence of newts and macrophytes. You'll need to add the `family = binomial` parameter to ensure that R carries out logistic regression:

```
> newt.glm = glm(presence ~ macro, data = gcn, family =
binomial)
```

4. Look at the result using the `summary()` command:

```
> summary(newt.glm)
Call:
glm(formula = presence ~ macro, family = binomial, data
= gcn)

Deviance Residuals:
    Min       1Q    Median       3Q      Max
-1.6042  -0.8209  -0.7464   1.2860   1.6817
Coefficients:
              Estimate Std. Error z value Pr(>|z|)
 (Intercept) -1.135512   0.217387  -5.223 1.76e-07 ***
macro         0.022095   0.005901   3.744 0.000181 ***
---
Signif. codes:  0 '***' 0.001 '**' 0.01 '*' 0.05 '.' 0.1
' ' 1
 (Dispersion parameter for binomial family taken to be 1)
      Null deviance: 261.37  on 199  degrees of freedom
```

```
Residual deviance: 246.55  on 198  degrees of freedom
AIC: 250.55
Number of Fisher Scoring iterations: 4
```

5. You can see that the *macro* variable is statistically significant ($p = 0.000181$).

In the preceding exercise you can see that the cover of macrophytes (the *macro* variable) is significant. What about R^2, the measure of "how good" the model fit is? This is not quite appropriate in logistic regression; however, you can determine a statistic that is broadly equivalent. This is often called D^2 as it relates to the deviance parameters. If you look at the bottom of the result you can see two lines headed *Null deviance* and *Residual deviance*. You can use the ratio of these values to determine D^2 and get something that broadly represents how good your model is.

To do this you need to evaluate the following:

$$1 - (\textit{Residual Deviance}/\textit{Null deviance})$$

If you do that for your model, you get:

```
> 1 - (newt.glm$dev / newt.glm$null)
[1] 0.05669383
```

The null deviance and deviance values are already stored for you (use the $ to extract them). Here you obtain a value of 0.0567, which is not that good. It tells you that there are probably a lot of other factors determining the presence or absence of newts.

You should draw some sort of graph to illustrate your results. There would not be a lot of point in plotting presence and absence against macrophyte cover because your response variable is either 0 or 1. What you can do is to show the predictive model where you have the probability of finding a newt against the macrophyte cover.

You'll need the `lines()` command to add the `fitted()` values to a plot and the `spline()` command to make a smooth curve in a similar manner to that used when you drew polynomial (Section 11.2.1) and logarithmic (Section 11.2.2) regressions.

To make a fairly good plot you need the following four command lines:

```
> plot(gcn$macro, fitted(newt.glm), type = "n",
xlab = "% Macrophyte cover", ylab = "Probability")
> lines(spline(gcn$macro, fitted(newt.glm)), lwd = 2)
> abline(v = mean(gcn$macro), lty = 3)
> abline(h = mean(fitted(newt.glm)), lty = 3)
```

Note that in the `plot()` command you use `type = "n"` so that nothing is plotted! The x-values are the cover of macrophytes and the result of the regression, the `fitted()` part, is used for the y-values. The `abline()` command is used simply to draw some straight lines on the plot.

The final logistic model plot looks like Figure 11.15. The regression line is a subtle S-shape.

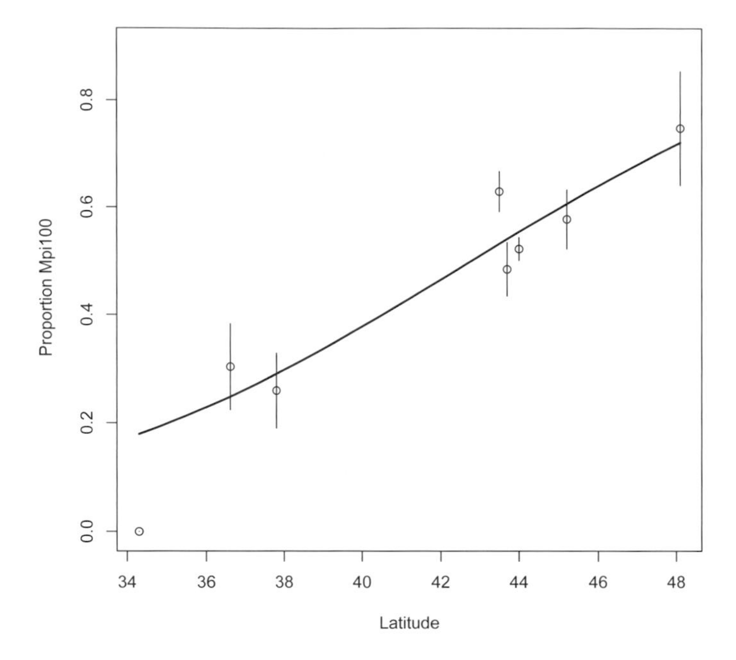

Figure 11.16 Proportion of Mpi100 allele with latitude for the Californian beach hopper *Megalorchestia californiana*. Data from McDonald, J.H. 1985 (*Heredity* 54: 359–366).

Predicting response values from your logistic regression

Earlier it was implied that you could use the results of your logistic regression to predict the chance of success. You can use the `predict()` command to do exactly that. The support website has some examples of how you can do this for the two examples you've seen here.

Go to the website for support material.

11.3.2 Multiple logistic regression

You can carry out logistic regression where you have one response variable and multiple predictor variables much as you can with regular multiple regression. You may simply add as many factors as you like to your regression model! Previously you saw how to undertake a stepwise regression (Section 11.1.2). You can do pretty much the same thing in logistic regression to build the "best" model.

The main difference is that when you use the `add1()` command you cannot use an *F*-test (it is not appropriate to logistic regression), and must use the `test = "Chisq"` parameter instead. Here is a quick reminder of the steps you need:

1. Make a model containing the response variable and an intercept only as a starting point, e.g.:

```
> mod = glm(presence ~ 1, data = gcn, family = binomial)
```

2. Now run the `add1()` command to see which factors are "important", e.g.:

```
> add1(mod, gcn, test = "Chisq")
```

3. Select the variable with the lowest *AIC* value, and which is significant. Add this to the original model in place of the 1, e.g.:

```
> mod = glm(presence ~ macro, data = gcn, family = binomial)
```

4. Run the `add1()` command again. If there are more significant terms, then choose the one with the lowest *AIC* value. Edit the model and add the term to the `formula`.
5. Repeat step 4 until you have no more significant variables to add.
6. Use the `summary()` command to see the final model summary.
7. Use the *Null* and *Residual Deviance* values to calculate a D^2 value.

You ought to draw a graph of your final "best" model but this is something of a problem as you have many factors. What you can do is show how "good" your overall model is. As part of the regression calculations, R determines the fit of the model and you can access this using the `fitted()` command, which you met before. The values that result are in order of the values of the samples. You want them to be in ascending order so you simply sort them using `sort()`. To plot your graph you type commands similar to the following:

```
> plot(sort(fitted(mod)), type = "l", lwd = 2, ylab = "Model
Probability", xlab = "Sorted sample number")
> abline(h = mean(fitted(mod)), lty= 2)
> abline(v= length(mod$fit)/2, lty= 2)
```

In the example *mod* is the name of the logistic regression model; the final graph resembles Figure 11.17.

The regression model in Figure 11.17 is based on the following:

```
> mod = glm(presence ~ macro + other.ponds + fish + shade +
water, data = gcn, family = binomial)
```

The main values are the fitted regression values: you sort them in order and therefore the *x*-axis is simply the samples sorted in ascending order of probability. It is helpful to have some guides on the plot, which is where the `abline()` commands come in. The horizontal line corresponds to the mean probability. The vertical line corresponds to the mean of the sample number, simply half-way along – because you had 200 samples, this is at 100. The `length()` command was used to get the number of samples and divide by 2.

In an ideal model you would expect the line to be a broad S-shape with the middle straight section at 45° and passing through the centre of the plot at a mean probability of 0.5. A very flat or very steep line would indicate a poor fit. Here your model is not too bad (although the D^2 value is low); it goes from close to 0 to about 0.9 and this is another good indication. A poor model would not cover the entire range of probabilities.

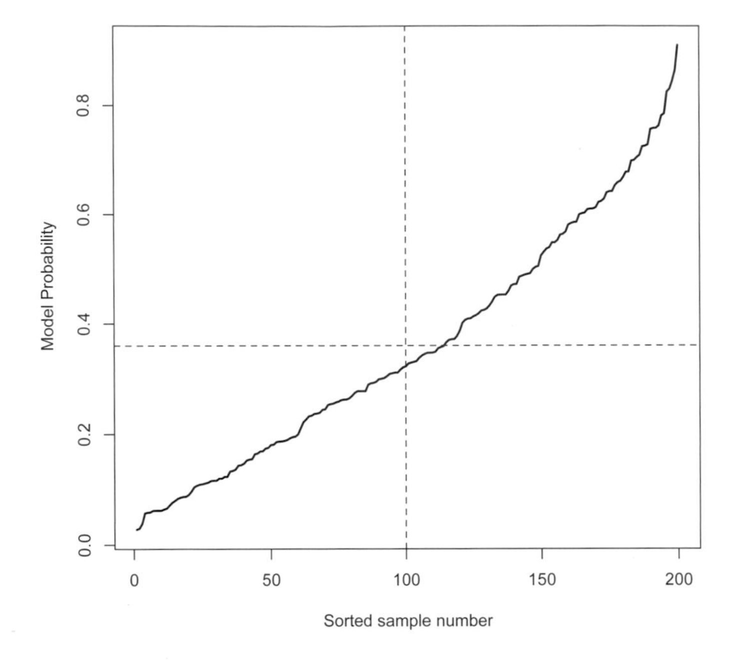

Figure 11.17 The success of the multiple logistic regression model for great crested newt presence/absence at ponds in Buckinghamshire, UK (see text for model parameters).

Note: Logistic regression model building

Go to the support website for an online exercise in model building using the newt data featured in the text.

Go to the website for support material.

EXERCISES

Answers to exercises can be found in the Appendix.

1. In multiple regression each variable has a separate coefficient, analogous to a slope. These coefficients (along with the intercept) describe the relationship between the response variable and the various predictors. TRUE or FALSE?
2. In a study of the number of species of alpine plants in Yellowstone National Park (data adapted from Alto, K. and Weaver, T. 2010. *Arctic, Antarctic and Alpine Research*, 40: 139) the results showed that the species richness (number of species) is related to rock cover and soil pH as follows: $rich = -0.152rock - 2.686ph$

+ 42.221. If the standard deviations are *rich* = 7.163, *rock* = 33.318 and *ph* = 0.549, what are the beta coefficients?

3. You've carried out a project exploring the abundance of a plant species in response to different soil pH. When you plot the results on a graph you see that the relationship appears to go upwards, then turn the corner and go downwards; in other words you have an inverted U shape. Which of the following mathematical relationships would you most likely need in a regression model?
   A. Logarithmic
   B. Logistic
   C. You cannot carry out regression
   D. Polynomial
   E. Linear
4. You are looking to build a regression model and have determined the correlation coefficients of four variables. Which one would be the best starting variable in your model?
   A. 0.576
   B. –0.477
   C. –0.787
   D. –0.645
5. Using R you can calculate theoretical values from a regression model using the ____ command. You can use these values to help draw a ____.

Chapter 11: Summary

Topic	Key points
Multiple regression	In regression you are looking for links between numeric variables (but see Logistic regression). The relationship between variables can be described by a mathematical formula ($y = mx + c$ for regular straight-line regression). The effect of each predictor variable is described by a coefficient (i.e. a slope). When you have more predictor variables you add more terms to your model, e.g. $y = m_1x_1 + m_2x_2 + c$. The term with the largest regression coefficient is the "most important" variable. The overall R^2 value tells you how "good" your model is (varies from 0 to 1). You can carry out regression in Excel using the `LINEST` function. The *Analysis ToolPak* can also do the calculations. In R the `lm()` command carries out regression.
Beta coefficients	Regular regression coefficients describe the mathematic formula that links the response and predictor variable(s). Beta coefficients are standardized against one another (units of standard deviation), which allows you to compare the magnitude of their effects directly.

```
> plrich = read.csv(file.choose())
```

2. Now look at the top few rows of the data:

```
> head(plrich)
  Site                Species
1  ML1 Achillea millefolium
2  ML1       Centaurea nigra
3  ML1    Lathyrus pratensis
4  ML1 Leucanthemum vulgare
5  ML1    Lotus corniculatus
6  ML1  Plantago lanceolata
```

3. Use the `table()` command to reassemble the data, and so get the frequency of species by site:

```
> table(plrich$Species, plrich$Site)
                        ML1 ML2 MU1 MU2 PL2 PU2 SL1 SL2 SU1 SU2
  Achillea millefolium    1   1   1   1   0   1   0   0   0   0
  Aegopodium podagraris   0   0   0   0   0   0   0   1   0   0
  Agrostis capillaris     0   1   1   1   1   0   1   1   1   0
```

4. The `ftable()` command allows you to use a formula, meaning you don't need to use the $ for the variable names:

```
> ftable(Site ~ Species, data = plrich)
                      Site ML1 ML2 MU1 MU2 PL2 PU2 SL1 SL2 SU1 SU2
Species
Achillea millefolium         1   1   1   1   0   1   0   0   0   0
Aegopodium podagraris        0   0   0   0   0   0   0   1   0   0
Agrostis capillaris          0   1   1   1   1   0   1   1   1   0
```

5. To get the species richness you need the column totals:

```
> colSums(table(plrich$Species, plrich$Site))
ML1 ML2 MU1 MU2 PL2 PU2 SL1 SL2 SU1 SU2
 15  16  21  14  13  11  16  24  27  26
```

You now have the species richness, which you can use in further analysis. It helps if you save the results to a named object, e.g.:

```
> ps = table(plrich$Species, plrich$Site)
> psr = colSums(ps)
> psr
ML1 ML2 MU1 MU2 PL2 PU2 SL1 SL2 SU1 SU2
 15  16  21  14  13  11  16  24  27  26
```

Once you have your species richness results you can use the data like any other response variable.

12.1.2 Diversity indices

Simply counting the number of different taxa in a sample gives the species richness, but this statistic will only give a simple picture, as a species only has to have one representative to be "counted". More sensitive measures of diversity take into account the relative abundance of species (Figure 12.1).

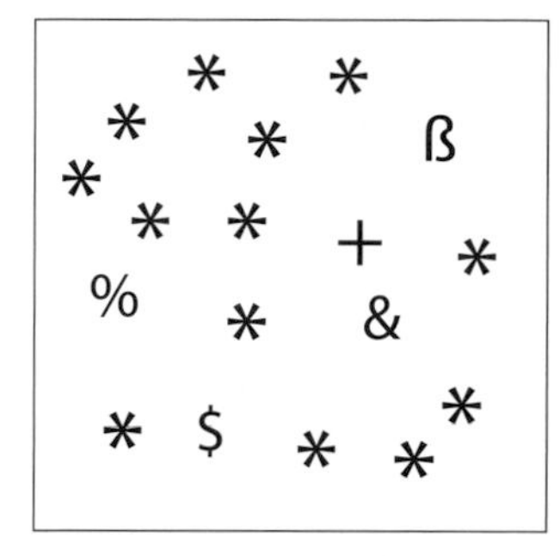
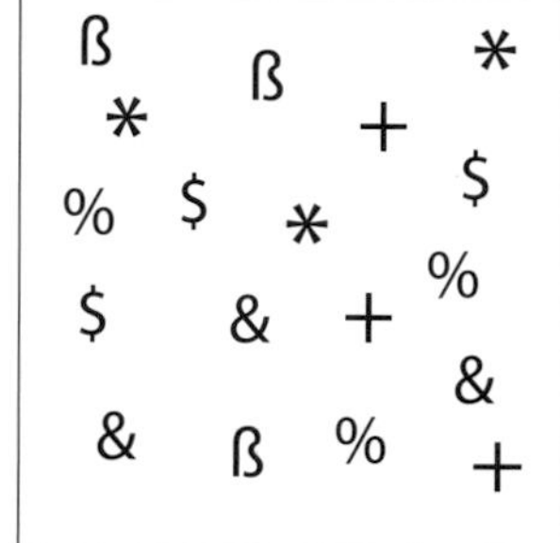

Figure 12.1 Representation of two samples, where each symbol represents a different species. Both samples have the same number of species (6) but the sample on the left is dominated by one species (*).

In Figure 12.1 you can see a representation of two samples, which both have the same species richness (6). However, the sample on the left is dominated by one species (*). The sample on the right shows that all species are equally dominant. It is reasonably evident that the diversity of the sample on the right should be greater than the sample on the left.

In order to take into account the different abundance of various species several diversity indices have been devised. The most commonly used are:

- *Simpson's index* – this index varies from 0 to 1, with the higher value indicating a more diverse sample.
- *Shannon–Wiener* (a.k.a. Shannon–Weaver) – this index varies from 0 with no upper limit (although values are usually <10). Higher values indicate higher diversity.

In both cases you need the abundance of the species involved. The measure of abundance that these indices were devised to utilize is a direct count of individuals but it is generally acceptable to use other relative measures, such as percentage cover, when counts of individuals are not possible. Once you have the abundance data the starting point is to determine the relative proportion that each species makes to the total.

Once you have the relative proportions you can work out the overall diversity index. The two most commonly used indices (Simpson and Shannon) are similar but subtly different, as you'll see next.

Simpson's diversity index

There are several versions of Simpson's diversity index; the most commonly used is Simpson's D index and the formula to calculate this is shown in Figure 12.2.

$$D = 1 - \sum \left(\frac{n}{N} \right)^2$$

Figure 12.2 The formula to calculate Simpson's *D* index of diversity. In the formula, *n* is the abundance of a taxon and *N* is the total abundance.

For each species (or taxon) you require the proportion that it makes towards the total, this is the n/N part of the formula (Figure 12.2). In order to compute the Simpson's index you'd follow these steps:

1. For each sample you need a column of values, where each value is the abundance of a different species (the n part of the formula).
2. Add up the separate abundance values to give a total abundance (the N part of the formula).
3. Determine n/N for each species you have. The sum of n/N should be 1.
4. Now square each of the individual n/N values.
5. Add together the values you calculated in step 4.
6. The final D index is 1 minus the value you get from step 5.

Simpson's D index varies from 0 to 1, with the higher values indicating higher diversity.

> **Note: Extremes of Simpson's diversity**
>
> Although Simpson's index can vary from 0 to 1, in practice you can never achieve the upper limit (you would need an infinite number of species, all with equal abundance). The lower limit can only be achieved if you have a single species.

Simpson's D index has an interesting property. The value of the index is "predictive": the index predicts the probability that if you sample randomly you'll get a different species from the previous time. For example, imagine you had an index value of 0.70. You then sample randomly and pick out a certain species. Now you sample again. The index predicts that you have a 70% chance of getting a different species.

Calculating Simpson's index using Excel is quite easy. In Table 12.1 you can see an example of how you might layout your calculations. The data are from samples of butterfly abundance that you used previously.

The species names are not essential but the column is useful for other row labels. You can use the `=SUM` function to add up the abundance values for the species. The proportions (n/N) are easily calculated; it is easiest to "fix" the cell reference for the N value (472 in Table 12.1) using $ in your formula. The sum of the proportions should be unity (a useful check).

Once you have the proportions you can square them. The sum of these squares, subtracted from 1, is the final diversity index.

There is a Simpson's *D* index calculator on the support website; the file is called *Diversity Simpson D.xls*. You can also use the *Freshwater invertebrates.xlsx* file for practice (you will meet these data shortly).

Table 12.1 How to layout data when calculating Simpson's diversity index using Excel.

	Count	P	P^2
M.bro	88	0.186	0.035
Or.tip	90	0.191	0.036
Paint.l	50	0.106	0.011
Pea	48	0.102	0.010
Red.ad	6	0.013	0.000
Ring	190	0.403	0.162
Sum	472	1.000	0.255
D	0.745		

You can calculate Simpson's D index easily using R. The easiest way is to use a package called *vegan* (short for *veg*etation *an*alysis), which is easily downloaded and installed in R. The *vegan* package contains many useful commands that are used in analysis of community data, including `diversity()`, which will calculate Simpson's (and Shannon's) diversity index.

The advantage of the *vegan* package is that the `diversity()` command can operate on data that are in different layouts. So, you can easily calculate diversity for a single sample or several at once. You need to get the *vegan* package installed to start with:

```
> install.packages("vegan")
```

Once you enter the command R will look to the Internet and get the package for you (recall Section 3.1.6). You'll most likely be asked to select a location to download from; select a location near you – this minimizes traffic on the websites and results in speedier downloads.

You load the library of routines as you need using the `library()` command:

```
> library(vegan)
```

Now the commands in the package are available to you. The `diversity()` command will compute Simpson's index and its general form is like so:

```
diversity(x, index = "shannon", MARGIN = 1)
```

`x`	A sample of data. This can be a single `vector` of values or a `matrix` containing several rows/columns.
`index = "shannon"`	The diversity index to use. The default is `"shannon"`; the alternative is `"simpson"`.
`MARGIN = 1`	If your data contains several samples you can compute the diversity by row or column. The default, `MARGIN = 1`, calculates diversity by row.

In the following exercise you can have a go at calculating Simpson's index for yourself.

Have a Go: Use R to calculate Simpson's index

You'll need the *Freshwater invertebrates.xlsx* file for this exercise. The data show the abundance of some freshwater invertebrates in a stream in Yorkshire.

Download material available from the support website.

1. Open the *Freshwater invertebrates.xlsx* file in Excel. You'll see that the first five columns show the taxonomy, with the fourth giving the family and the fifth a common name. The last column gives the abundance (as a count) from combined sampling.
2. Copy the abundance values to the clipboard.
3. Now switch to R. You will use the `scan()` command to read the data from the clipboard. Assign the result a name:

```
> fwi = scan()
```

4. Once you press *Enter*, R will wait for the data. Paste the data from the clipboard (use *Ctrl+V* or the *Edit > Paste* command from the R console).
5. There are 21 data items. R will wait at this point (you can paste multiple times should you need). Press the *Enter* key to complete the `scan()` process. R will tell you that it has read 21 items.
6. Now use the `diversity()` command to calculate the Simpson's diversity index for the invertebrate sample:

```
> diversity(fwi, index = "simpson")
[1] 0.7325558
```

You only have a single sample here. Shortly you'll see how you can obtain the diversity index for several samples (in the next section, the Shannon index).

Shannon diversity index

The Shannon diversity index is broadly similar to Simpson's index, in that you start with the abundance of each species and work out their proportions towards the total for your sample. In Simpson's index the proportions (*n*/*N*, Figure 12.2) are squared but in the Shannon index they are dealt with slightly differently (Figure 12.3).

$$H = -\sum\left(\frac{n}{N}\right)\text{Ln}\left(\frac{n}{N}\right)$$

Figure 12.3 The formula to calculate the Shannon diversity index. In the formula, *n* is the abundance of a taxon and *N* is the total abundance.

In the formula n/N is the proportion of each species' abundance in the sample. The *Ln* part indicates that you take the natural log of the proportion. If you want to compute the Shannon diversity index you'd need to follow these steps:

1. For each sample you need a column of values, where each value is the abundance of a different species (the n part of the formula).
2. Add up the separate abundance values to give a total abundance (the N part of the formula).
3. Determine n/N for each species you have. The sum of n/N should be 1.
4. Now for each n/N value take the natural logarithm.
5. Multiply each n/N value by the natural log you calculated in step 4.
6. Add up all the values from step 5. This will be a negative number.
7. Change the sign of the value from step 6 to get the final Shannon index.

The Shannon index can take more or less any value from zero upwards. Generally the more species you have the larger the value of the index (and the higher the diversity). In practice you are unlikely to get values >10.

Note: Logarithms in the Shannon diversity index

It is usual to use the natural logarithm when calculating the Shannon index. However, in theory you can use any logarithm. Mathematically speaking you can make a good case for using log base 2. In practice it best to stick to the natural log, as this allows others to compare results more easily.

Because the upper limit of the Shannon index is not 1 it makes comparisons between samples a bit easier than Simpson's index.

Calculating the Shannon index is fairly easy in Excel. In Table 12.2 you can see an example of how you might layout your calculations. The data are from samples of butterfly abundance that you used previously.

Table 12.2 How to layout data when calculating the Shannon diversity index using Excel.

	Count	P	P.Ln(*P*)
M.bro	88	0.186	–0.313
Or.tip	90	0.191	–0.316
Paint.l	50	0.106	–0.238
Pea	48	0.102	–0.232
Red.ad	6	0.013	–0.055
Ring	190	0.403	–0.366
Sum	472	1.000	–1.521
H	1.521		

As with the calculation of Simpson's index (Table 12.1) the species names are not essential. Calculating the proportions is the same as before (use the $ to "fix" the cell reference for the total abundance). You can use the `=LN` function to calculate the natural logarithm. In Table 12.2 you can see that the final column contains the proportions multiplied by their log in "one go"; if you prefer to split the calculation, with a column for the log and one for the multiplication, that's fine.

The sum of the final column is the value of the Shannon diversity index, but with the opposite sign (logs of numbers <1 are negative). The final row simply displays the result with the opposite sign.

If you have blank rows or abundance values of zero then you'll get errors if you try to take the logarithm. You can get around this by using an `=IF` function to "take care" of potential blank rows. This allows you to make a template spreadsheet that you can paste values into.

Tip: Logarithms and zero or missing values

Instead of using an `=IF` function to "take care" of blank data you can add a tiny amount to the value to be logged. For example instead of using `=C2*LN(C2)` use `=C2*LN(C2 + 1E-10)`. Typing `1E-10` is equivalent to 1×10^{-10}, a very small value (note that you can use lower-case `e` instead of `E`).

There is a Shannon diversity index calculator on the support website: the file is called *Diversity Shannon.xls*. You'll see how the `=IF` function is used in this spreadsheet.

Calculating the Shannon diversity index is easy using R, especially if you use the *vegan* package (as described earlier in the section on Simpson's index). The `diversity()` command in the *vegan* package can calculate the index for a single sample or several at once. By default the `diversity()` command calculates the Shannon index, you can tell it to do so explicitly by adding `index = "shannon"` as a parameter (as you will see shortly).

You saw how to use the `diversity()` command to calculate an index for a single sample earlier (Simpson's index); in the following exercise you can have a go at using R to calculate an index for a dataset containing several samples.

Have a Go: Use R to calculate the Shannon diversity index

You'll need the data file *Plant species abundance.csv* for this exercise. The data show the results of a vascular plant survey by students at a field centre in Shropshire, UK. The samples were collected from observations in quadrats measuring 2 m × 2 m. These are the same data you used when looking at species richness (the abundance information was removed from the earlier dataset). The abundance data is a combination of percentage cover and frequency.

Download material available from the support website.

1. Open the CSV file and have a look (the file should open in Excel). You can see that there are columns for each sample and rows for the species names. You do not really want the column of species names to be part of the "data". You'll take care of this when importing the dataset into R.
2. Close the spreadsheet and open R. Use the `read.csv()` command to import the datafile. Set the first column (the species names) to act as row names:

```
> psa = read.csv(file.choose(), row.names = 1)
```

3. Look at the top few rows of the data:

```
> head(psa)
                        ML1 ML2 MU1 MU2 PL2 PU2 SL1 SL2 SU1 SU2
Achillea millefolium      6   3   5 4.0   0 3.2   0 0.0 0.0 0.0
Aegopodium podagraris     0   0   0 0.0   0 0.0   0 1.6 0.0 0.0
Agrostis capillaris       0   8   8 5.6   8 0.0   5 2.0 3.2 0.0
Agrostis stolonifera      0   0   0 0.0   0 5.0   0 0.4 0.0 0.0
Anthriscus sylvestris     0   0   0 0.0   0 0.0   0 0.0 1.2 3.0
Arctium minus             0   0   0 0.0   0 0.0   0 0.0 0.0 0.4
```

4. Make sure the *vegan* package is loaded and ready to use:

```
> library(vegan)
```

5. Now use the `diversity()` command to get the Shannon diversity index for the ten samples (the columns):

```
> diversity(psa, index = "shannon", MARGIN = 2)
     ML1      ML2      MU1      MU2      PL2      PU2
SL1      SL2
2.598084 2.421995 2.626107 2.368889 2.240545 2.095538
2.385839 2.763634
     SU1      SU2
2.731173 2.676890
```

Note that the `diversity()` command computes the Shannon index by default so you did not need to type the `index` part. The command generally expects the rows to be the samples (`MARGIN = 1`), so you did need the last part of the command.

In the preceding exercise you calculated the Shannon diversity index for the ten samples. Use the same data and look at the Simpson's index for comparison.

12.1.3 Diversity and scale

Any measure of diversity you use is going to be influenced by scale, so you should think carefully about the way you collect your data. Diversity is generally regarded at three levels (or scales).

- *Alpha* diversity – this is the smallest scale or measurement unit, most likely from a single sample in a habitat.
- *Gamma* diversity – this is the largest scale of measurement, where you combine your measurements of alpha diversity into one large sample (think landscape or ecosystem diversity).
- *Beta* diversity – this is the intermediate scale and links the *alpha* and *gamma* levels of diversity. You can think of *beta* diversity as the change in diversity between samples (at *alpha* level).

The relationship between the scales of diversity can be complicated! In general the relationship is:

$$\alpha \times \beta = \gamma$$

However, this relationship does not hold true for the basic Simpson's or Shannon diversity indices (it becomes additive: $\alpha + \beta = \gamma$). Details of diversity and scale are beyond the remit of this book; for a fuller treatment see: *Community Ecology: Analytical Methods using R and Excel* (Gardener 2014).

What you do need to think about carefully is your own scale of measurement when you collect data for diversity analysis (most likely at *alpha* level). Is it really appropriate to determine diversity of a meadow using 1 m^2 quadrats, for example?

12.1.4 Comparing diversity

When you set out to look at differences between "things" you usually collect replicated data, i.e. repeated measurements as a sample. You then use these repeat measurements to get an average (e.g. mean or median) and a measure of variability. Your averages and measures of variability allow you to carry out a test of differences between two (or more) samples (recall Chapters 7 and 10).

If you are looking at diversity you can more or less take the same approach. You'll need to collect repeated measurements of diversity so that you can obtain an average for each of your predictor variables (e.g. different habitats or locations). When you have simple species richness things are fine; you can use the richness like any other response variable and use "classic" hypothesis tests.

The way that other diversity indices are calculated makes it slightly less appropriate to use regular hypothesis tests. However, it is possible to compare a diversity index if you have repeated measurements. You just use the diversity index like any other response variable. Start by checking the distribution (normal or non-parametric) and go from there (see Table 5.1).

If you have only got a single measurement of diversity from each of your predictor variables then things are a little harder. The basic indices of diversity (Simpson's and Shannon) have received some attention in this area and there are versions of the *t*-test that allow you to compare two samples. However, there are some problems with these methods and researchers are turning more towards randomization methods for this kind of calculation. Details of these methods are beyond the remit of this book; for a fuller treatment see: *Community Ecology: Analytical Methods using R and Excel* (Gardener 2014).

Note: The Hutcheson *t*-test for comparing Shannon diversity index

The Hutcheson *t*-test is a modification of the classic *t*-test devised to allow a test of significance of the difference between the Shannon diversity of two samples. Go to the support website for some additional notes (and a spreadsheet calculator, *Shannon Diversity t-test calculator.xlsx*) about this analytical method.

Download material available on the support website.

If you are looking at communities and simply comparing indices of diversity you may well be missing important elements in the differences between communities. Look at Table 12.3, which shows two (theoretical) communities.

Table 12.3 Two theoretical communities. The diversity of the two samples is virtually the same yet the make-up of the communities is clearly completely different, as there is only one species common to both.

Species	Site A	Site B
A	23	0
B	17	0
C	2	0
D	6	0
F	12	11
G	0	8
H	0	2
I	0	19
J	0	25

If you work out the diversity of the two samples from Table 12.3 you would find them virtually the same. Yet the communities are clearly quite different, only one species is common to the samples. An alternative community approach to an index of diversity is to compare samples according to the species that are present. This allows you to see which samples are most similar to one another and which are most dissimilar. This is the topic covered in the next section.

12.2 Similarity

The analysis of similarity does essentially what the name suggests; it compares samples and allows you to see which are most similar to one another. Analysis of similarity is a useful tool when looking at communities; it may be more useful in many cases than simply looking at diversity.

In analysis of similarity you calculate an index of similarity for every pair of samples you have. There are several methods, as you will see shortly. After you have created your matrix of similarities you can go on to visualize them using a special sort of diagram, a dendrogram (see Section 6.6).

The way you calculate your similarity index depends to a large extent on the kind of data you've got to start with. There are two sorts of input data you can deal with:

- *Presence–absence* – where you do not know the abundance of species, you only know if they were present or not.
- *Abundance* – where you know the abundance of the species. The abundance data can be in any (numerical) form (e.g. percentage cover, count, frequency).

The first case (presence–absence) is the simplest and uses the number of shared species as a key element in the calculation.When you have abundance information you can use more "sensitive" measures of similarity.

Many indices of similarity (and the corollary, dissimilarity) have been devised; in the following sections you'll see a few of the more commonly used measures.

12.2.1 Presence–absence data and similarity

When you know the presence or absence of species in samples you have information, albeit limited, to work with. In order to calculate how similar two samples are you need to look at three basic things:

- How many species are in sample 1 – this is usually designated *A*.
- How many species are in sample 2 – this is usually designated *B*.
- How many species are common to both – this is usually designated *J*.

The three values, *A*, *B* and *J*, are used to compute an index of similarity. Look at Table 12.4, which shows a trivial example of the presence of some species in two samples.

Table 12.4 A hypothetical example of species' presence–absence. Each sample has four species ($A = 4$, $B = 4$), but there is only one species in common to both ($J = 1$).

Species	Site A	Site B
a	1	0
b	1	0
c	1	0
d	1	1
e	0	1
f	0	1
g	0	1

It is easy to see in the example from Table 12.4 that there are four species in each sample (so A and B are 4) and that there is only one species in common ($J = 1$).

In order to calculate how similar these two samples are you need to formulate the A, B and J values appropriately. Many indices of similarity have been devised; two commonly used ones are:

- The *Jaccard* index.
- The *Sørensen* index (this is sometimes called Bray–Curtis).

Both these indices are easily computed, as you will see shortly. Each index can have values ranging from 0 (no species in common) to 1 (all species common to both).

The Jaccard index of similarity

The Jaccard similarity index is calculated using the formula shown in Figure 12.4.

$$Jaccard = \frac{J}{A + B - J}$$

Figure 12.4 The formula for the Jaccard index of similarity. In the formula A = number of species in first sample, B = number of species in second sample and J = number of shared species.

The higher the index the more similar the two samples (the index can have values between 0 and 1). If you calculate the Jaccard index for the example in Table 12.4 you get:

$$\begin{aligned} &A = 4,\ B = 4,\ J = 1 \\ &\text{Jaccard} = J/(A + B - J) \\ &\qquad = 1/(4 + 4 - 1) \\ &\qquad = 1/7 \\ &\qquad = 0.143 \end{aligned}$$

When you have several samples the index of similarity can show you which communities are most similar. You can also visualize the situation using a dendrogram (this looks a bit like a family tree).

Excel can calculate the Jaccard index quite easily and you can use the `=SUMPRODUCT` function to help you work out how many species are common to a pair of samples. The function takes two columns of numbers; for each row it multiplies one value by the other, and then it sums the totals for all the rows. You will have a 1 for the presence and 0 for the absence, so a species that is shared will give 1 × 1 = 1. A species that is not shared will give 1 × 0 = 0 (or 0 × 1 = 0). In the following exercise you can have a go at a fairly simple example using three samples.

There are two main R functions for calculating similarity indices:

- The `dist()` command, which will calculate several similarity indices (including Jaccard). This is built into the basic R package.
- The *vegan* package, which you encountered earlier, contains the `vegdist()` command. This calculates several other similarity indices (including Jaccard).

Have a Go: Use Excel to calculate Jaccard similarity

You will need the *hornbill.csv* file for this exercise. The data show the presence of different fruits in the diet of three species of hornbill in India (data adapted from Datta, A. and Rawat, G.S. 2003. *Biotropica* 35: 208).

Download material available on the support website.

1. Open the *hornbill.csv* file. You'll see a column labeled *Fruit*, which gives an abbreviated name of the fruit and a column for each of the three bird species. The presence of a fruit is given by a 1 and its absence by 0.
2. In cell A39 type a label `Total` to indicate the total number of fruits in the diet of each hornbill. Now in B39 type the formula to add up the number of fruit species: `=SUM(B2:B38)`. Copy the formula to cells C39 and D39 so you have totals for all the birds (14, 29 and 11).
3. In cells B41:D41 type the names of the bird species (`GH`, `WH` and `OPH`). Type the names again in cells A42:A49. You now have an empty "table" that you can fill in with the pair-by-pair *J* values.
4. In cell B43 type a formula that gives the number of shared species between the *GH* and *WH* bird species: `=SUMPRODUCT(B2:B38,C2:C38)`, you should get 11.
5. Now use a similar approach in cell B44 to get the shared fruit species for the *GH* and *OPH* hornbills: `=SUMPRODUCT(B2:B38,D2:D38)`, you should get 6.
6. In cell C44 calculate the *J* value for the *WH* and *OPH* pairing: `=SUMPRODUCT(C2:C38,D2:D38)`, you should get 6.
7. Copy the bird species name labels into cells B46:D46 and A47:A49, so you now have a table ready for the similarity index values.
8. In B48 type a formula to calculate the Jaccard index for the *GH, WH* bird samples: `=B43/(B39+C39-B43)`, you should get 0.344.
9. Repeat step 8 for the other two pairs of samples, e.g. *GH/OPH*: `=B44/(B39+D39-B44)` gives 0.316 and *WH/OPH*: `=C44/(D39+C39-C44)` gives 0.176.

You should now have three values representing the Jaccard similarity between the three pairs of samples of bird fruit diet (Table 12.5).

Table 12.5 Jaccard similarity in fruit diet of three species of hornbill in India.

	GH	WH
WH	0.344	
OPH	0.316	0.176

You can see that the most similar are the *GH* (great hornbill) and *WH* (wreathed hornbill) species. The *GH* and *OPH* (oriental pied hornbill) species are next in similarity, whilst the *WH* and *OPH* species are the least similar.

Whilst the calculations are easy to perform in Excel you can see that it would rapidly become quite tedious if you have a lot of samples to compare. The biggest problem is creating the triangular matrix for the pair-by-pair comparisons. Excel is also not suitable for drawing dendrograms (which you will see later), so for comparing multiple samples you may prefer to turn to R, which can carry out all the calculations with ease.

Both these commands actually calculate an index of dissimilarity (essentially 1 – *D*, where *D* is the index of similarity). A dissimilarity is more generally useful, as it makes it slightly easier to draw a dendrogram. The commands operate in a similar fashion:

```
dist(x, method = "euclidean")
vegdist(x, method =  "bray")
```

In both commands, `x` is your data, which is expected to be one where the rows are the samples and the columns are the species. In the previous example your data were arranged with the columns as the samples. This is not a problem, as you can transpose the data "on the fly". Both commands can use several methods, the ones shown here are the defaults.

You'll see how to use R to calculate indices of similarity shortly; but first you'll see the Sørensen index of similarity.

The Sørensen index of similarity

The Sørensen index is also sometimes known as the Bray–Curtis index. It is calculated from presence–absence data using the *A*, *B* and *J* values you've already seen (Figure 12.5).

$$Sorensen = \frac{2J}{A + B}$$

Figure 12.5 The formula for the Sørensen index of similarity (also known as Bray–Curtis). In the formula *A* = number of species in first sample, *B* = number of species in second sample and *J* = number of shared species.

If you look at the trivial example in Table 12.4 you can work out the Sørensen index like so:

$$A = 4,\ B = 4,\ J = 1$$

$$\begin{aligned} \text{Sørensen} &= 2J/(A + B) \\ &= (2 \times 1)/(4 + 4) \\ &= 2/8 \\ &= 0.25 \end{aligned}$$

You saw earlier how to use Excel to calculate the *J* values (shared species) with the `=SUMPRODUCT` function. The same applies here because you are still dealing with presence–absence data. The only difference is that the final formula is slightly different. You could look back to the previous exercise and recalculate using the Sørensen index; your results should resemble Table 12.6.

Table 12.6 Sørensen similarity in fruit diet of three species of hornbill in India.

	GH	WH
WH	0.512	
OPH	0.480	0.300

If you compare this to Table 12.5 you can see that, although the actual values are different, your conclusions are the same.

You can use R to calculate the similarity using the `dist()` or `vegdist()` commands (the latter is in the *vegan* package). However, the `dist()` command cannot calculate the Sørensen index (only the Jaccard index for presence–absence data). In the following exercise you can have a go at calculating similarity indices for the hornbill fruit data you met previously.

Have a Go: Use R to calculate the Sørensen index of similarity

You'll need the hornbill data for this exercise. The data are already imported into the *S4E2e.RData* file, but you can also import the *hornbill.csv* file yourself if you like.

Download material available on the support website.

1. If you want to import the data for yourself then use the following; otherwise just go to step 2:

```
> hornbill = read.csv(file.choose(), row.names = 1)
```

2. Look at the data by typing the name, `hornbill`. Note that the fruit species names (as abbreviated scientific names) are set to be the row names. There are three columns of data, one for each hornbill species.

```
> head(hornbill)
          GH WH OPH
Acti.obov  0  1   0
Acti.angu  0  1   0
Alse.pedu  1  1   0
Amoo.wall  1  1   1
Apha.poly  1  1   0
Beil.assa  1  1   0
```

3. You need the *vegan* package to be loaded, so ensure it is ready:

```
> library(vegan)
```

4. The commands that compute dissimilarity require the data samples to be the rows. You can transpose the data using the `t()` command. In this exercise you will make a copy of the data in transposed form:

```
> fruit = t(hornbill)
```

5. Calculate the Sørensen similarity using the `vegdist()` command. Remember that the default is to compute the dissimilarity:

```
> 1 - vegdist(fruit, method = "bray")
           GH        WH
WH  0.5116279
OPH 0.4800000 0.3000000
```

6. Now calculate the Jaccard index:

```
> 1 - vegdist(fruit, method = "jaccard")
           GH        WH
WH  0.3437500
OPH 0.3157895 0.1764706
```

7. Now use the `dist()` command to work out the Jaccard index:

```
> 1 - dist(fruit, method = "binary")
           GH        WH
WH  0.3437500
OPH 0.3157895 0.1764706
```

In the preceding exercise you computed the similarity by subtracting the result from unity. In the main it is more useful to calculate indices of dissimilarity, as this makes it easier to make a visual representation of the relationship between samples.

Visualizing the similarity between community samples

Once you have at least three samples you can draw a diagram to represent the relationships (in terms of the similarity) between them. The dendrogram is the diagram of choice (recall Section 6.6) for displaying dissimilarities between multiple samples in a meaningful manner.

Excel is not designed to draw dendrograms so you'll need a different approach. You have three main options:

- Use a pencil and paper!
- Use Excel but use the drawing tools.
- Use R, which has tools to create and plot dendrograms.

The first two methods are obviously not "computer led" so it is important to know how to construct a dendrogram from a matrix of dissimilarity values.

1. Make your matrix of dissimilarity index values.
2. Look for the smallest dissimilarity. Join those items. Make the crossbar at the height of the dissimilarity (draw a *y*-axis to represent the dissimilarity).
3. Look for next largest dissimilarity. If this is a "new pair" then join them separately. If this involves an item already joined then join to that. Make the crossbar at the height of the dissimilarity.
4. Repeat step 3.

Drawing a dendrogram by hand can pose a few difficulties, particularly when you have more than four or five samples. You'll often have to juggle items to get others to "fit". If you are drawing by hand then try some preliminary sketches first. If you are using the drawing tools in Excel then you can drag items around to make the space you need.

There is an additional exercise on the support website that shows you how to build a dendrogram from a dissimilarity matrix.

Support material available on the website.

Tip: Using Excel drawing tools for dendrograms

You can produce a good dendrogram using the drawing tools in Excel. Use basic boxes for the sample names (right click to add text). You can find the drawing tools in the *Insert > Shapes* menu. The elbow connectors can join items but you will need a shape (a small square works well) at the nodes (where a joining line joins to another line).

The R program can produce a dendrogram with ease. The `hclust()` command takes a dissimilarity matrix and works out how to join the items into a dendrogram. Once you have done that you can use the `plot()` command to draw the dendrogram. In the following exercise you can have a go at making a dendrogram for yourself.

Have a Go: Use R to make a dendrogram

You'll need the plant presence–absence data for this exercise. The data are part of the *S4E2e.RData* file, as a data object called *ps*. You may also have created the data item as part of the exercise using the *Plant species lists.csv* file.

Download available from the support website.

1. Open R and look at the first few rows of the *ps* data object:

```
> head(ps)
                        ML1 ML2 MU1 MU2 PL2 PU2 SL1 SL2 SU1 SU2
Achillea millefolium     1   1   1   1   0   1   0   0   0   0
Aegopodium podagraris    0   0   0   0   0   0   0   1   0   0
Agrostis capillaris      0   1   1   1   1   0   1   1   1   0
Agrostis stolonifera     0   0   0   0   0   1   0   1   0   0
Anthriscus sylvestris    0   0   0   0   0   0   0   0   1   1
Arctium minus            0   0   0   0   0   0   0   0   0   1
```

2. Use the `dist()` command to make a Jaccard dissimilarity matrix; the rows need to be the samples so you will need to transpose the data:

```
> ps.jac = dist(t(ps), method = "binary")
```

3. Now make the hierarchical cluster object, which will be the basis of the dendrogram:

```
> ps.hc = hclust(ps.jac)
```

4. Finally you can draw the dendrogram (Figure 12.6):

```
> ps.hc = hclust(ps.jac)
```

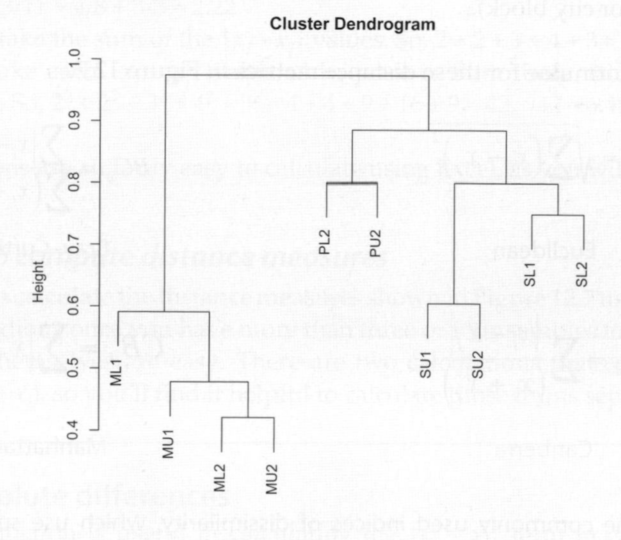

Figure 12.6 Jaccard dissimilarity for ten samples of plant species from sites in Shropshire.

In the preceding exercise you used the `dist()` command to make a matrix of dissimilarity (Jaccard). The command can produce other dissimilarities, as can the `vegdist()` command in the *vegan* package. As long as you have a matrix of dissimilarity you can use the `hclust()` and `plot()` commands to make and draw a dendrogram.

Chapter 12: Summary – *continued*

Topic	Key points
Comparing diversity	Richness can be used like a regular response variable as long as samples are collected with the same sampling effort (e.g. same quadrat size). Diversity indices can be compared to a response variable if the sampling effort is the same and you have replicated data. For single samples you can use a modified version of the *t*-test (the Hutcheson *t*-test) to compare Shannon indices.
Similarity and dissimilarity	Similarity is a measure of how similar two samples are (indices range from 0 to 1). When you have presence–absence data you use the number of shared species to help calculate an index of similarity. Common indices are Jaccard and Sørensen. When you have abundance information you can use a range of metrics to calculate a similarity index. Common indices are: Euclidean, Canberra, Bray–Curtis and Manhattan (city block). It is common to calculate an index of dissimilarity (1 – *d*), as this is easier to visualize. You can calculate indices using Excel but it can be tedious if there are several samples. The `dist()` and `vegdist()` commands (the latter in the *vegan* package) can be used to compute a range of (dis)similarity indices.
Visualizing community similarity	The dendrogram is the chart of choice to use to visualize the similarity (dissimilarity) between community samples. The dendrogram is like a family tree and shows the most similar samples nearest to one another. You can use drawing tools in Excel to make a dendrogram but there are no set chart types that you can press into service. You can use the `hclust()` command in R to make a cluster result from a dissimilarity matrix. The `plot()` command will then produce a dendrogram.

13. Reporting results

There is undoubtedly some personal satisfaction in planning a project, collecting the data and exploring the results; however, if you do not convey these results to the wider world then the work might as well not exist.

What you will learn in this chapter

- » Different ways to present your findings
- » How to present results of statistical analyses
- » How to present the best graphs
- » The elements of a scientific paper or report
- » How to use references
- » How to make best use of PowerPoint

The presentation of your work is an important stage in the scientific process. It helps you to move forwards and to determine "what next?" as well as adding to the body of scientific knowledge and helping other researchers in the future. Your work may be presented to the general public, the scientific community, a client or simply your tutor. There are various elements in presenting results and various ways to do the actual presenting.

13.1 Presenting findings

In general there is no point doing some scientific study if you aren't going to tell anyone what you have done. The three main ways to get your message out into the world are:

- Publishing a written article (e.g. paper, report, Internet, newsletter).
- Giving a talk at a conference or meeting.
- Presenting a poster at a conference or meeting.

This is the final stage in the loop that started with the planning process. You might even have an idea of where your work will be presented right at the planning stage. As you move through the process to recording and analysis you end up with the business of reporting.

following manner: $*p < 0.05$, $**p < 0.01$, $***p < 0.001$. Note that the 2% level ($p < 0.02$) does not have a star rating. Although it is a fairly standard convention you might also include the legend in the caption saying how many stars equate to which level of significance.

13.4 Graphs

A graph should convey as much information as possible in the simplest manner. You need to present a summary (i.e. not raw data) and of course follow the same rules as before (recall Chapter 4) showing centrality and spread. You can produce very good graphs in Excel but sometimes the default settings need to be tweaked to make a really good figure (recall Chapter 6).

You saw how to produce graphs in Chapter 6 (with additional examples elsewhere). Good graphs have the following attributes:

- Clear and not too cluttered. If you have too many bars or lines then consider making several graphs rather than one.
- No fancy fonts or effects. A basic 2D graph is perfectly sufficient and 3D bar charts with multi-coloured backgrounds are not essential to delivering the message (and may be misleading or hard to read).
- Labels. Axes need to be clearly labelled and include units where appropriate. Make sure the labels are not too small (or too large).
- A caption. The caption is an important thing, as it tells the reader what they are looking at. In general you do not need to use the title facility in Excel (or any other graphical program) and it is better to use a caption generated by your word processor. Captions for figures go below the figure (table captions go above). The caption allows you to cite the figure or table from the text and so has a second, useful function.
- The plot area needs to be a good size in order to display the relevant information (e.g. mean, standard deviation).

Here are some examples to illustrate how you can maximize the usefulness of your graphs. The first one is a bar chart (Figure 13.1). This sort of graph is useful for showing differences between samples (two or more, recall Section 6.3).

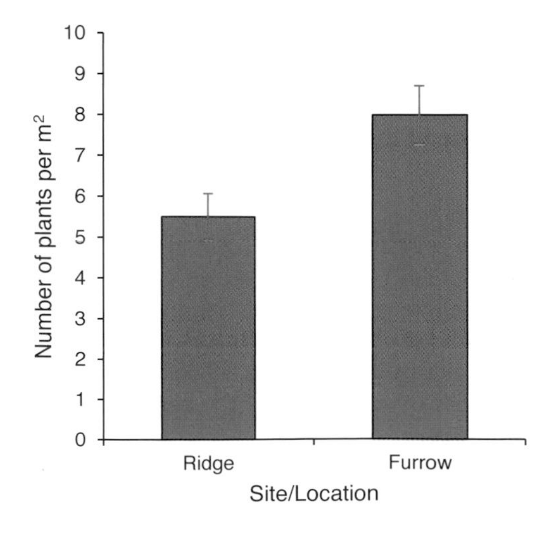

Figure 13.1 Abundance of *Ranunculus repens* per m^2 in ridges and furrows of a medieval field in Buckinghamshire. Bars show mean and error bars represent standard error.

Here the caption says pretty much all you need. The reader can get quite a lot out of this without even reading the text (a most important attribute). The axes are labelled appropriately (including units). The main bars show the mean (so you already know the data are normally distributed) and the error bars are standard error (which is standard deviation divided by the square root of *n*, Section 4.5.1). Error bars can be added reasonably easily in Excel or R (see Section 6.3.1).

Sometimes it is useful to convey the statistical result right on the graph and this can be done with a text annotation. The following graph (Figure 13.2) shows an example of this in a box–whisker plot.

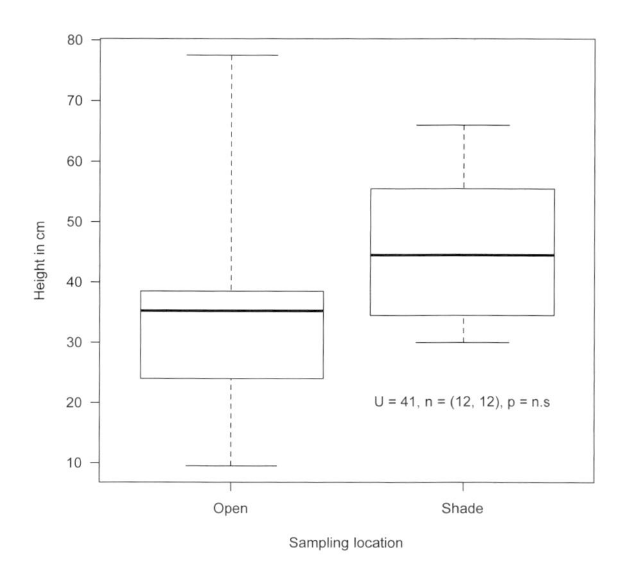

Figure 13.2 Height in centimetres of stinging nettles (*Urtica dioica*) in samples with differing exposure to sunshine. Stripe shows median, boxes show inter-quartiles and whiskers show range.

In this graph (Figure 13.2) you can see immediately that the data are non-parametric: the plots are not symmetrical around the stripes. You could add a few words in the caption to describe the stats results but in this case the results are added to the chart. It is evident from reading the graph that you performed a *U*-test and had 12 replicates in each sample (there is no need to start your caption with "graph to show...", that should be pretty obvious). The result is not statistically significant ($p = 0.078$).

When you are showing the result of a correlation, you present a more appropriate graph, a scatter plot. The following graph (Figure 13.3) shows the relationship between water speed and abundance of a freshwater invertebrate. In Figure 13.3 you see a scatter plot with the dependent variable labelled as abundance and the independent axis as speed (with the units). This time the stats result is given in the caption. A value for *n* is included even though you can see the points on the graph; some may overlap and if there are a lot it would be hard to count. Let's be kind to the reader and put the information in the caption (and it is good practice to include it every time).

You usually want to try to get the points to fill the chart area as much as possible. This can mean altering the axes so that they do not include zero. In Figure 13.3 the *x*-axis might be better rescaled to start from 2, rather than zero.

Remember that with basic correlation you do not attempt to produce a line of best fit! You are looking to determine the strength of the relationship and this may not be in the form of a straight line (although this looks pretty good). It is acceptable to include a

Table 13.2 The elements of a scientific paper.

Element	Purpose
Title	The title is the first thing that is generally seen so it needs to be clear and succinct and yet make the reader want to dive in.
Contact details	This is not just advertising; other researchers may want to contact you to discuss your work and perhaps to collaborate in further research. Include your institution, address and email.
Abstract	This is a complete précis of the work, including stats results. If you have a lot of results you can include only the most important ones. This needs to be brief but remember that this may be the only thing that is read. Don't include citations.
Keywords	Usually the journal offers you the chance to add a few keywords. If your paper ends up being added to a database, these may be the search items. In addition, the reader can see what you think are the key elements. You can add the species name for example.
Introduction	This is where you set the background to the subject area. It is also the place where you show off how much background reading you have done. Lead the reader into the subject in general and hit them with your specific aims at the end. In general, detailed hypotheses are not given but in student works they might be expected.
Methods	This is where you describe what you did and where. You should include site descriptions and possibly even species details (sometimes they are better in the introduction). The point is that a reader should be able to repeat your work following the instructions you provide. Avoid equipment lists and bullet points.
Results	This section details what you found. Raw data are not generally included although species lists could go in as long as they are not too long. The point here is to summarize: present averages, dispersion and how big the samples are. Give stats results too. Say everything in words as well as referring to graphs (or tables).
Discussion	This is where you say what you think it all means and to summarize the findings. Try not to bring in too much that is new (the bulk of your citations go in the introduction).
Acknowledgements	This is where you say thank you. Avoid oversentimentality but do put in people like landowners and anyone who put up money.
References	Everything you cite in the text goes in the reference list. Check out the formatting requirements of the journal; a good start is to use the standard Harvard layout.
Appendix	This is for stuff that you really don't want anyone to look at but which really dedicated obsessives might want to see. Long, complicated species lists and tables of raw data might be things you'd include for example. However, you should not just dump all your results here; use it as a repository for useful items that would otherwise get in the way of the main report.

are sending to a particular journal, check their format carefully because they will have slightly different rules to others even though the basic format remains the same. Table 13.2 describes the basic elements of a scientific paper.

The various elements shown in Table 13.2 form a basic framework that applies to more or less all scientific presentations, regardless of the audience. You may place different emphasis on certain elements (or omit them entirely) but you should always keep the basic framework in mind. The structure is important because readers need to know where to find certain pieces of information. Remember that you are trying to tell a coherent story.

13.5.2 Writing style

The idea behind writing a paper is to get across the main elements of your work as succinctly and clearly as possible. Different branches of science have subtly different styles but there are many elements in common. It is most common to use a passive style in reporting what was done. In other words you say, "the vegetation was sampled" rather than "I sampled the vegetation". This leads to a rather impersonal style and this is exactly the point. A research paper should be about the work that was carried out, not about how great you are. If you turn out a good paper, people will realize how great you are!

The passive style forces you to think carefully about how you phrase your sentences and leads to clearer work. Try to use short words and avoid jargon wherever possible, unless of course the word you want is specific to your field and is the best available. Your aim is to aid understanding, so use the shortest and simplest way to get across each point and avoid ambiguity.

13.6 Plagiarism

Plagiarism is a form of stealing. Essentially it involves you setting forth someone's work and passing it off as if it were your own. Of course you need to use previous knowledge in your work but you need to acknowledge where your knowledge/information came from.

There are three types of knowledge:

- Common knowledge. This is stuff that everyone knows that does not need to be referenced, e.g. "the fox is a mammal", "France is in Europe". These are generally understood by everyone to be true and self-evident. They are not likely to be central to your hypothesis.
- Knowledge and opinions that belong to someone else. For example, "IUCN consider this species to be seriously endangered". You should not state, "This species is seriously endangered" without referring to the original authors. The conclusion that this species is endangered came from someone else's work, so you refer to them. This sort of fact tends to be in your introductory material.
- Knowledge and opinions that you have developed yourself. These are generally in your results and conclusions. They stem from your research so there is no need to reference them. Of course if you are referring back to some previous work that you prepared, naturally you should refer to this as if it were someone else's. The key to avoiding plagiarism is to know how and when to cite references.

13.7 References

References come in two parts. There is a bit in the text that essentially says "look at this for information" and a list at the end that gives the original sources. The first part is called the *citation* and you generally quote the source as (author, date) in brackets. The citation could point to a figure or table (where you will put a caption, see Section 13.4). The second part is the references themselves. If you do not include citations in the text, the list of sources is more properly called a bibliography (essentially a list of the material you used in preparing the work).

There are two main ways to indicate that you wish to refer to some other work: you can give the citation in brackets like so (Gardener, 2010) or (Figure 23), or you can use numerical form (number citations as superscripts[1] or [1] are the two most common). When you use citations as (name, date), your reference list is sorted in alphabetical order. When you use a number, you sort the references by the order in which they appear (i.e. numerically). Generally, the (name, date) approach is preferred. Once you are familiar with a subject, you tend to recognize names of authors and a name means something whereas a number is meaningless without the corresponding reference. The number style tends to be adopted in more "popular" works and in places where you do not wish to interrupt the narrative. For most scientific publishing, the (name, date) style is most appropriate.

> **Note: References in web articles**
>
> If you are writing a report for the Internet your citations can be made into hyperlinks, which can jump directly to the target. For most purposes the (name, date) style of citation is still the most useful.

There are two main ways to use a citation. You usually refer to an area of study and give examples of the research, e.g. "Studies have shown that this plant is common in this type of habitat (Crawley, 1999)". On the other hand you may refer to the individual directly, e.g. "Crawley (1999) showed that this plant is common in this habitat". In the second form, you do not need to repeat the author in the brackets. As a general rule it is good to get your citations at the end of sentences so the more impersonal first example would be preferred. Where you have more than one author, your approach depends on how many there are. If there are only two, you give both, e.g. "(Gardener & Gillman, 2002)". Where there are more than two, you present only the first author and add *et al.* to the name, e.g. "(Chamberlain *et al.*, 2010)". The *et al.* part is Latin for "and others" (it is short for *et alia* so the *al.* part has a full stop) and is usually in italics. Some journals prefer you to give up to three authors names so you should check before you write too much. Where you have two different references by the same author(s) and in the same year, you append letters, e.g. "(Gardener & Gillman, 2001a, 2001b)".

Where you need to refer to a specific part of a larger work, perhaps a figure or table, you can append this to the citation, e.g. "(Smith, 2008, figure 2.1 p. 87)". This allows you to refer to another part of the work later but still only include a single entry in your final reference in the list.

The reference list itself is an important part of the report. This is where you demonstrate where all your sources came from. There are many different formats of setting out reference lists and different journals require different layouts; however, the underlying set-up is the same and is based on the Harvard system. A reference needs to contain a variety of information:

- Author name(s). You give all the names along with the initials (note that the initials do not appear in the citation).
- Year of publication. Only the year is given although there are some exceptions so it is a good idea to check the rules for the publication you are aiming at.
- Title of the work. For most published works, this is not a problem but if you are referring to a website then it can be tricky. If you cannot find an official title, you may have to make one up ("homepage" is a good standby).
- The place it was published. This is a broad category as the exact information depends on the type of publication. For a journal article you give: journal, issue and pages. For a book you give the publisher and the city of publication. For a website you give the URL.

When you refer to a website and therefore present a URL, you add an additional piece of information, [date accessed]. This is because websites can be changed.

All references need an author, date, title and the place published. You can see that this is not so different from the basic tenets of biological recording: who, what, where, when. Before you write too much, you should check out the exact style required.

13.8 Poster presentations

If you give a talk you may be able to speak to fifty people in a room. At a large conference there are usually many talks being given simultaneously and there may be people who were unable to get to your talk. The poster session allows you to make a presentation, which is left on display for hours, sometimes days. It can potentially reach hundreds of people at the meeting because it is hanging around for so long. Generally, there is a set session where you stand by your poster and present it to anyone who expresses an interest; otherwise it stands alone.

Your poster needs to have impact; it may be glimpsed from some distance away so make text large. Do not put too much detail. You are attempting to summarize your work in a single page, albeit a big one. You need to reduce the information to the absolute minimum to get your message over. All the major elements need to be present: title, introduction, methods, results, discussion etc., which can be a challenge.

There is no set layout for a poster; you may have the background information at the top and follow down the page or you may use a more radical layout – as long as the reader can follow what you have done that is okay. Many posters are created from a single sheet of paper (A1 or A0 for example) but you can make perfectly good posters from separate sheets of A3. Each conference will have a slightly different area available.

- Use big fonts. Can you read the headings from 2–3 m away?
- Reduce clutter, use bullet points and keep text short and snappy.

It can be useful to have a supply of A4 or A5 sized posters as handouts that people can take away (you can pin a container under the poster). Make sure that you can read it at this size without a magnifying glass!

13.9 Giving a talk (PowerPoint)

PowerPoint is a great tool for presenting your findings to an audience in a talk setting; however, it can also be something of a distraction so here are a few key things to remember about using PowerPoint:

- Make text large.
- Avoid long sentences.
- Use bullet points as a memory aid for yourself and don't type up your talk onto the screen.
- Make graphs and images large and clear.
- Avoid fancy animation effects, they merely distract from your main points.
- Try to bring in items as bullet points one at a time rather than all at once; the audience will start to read your items before you get to them and will therefore not be listening.
- Set out your talk/slides exactly like a paper, i.e. title, introduction, aims, methods, results, discussion, final conclusion(s), acknowledgements.
- Have a blank slide at the end.

The acknowledgements slide is very useful as it signals the end of your talk and you can end with a statement like, "Finally I would like to thank the following people, thank you for listening". This clearly signals the end and people can applaud (or throw things as appropriate) and then you can be ready to accept questions.

EXERCISES

Answers to exercises can be found in the Appendix.

1. Which of the following results is most acceptable to present in a report?
   A. ... $t = 2.61$, $df = 14$, $p = 0.0204$
   B. ... $t = 2.61$, $p < 0.05$
   C. ... $t_{14} = 2.61$, $p = 0.02$
   D. ... $t_{14} = 2.61$, $p < 0.05$
2. You should use visual summaries in your reports. Place a ____ in the text that points to your ____.
3. You only need to include an acknowledgements section if someone gave you money to help carry out the research. TRUE or FALSE?
4. If you are using a poster as a presentation tool you need to make text ____ and keep things ____. It can be helpful to have a ____ for people to take away.
5. The appendix is the place to put stuff that was too long or boring to go in the main report. TRUE or FALSE?

Chapter 13: Summary

Topic	Key points
Publishing and audience	There are many ways to present your work to the wider world. The basic elements of a scientific report will be the same but different publishing methods require different emphasis on the different elements. Target your report to your particular audience.
Reporting results	It's all about the results. All numerical data need summarizing: give an average and a measure of dispersion as well as sample size. Write the summary in the text and then cite the figure (or table) that visualizes the results. Results of analyses (e.g. statistical tests) should be reported in a few words in the text followed by a citation that points to an appropriate figure. Each stats test has its own symbol. Generally you need to state the test result (e.g. t-value) and the sample sizes (or degrees of freedom). The significance should be stated clearly. Use $p < 0.05$ rather than an exact p-value. Critical values may be given (student reports are more likely to include them). If results are not significant then say "n.s." rather than $p > 0.05$ or the exact p-value.
Tables	ANOVA and regression results are often best given in a table (which should be cited in the text). Avoid large tables of results; present the key elements and give longer tables as an appendix if required. Tables need a caption to describe their contents. The caption should go above the table.
Using graphs	Always aim to present visual summaries of your results. Make sure graphs are clear and not cluttered; present several simple graphs rather than one complicated one. Choose the best type of chart for your requirements. Label axes and units and give a legend if required. Titles are not required in most cases. Use a caption under the figure to describe the figure. Make captions "standalone" as far as you can.
Elements of scientific reports	All scientific reports (however informal) should contain the same basic elements. The emphasis of the different elements will vary, according to the style of report (some elements may even be missing). In general the elements are: title, abstract, introduction, methods, results, conclusions, references, acknowledgements. Each element has a specific purpose, which helps you to deliver a coherent report and helps the audience understand it!
Referencing	Referencing is important (think of it like networking). References come in two parts. The citation is the part in the main text that refers to something else. Citations can be plain numbers but most often are (name, date) style. The citation points to an item in your reference list, or to a figure or table in your report. The reference list gives the sources you referred to (cited) in the text. All references have the same basic elements: name, data, title, source. The source could be a scientific paper, a book, a web article or something else. There are different styles of reference but the elements are the same.

Chapter 13: Summary – *continued*

Topic	Key points
Posters	A poster is essentially a large piece of paper where you can illustrate your report. The elements of a regular report should be there but you need to ensure that the words can be read from some distance! Generally you will write fewer words (use bullet points) and have large clear figures.
Talks	A talk usually involves PowerPoint. You need the same elements as for a regular report. Avoid writing what you are going to say on the slides. Use bullet points as "key points", and bring them in one at a time, but avoid fancy animation! Make text large so the audience at the back of the room can see. End with an acknowledgements slide (but have a blank afterwards in reserve) so you have a definite "end".

14. Summary

By the time you reach this point, you will have realized that there is more to data analysis than simply looking at averages. The process should begin before you even have any data to analyse. Start by planning what it is you want to do and decide upon the best way to collect the data you need. The elements of planning include research about the species you are looking at, as well as methods of sampling and how you are going to record the data once you have it. You should have a good idea of what analytical approach you are going to use before you get anywhere near using a calculator.

A pilot study may well be a useful stage in the planning process and the experience may help you to refine your original ideas and hypotheses, and enable you to collect more meaningful data. This refinement is part of the scientific method, continuous evaluation and re-evaluation. At each stage you should evaluate your work and ensure that you have done the best job before proceeding to the next stage.

Writing down the data is a stage in itself and should not be taken lightly. The way you record your data can aid your analysis later so it is important to get this right. If you record your data appropriately then it may be used easily with little further processing. If, however, you record your data in an inappropriate manner, you may require a lot of extra time and effort in rearranging your data.

Once the data are collected, the main analytical processes can begin. This will usually start with summary data, for example averages. You should produce graphical summaries to help you visualize the data and confirm the approach for the next stage. These graphical summaries will often include checking the distribution of the data, for example using histograms. Once you have affirmed your analytical approach, you will move on to undertaking the actual analyses. There are two stages in this: one is doing the mathematical part and the other is preparing summaries of those results (graphs and/or tables). A final (but often overlooked) part of the analysis is to make some sense of results in the context of the biological system in which they were collected. You will need to do this as part of the presentation of the work.

The presentation and interpretation of the work is the final part of the process and closes the loop of the scientific process that began with the planning at the beginning. The presentation of your work may be quite short and informal, perhaps a simple group meeting involving a few PowerPoint slides. At the other end of the scale, your presentation may be a lot more complex and involve a dissertation running to many pages. Whatever the type of presentation, there are elements in common to them all. The structure for all presentations should include: an introduction, methods, results and some interpretation. Within that structure you have a lot of freedom; some presentations will focus more on interpretation and others will focus more on actual results.

The stages summarized here form the basis for the scientific method. You get an idea, work out how to tackle the problem, collect appropriate data to test your idea, and then you carry out some analysis and report on your results. Your original idea may be supported or not supported. Science thus moves on in this looping process, each statement is treated sceptically until unequivocal empirical evidence is produced. All good scientists, from the great and lofty to the humble student, use the same process to advance (so there is hope for us all). This quote from Terry Pratchett seems appropriate:

> "Science is not about building a body of facts. It is a method for asking awkward questions and subjecting them to a reality-check, thus avoiding the human tendency to believe whatever makes us feel good."

Glossary

alternative hypothesis Often written as H1. This is what you expect to find; however, it is not easy to prove mathematically so the concept that is usually tested is the null hypothesis (H0).

analysis of variance (ANOVA) Analysis of variance. A statistical process that allows comparison of more than two samples. Data should be normally distributed.

association A way of linking together data. When data are in categories, the links between categories are called associations. *See* chi-squared.

average A measure of the central tendency of a sample of numerical data. *See* mean, median, mode.

bar chart A graphical method of displaying values for various categories. *See also* histogram.

box–whisker plot A graphical method of displaying numerical sample data. The plot shows five quartile values.

Analysis ToolPak An add-in for Excel that permits a range of statistical analyses to be conducted.

bibliography A list of sources that were used in the preparation of a document. They are not referred to explicitly in the text of the document (i.e. they are not cited), compare to References.

Braun-Blanquet Josias Braun-Blanquet, a Swiss botanist who developed methods of examining plant communities. The Braun-Blanquet scale is a simplified abundance scale (*see also* Domin scale). There are several variants but the main scale runs thus: + = <1%, 1 = 1–5%, 2 = 5–25%, 3 = 25–50%, 4 = 51–75%, 5 = >75%.

Bray–Curtis index A measure of how similar (or dissimilar) two ecological communities are (*see* similarity, dissimilarity).

Canberra distance A measure of how similar (or dissimilar) two ecological communities are (*see* similarity, dissimilarity).

chi-squared A statistical method for examining the association between categorical factors, developed by Pearson.

citation A reference to a source of information within the text of a document. The citation itself is brief (name, date) and the full source is listed in the reference section.

coefficient Usually refers to multiple regression, each factor has a coefficient that is analogous to the slope in a simple straight-line relationship. *See* multiple regression.

confidence interval A measure of the variability of a numerical sample. Expressed as a value and a percentage (or proportion), the percentage of the data that lie within a certain distance from the mean: e.g. $CI_{0.95} = 1.5$ indicates that 95% of the data lie within 1.5 units of the mean for the sample.

constancy A term used in the National Vegetation Classification system. Five quadrats are used in the NVC system and constancy refers to how many of the five a species occurs in (so it is a measure of frequency). Usually written as a Roman numeral I–V.

contingency table A table of observed values for observations in various categories. Used in the chi-squared test for association, e.g. categories could be habitat and invertebrate order.

correlation A link between two variables, e.g. stream speed and mayfly abundance. If one value (e.g. speed) increases and the other (e.g. abundance) decreases the correlation is negative. If both factors change in the same direction then the correlation is positive. *See* Spearman's rank and Pearson's product moment.

critical value A value regarded as the cut-off point for a statistical test. In some tests when this value is exceeded the result is regarded as significant. For other tests the result is regarded as significant if the calculated value is less than the critical value.

DAFOR scale A relative abundance scale (therefore ordinal): D = dominant, A = abundant, F = frequent, O = occasional, R = rare. Can be applied to any organism and is defined by the user for convenience. Often the letters are converted to numerical values e.g. D = 5, R = 1.

degrees of freedom Related to the sample size of data. Usually the degrees of freedom is the sample size –1 but there are variations according to the statistical test being applied.

Dependent variable In a correlation or regression this is the variable that is thought to be affected by others. *See* response variable.

density plot A method of displaying the distribution of a numerical sample as a continuous line. An alternative approach to a histogram (*see also* tally plot and stem–leaf).

dissimilarity Used to describe how (dis)similar two ecological communities are. There are several indices of dissimilarity (*see also* similarity), e.g. Jaccard, Sørensen, Euclidean, Bray–Curtis, Manhattan, Canberra.

dispersion A term used in relation to the spread of data in a sample, e.g. standard deviation, range.

distance measure A measure of how similar (or dissimilar) two ecological communities are (*see* similarity, dissimilarity).

diversity A term used to describe the number of different species in an ecological community. There are different methods to describe diversity: species richness and diversity indices (e.g. Shannon, Simpson's index).

diversity index Describes the number of species in an ecological community and their relative abundance. Commonly used indices are Shannon and Simpson's.

Domin scale An abundance scale used for determining plant percentage cover. Named after a Czech botanist, the scale goes from 0 to 10 like so: 1 = <4% with 1–3 individuals, 2 = <4% with 4–10 individuals, 3 = <4% with >10 individuals, 4 = 4–10%, 5 = 11–25%, 6 = 26–33%, 7 = 34–50%, 8 = 51–75%, 9 = 76–90%, 10 = 9–100%.

error bars A way of illustrating graphically a measure of the spread of the data. For example in a bar chart the bars may represent the mean values whilst error bars could represent standard deviation, standard error or confidence interval.

Euclidean distance A measure of how similar (or dissimilar) two ecological communities are (*see* similarity, dissimilarity).

goodness of fit A type of statistical analysis that compares the observed frequencies in a number of categories to be compared to theoretical frequencies. Similar in approach to the chi-squared test.

histogram A graph to show frequency distribution of a sample. Like a bar chart except that the bars should touch, i.e. they show a continuous range of values split into convenient categories (or bins).

hypothesis Something that you are trying to test. Usually this is a single thing that might be proven. *See* null hypothesis and alternative hypothesis.

independent variable In a correlation or regression this is the variable that controls the level of the dependent variable; however, you should beware of cause and effect. *See* predictor variable.

intercept A straight-line relationship between two numerical variables can be represented by the equation $y = mx + c$. The intercept is where the straight line crosses the y-axis (c in the equation). *See* multiple regression.

Jaccard index A measure of how similar (or dissimilar) two ecological communities are (*see* similarity, dissimilarity).

Kruskal–Wallis A statistical approach that allows comparison of more than two samples when the data are not normally distributed.

logistic regression A way of examining relationships between numerical variables where the dependent factor is binary, i.e. is either 0 or 1 (presence or absence) or has two alternatives.

Manhattan index A measure of how similar (or dissimilar) two ecological communities are (*see* similarity, dissimilarity).

mean A measure of the central tendency of a numerical sample. Calculated as the sum of the values in the sample divided by the number of observations. Used for normally distributed data. *See* parametric.

median	A measure of the central tendency of a numerical sample. Calculated as the middle value when they are ranked in ascending numerical order. Used for non-normally distributed data.
mode	A measure of the central tendency of a numerical sample. Calculated as the most frequent value. Usually only used for very large samples and rarely used in statistical testing.
multiple regression	A statistical method that examines the link between a single dependent variable and several independent variables. The relationship between the dependent and independent variables is assumed to be a straight line of the form $y = mx + c$.
non-parametric	When a sample of data is normally distributed (i.e. is skewed) the data are described as non-parametric.
null hypothesis	Often written as H0. This is a hypothesis that you disprove mathematically in order to support your alternative hypothesis (H1), which is what you expect to find. Think of it as the dull hypothesis, e.g. there is no difference; there is no correlation; there is no association.
NVC	National Vegetation Classification. This is a standardized method of surveying plant communities.
***p*-value**	A measure of the likelihood of a result happening by random chance. In statistics, a value of $p = 0.05$ is taken as significant, i.e. there is only 5% chance that the result could have occurred randomly.
parametric	Normal distribution. Describes the frequency distribution of a sample where the data are symmetrical about the middle and form a bell-shaped curve.
Pearson's product moment	A statistical method for determining the relationship (correlation) between two factors that are normally distributed. The relationship is assumed to be a straight line in the form of $y = mx + c$.
pie chart	A graphical method of displaying data as a proportion. Each slice of the pie represents the proportion towards the total. Pie charts can always be represented as a bar chart instead.
pivot table	A summary table generated by Excel. Pivot tables are a useful way to rearrange and summarize data, especially when the data is in the form of biological records.
post-hoc test	Literally "after this". A method of comparing samples pair by pair after a multi-sample test is applied. *See* Tukey HSD test.
predictor variable	A predictor variable is one that has some effect on another variable (directly or indirectly). In a scatter plot this would be represented on the x-axis. Effectively you are saying that levels of this variable help you to predict levels in the response variable.
presence–absence	A form of binary data where you have the presence or absence of a species in a sample but no information on the abundance (*see also* similarity, diversity).

quartile A value half-way between two extremes. The main quartiles lie 1 and 3 quarters of the way along a series of values laid out in size order (i.e. ranked), so are half-way between the middle and one end.

R program A statistical programming language used for many types of analysis. If is free and open source.

range The difference between the maximum and minimum values in a sample of numerical data. Equates to the 0th and 4th quartiles. *See* box–whisker plot.

references A list of sources used in the preparation of a document. All sources listed are cited in the text, usually as (name, date), compare to Bibliography.

residual A measure of how far away from the line of best fit a datum is.

response variable This is the variable that is of interest to the researcher. It is the variable that is affected by changes in the predictor variable(s). In a scatter plot, this would be represented on the y-axis.

running mean A cumulative average (median might also be used). The mean value is calculated each time a new observation is made. This may be used to help determine when enough data has been collected.

scatter plot A graphical method of displaying the relationship between two numerical values. *See* correlation.

similarity Used to describe how similar two ecological communities are. There are several indices of similarity (*see also* dissimilarity), e.g. Jaccard, Sørensen, Euclidean, Bray–Curtis, Manhattan, Canberra.

Simpson's index An index of diversity that takes into account the number of species and their relative abundance.

Shannon index An index of diversity that takes into account the number of species and their relative abundance.

skewed Usually refers to not normally distributed data (i.e. non-parametric), where the frequency distribution has its highest point skewed from the middle (in normal distribution the highest point would be in the middle of the distribution).

slope A straight-line relationship between two numerical variables can be represented by the equation $y = mx + c$. The slope is a measure of how steep the straight line is (m in the equation). *See* multiple regression.

Sørensen index A measure of how similar (or dissimilar) two ecological communities are (*see* similarity, dissimilarity).

Spearman's rank A statistical method for determining the relationship (correlation) between two factors. These factors do not need to be normally distributed and the relationship does not have to be linear (although it should not be a U or inverted U-shape).

species richness The number of different species in a sample. *See also* diversity.

standard deviation A measure of the spread of data from the middle (mean) in normally distributed samples.

standard error A measure of the spread of data from the middle in normally distributed data. It is calculated as the standard deviation divided by the square root of the sample size.

stem and leaf A type of frequency distribution graph. The values in a sample are rewritten so that they appear to form a tally plot. The advantage over the tally plot is that the original data values can be reconstructed from the graph.

***t*-test** Usually called the Student's *t*-test after the pen name of the original author. The *t*-test is a statistical method for determining the difference between two samples of normally distributed data.

tally plot A simple form of frequency distribution where simple tally marks are placed against categorical bins representing size class.

Tukey HSD test A form of post-hoc test. Tukey's honest significant different test is carried out after analysis of variance to look at pairwise comparisons.

***U*-test** The *U*-test is a statistical test for examining the difference between two samples when the data are not normally distributed. It is often called the Mann–Whitney *U*-test although Wilcoxon if often attributed as the author.

variance A measure of the spread of data (*see* dispersion). Variance is the standard deviation squared.

***z*-test** A version of the *t*-test for determining the statistical difference between two samples of normally distributed data. The *z*-test is essentially a *t*-test with large sample sizes (>25).

Appendix 1
Answers to exercises

The answers to the end-of-chapter exercises are set out in the following sections.

Chapter 1

1. In a project looking for links between things you would be looking for Correlation or Association depending on the kind of data.
2. Categorical > Ordinal > Interval. With categories you cannot say which is "biggest" only that categories are different. With ordinal data you can say that one item is larger than another but not by how much (e.g. *Dominant* is bigger than *Rare*). Interval data is the most sensitive because you have more "exact" measurements.
3. Domin, DAFOR and Braun-Blanquet are all examples of Ordinal scales, whereas *red*, *blue* and *yellow* would be examples of Categories.
4. D. Random is not a kind of transect. You may choose to measure at random intervals along a transect but the transect itself would be described as one of the other categories.
5. TRUE: one of the main aims of a sampling method is to eliminate bias (the other is to be representative).

Chapter 2

1. A, B and E. Maintaining good records (D) will allow you proper credit but that is not a good reason. Knowing where you've been (C) is important but this is because it allows repeatability and verification.
2. Sample layout (each column a separate sample) is one form. Recording (or scientific recording) layout (where each column is a separate variable) is the other.
3. C. A sample should not generally be a row in your dataset. Each row should be a separate observation/record/replicate. You may rearrange your data in a form where the rows are samples, for example if you are looking at community data.
4. FALSE. The sample layout is the least flexible. The recording layout allows you greater flexibility to add variables and records and to rearrange data.
5. In your dataset each column should represent a separate variable.

Chapter 3

1. D. Names of R objects must start with a letter. Names are case sensitive and can include period or underscore characters.
2. C. The `scan()` command can read data from the clipboard. You press *Enter* on a blank line to finish the operation.
3. If you had data in DAFOR format and needed to convert to a numerical format you could use a lookup table with the VLOOKUP or HLOOKUP function.
4. TRUE. A pivot table allows you to manage and rearrange your data. To go from recording to sample format you'd need to have a index (a simple observation number is sufficient, see the exercise in Chapter 3).
5. You can use the read.csv() command to import data from Excel into R.

Chapter 4

1. The median is the best measure of central tendency (average) to use when your data are non-parametric.
2. D. The variance and standard error are both suitable measures of dispersion for use with normally distributed data (standard deviation and confidence interval are also fine). The inter-quartile range and range measures are more suitable for non-parametric data. The mean is a measure of centrality, not of dispersion.
3. The histogram and tally plot are forms of chart that allow you to visualize data distribution. Both use the frequency of data in size classes called bins.
4. FALSE. The logarithmic transformation is probably the best choice. The arcsine transformation is better when you have data with defined boundaries (such as percentage).
5. Yes, the data are parametric. You can draw a histogram or a tally plot to look at the data visually (try between 5 and 9 bins). Using R the `shapiro.test()` command would confirm that there is no significant departure from normality:

```
> q5
 [1] 28 26 28 27 29 28 24 26 28 32
> shapiro.test(q5)
   Shapiro-Wilk normality test
data:  q5
W = 0.9295, p-value = 0.4435
```

Chapter 5

1. There are two forms of project that explore links between things; these are correlation (or regression) and association. The difference between these projects is the kind of data. In tests of association your data are categorical, whereas in correlation and regression you have ordinal or interval data.
2. FALSE. It is common to have a biological variable and an environmental one but this is not mandatory. In fact it can be a useful exercise to explore correlations between environmental variables.
3. C (or H) and E. Whilst A is okay as a hypothesis you really ought to make a more definite prediction (so C or H are better choices). The H0 is not simply the opposite. B (or D) and F. Similarly you should make a prediction about the difference (either B or D are acceptable).
4. FALSE. The *t*-test is for differences, but between only two samples. If you have more than two samples of parametric data you use ANOVA.
5. C. In this case you are looking for links (between bee species and flower type). Your data are categorical (categories of bee and categories of flower) so you need a test of *association* (chi-squared).

Chapter 6

1. C or E. The bar chart or box–whisker plot are best here. You would determine the average (mean or median) and a measure of dispersion for each sample.
2. The box–whisker plot usually displays non-parametric data, i.e. the median, inter-quartiles and range.

3. FALSE. A scatter plot is required for this, in which case the response variable would indeed go along the y-axis.
4. FALSE. You can add a joining line but that does not make a "proper" line plot. In a true line plot the interval between items on the x-axis is fixed (the items are treated as categories). In a scatter plot the x-axis is a continuous numerical scale.
5. When you have compositional data you can use a pie or a bar chart to display the results but only a bar chart allows you to display multiple categories.

Chapter 7

1. E. In a *U*-test you rank the data as if they were all one sample. You use the sum of the ranks from one sample and the sum of the ranks from the other. You need the number of replicates in the *U*-test formula. So, you do not need any of the summary statistics to carry out the test. You would need the median and range if you wanted to produce a graph (and probably also the inter-quartiles).
2. First of all you need the variance, which is the standard deviation squared:

   Field: $1.41^2 = 1.99$
   Path: $1.67^2 = 2.79$

   Now you can substitute the values into the formula for the t-test (Figure 7.4):

   $t = |4 - 5.75| \div \sqrt{(1.99/8) + (2.79/8)}$
   $t = 1.75 \div \sqrt{(0.25 + 0.35)}$
   $t = 1.75 \div 0.77$
   $t = 2.26$

   The critical value for t at 14 degrees of freedom $(8 - 1 + 8 - 1)$ is 2.14. Your calculated value exceeds this so the result is statistically significant at $p < 0.05$.
3. The ranks are: 2 4.5 2 7 8 2 4.5 6. There are three 0 values, these occupy ranks 1, 2 and 3. The average rank is therefore 2. There are two 1s, which will occupy ranks 4 and 5. Their ranks will therefore be 4.5. The next available rank is 6, which goes to the number 3. The value 4 gets rank 7 and the top value 5 gets rank 8.
4. TRUE. There are versions of both t-test and *U*-test.
5. You can rank data in Excel using the RANK.AVG function.

Chapter 8

1. Correlation describes the strength and direction of the link/relationship between two variables. If your data are normally distributed you can use Pearson's product moment. If data are non-parametric then you use Spearman's rank correlation.
2. First you need to rank the data (each variable is ranked separately). The ranks are shown in the following table.

A	B	$Rank_A$	$Rank_B$	D_{A-B}	D^2
4	4	1	1	0	0
6	12	2	2.5	–0.5	0.25

12	23	3	5	–2	4
23	18	5	4	1	1
18	12	4	2.5	1.5	2.25
35	35	6	6	0	0

Now you evaluate the difference between each pair of ranks (column 5); the sum should come to zero.
Take each of the differences and square them; this removes any negative values. The sum of the squares should be 7.5.
Now you can substitute values into the Spearman formula (Figure 8.5). The values are shown in the following table.

$6\sum D^2$	45
$n(n^2 - 1)$	210
r_s	0.786

The critical value for six pairs of observations is 0.866 (Table 8.5). So the result is not statistically significant.

3. A, B and D are all significant. The degrees of freedom are the number of pairs of observations minus two, so $24 - 2 = 22$. Ignore the sign when comparing to the critical value (Table 8.6). The results A and D show negative correlations, whilst B is positive (C is positive but not significant).
4. The correlation coefficient and the t-statistic are linked by a simple formula:

   $t = \sqrt{(r^2 \times df \div 1 - r^2)}$ (Figure 8.12)

   Once you've calculated a value for t you can use `TINV` to get a critical value.
5. This would be TRUE if the relationship were described by a mathematical formula (such as in Pearson's correlation), where data are normally distributed. This would be FALSE if data were non-parametric and you used Spearman's rank correlation (you should not add the trendline to the scatter plot).

Chapter 9

1. Association analysis is used to explore links/association between the frequency of items in categories.
2. First of all you need to get the row and column totals (and the grand total). Then you can work out the expected values: *row total* × *column total* ÷ *grand total*. The expected values are shown in the following table:

76	61
46	37

Then you'll need the observed – expected values. Because you have a 2 × 2 contingency table it is usual to apply Yates' correction, which reduces the values by 0.5. The results are shown in the following table:

19.5	–19.5
–19.5	19.5

Now determine the chi-squared values and add them together for a total, which comes to 29.9 (see following table).

5.0	6.2
8.3	10.3

Compare to the critical values (Table 9.6) and you'll see the result is statistically significant.

3. D. Degrees of freedom are worked out from the number of rows and columns like so:

   df = rows – 1 × columns – 1
   df = 5 – 1 × 5 – 1 = 16

   Table 9.6 does not include critical values for that level of degrees of freedom so you'll need to use `=CHIINV(0.05, 16)` in Excel to get a critical value (26.296).
4. FALSE. The result shows that there is no significant difference between your study and the theoretical ratio of colours.
5. B. A bar chart of the Pearson residuals would be the most useful graph. You could show all the results in "one go" or use several charts. A pie chart can only show the original compositional data (and only from one category at a time).

Chapter 10

1. Firstly you need to work out are the degrees of freedom. There are six treatments so the df for the predictor is 6 – 1 = 5. There are 6 × 12 = 72 observations in total. So, the df for the error (residuals) is 72 – 5 – 1 = 66.
   The critical value can be looked up in a table but Table 10.1 does not go higher than 30. So use the `FINV` function in Excel, `=FINV(0.05, 5, 66)` to give the result, 2.35.
   Now you can get the mean squares (MS) by dividing SS by df.
   The F-value is the $MS_{predictor} \div MS_{Error}$.
   The results are summarized in the following table.

	df	*SS*	*MS*	*F*	Crit
Spray	5	2669	534	34.71	2.35
Error	66	1015	15		

2. In a two-way analysis of variance you have two predictor variables. These can act independently or form interactions, which need to be accounted for.
3. FALSE. The Kruskal–Wallis is only an alternative for a one-way ANOVA (i.e. when you have a single predictor variable).
4. After you have carried out the main analysis you must run post-hoc tests, which are more in-depth pairwise comparisons.
5. A and E. A bar chart would be fine as long as you had error bars. A line plot is not correct. Options B, C and D would all be fine.

Chapter 11

1. TRUE. The coefficients and intercept define the mathematical relationship between variables. In polynomial regression there are two coefficients for the main variable, x and x^2, but the statement is still broadly true.
2. You can calculate a beta coefficient using the regular coefficient and the standard deviations like so: $b' = b \times sx \div sy$ (Figure 11.6). Substituting the coefficients from the equation gives:

$$b'_{rock} = -0.152 \times 33.318 \div 7.163 = -0.707$$
$$b'_{pH} = -2.686 \times 0.549 \div 7.163 = -0.206$$

3. D. The inverted U shape is encountered with polynomial relationships ($y = ax + bx^2 + c$). Technically this is still a linear relationship so E is marginally correct too.
4. C. The correlation coefficient with the highest absolute value is the best starting point for a regression model.
5. Using R you can calculate theoretical values from a regression model using the fitted() command. You can use these values to draw a trendline/best-fit line.

Chapter 12

1. Species richness is the simplest measure of biodiversity. However, comparisons between samples can only be made if sampling effort is consistent.
2. TRUE. A Simpson's index of 0.75 means that you are likely to have a 75% chance of getting a different species if you sampled twice.
3. D. The Jaccard similarity is calculated as J ÷ A + B – J. Substituting values gives: 10 ÷ 20 + 25 – 10 = 10 ÷ 35 = 0.29. Answer A is the Sørensen similarity. Answers B and C are the *dissimilarity* for Jaccard and Sørensen respectively.
4. The Euclidean metric can be used as a measure of similarity (or dissimilarity) when your samples include abundance measurements.
5. This is broadly TRUE but it is the `plot()` command that actually does the drawing. The `hclust()` command creates the hierarchical structure from a (dis)similarity matrix.

Chapter 13

1. D. Answers A and C give an exact p-value. B lacks the sample size(s).
2. You should use visual summaries in your reports. Place a citation in the text that points to your graph/figure.
3. FALSE. There is always someone who helped you out, but don't thank your personal deity or parents (unless your folks are professors of ecology).
4. If you are using a poster as a presentation tool you need to make text large/big and keep things simple. It can be helpful to have a summary for people to take away.
5. FALSE. The appendix ought to contain useful and relevant items. However, the items in the appendix would be considered too long or distracting to be in the main report. Student reports often include raw data in appendix so their professor can check the

results.

Appendix 2
Tables of critical values

These tables are copies of the tables in the main text, reproduced for ease of reference.

Table 7.1 Critical values for the Student's *t*-test. Reject the null hypothesis if your calculated value is greater than the tabulated value.

Degrees of freedom	Significance level			
	5%	2%	1%	0.1%
1	12.706	31.821	63.657	636.62
2	4.303	6.965	9.925	31.598
3	3.182	4.541	5.841	12.941
4	2.776	3.747	4.604	8.610
5	2.571	3.365	4.032	6.859
6	2.447	3.143	3.707	5.959
7	2.365	2.998	3.499	5.405
8	2.306	2.896	3.355	5.041
9	2.262	2.821	3.250	4.781
10	2.228	2.764	3.169	4.587
11	2.201	2.718	3.106	4.437
12	2.179	2.681	3.055	4.318
13	2.160	2.650	3.012	4.221
14	2.145	2.624	2.977	4.140
15	2.131	2.602	2.947	4.073

Table 7.9 Critical values for *U*, the Mann–Whitney *U*-test at 5% significance level. Reject the null hypothesis if your value of *U* is equal or less than the tabulated value.

n	1	2	3	4	5	6	7	8	9	10
1	–	–	–	–	–	–	–	–	–	–
2	–	–	–	–	–	–	–	0	0	0
3	–	–	–	–	0	1	1	2	2	3
4	–	–	–	0	1	2	3	4	4	5
5	–	–	0	1	2	3	5	6	7	8
6	–	–	1	2	3	5	6	8	10	11
7	–	–	1	3	5	6	8	10	12	14
8	–	0	2	4	6	8	10	13	15	17
9	–	0	2	4	7	10	12	15	17	20
10	–	0	3	5	8	11	14	17	20	23

Table 7.13 Critical values for Wilcoxon matched pairs. Compare your lowest rank sum to the tabulated value for the appropriate number of non-zero differences. Reject the null hypothesis if your value is equal to or less than the tabulated value.

N_D	Significance level		
	5%	2%	1%
5	–	–	–
6	0	–	–
7	2	0	–
8	3	1	0
9	5	3	1
10	8	5	3
11	10	7	5
12	13	9	7
13	17	12	9
14	21	15	12
15	25	19	15
16	29	23	19
17	34	27	23
18	40	32	27
19	46	37	32
20	52	43	37
25	89	76	68
30	137	120	109

Table 8.5 Critical values for r_s the Spearman's rank correlation coefficient. Reject the null hypothesis if your calculated value is equal to or greater than the tabulated value.

No. of pairs,	Significance level		
n	5%	2%	1%
5	1.000	1.000	–
6	0.886	0.943	1.000
7	0.786	0.893	0.929
8	0.738	0.833	0.881
9	0.683	0.783	0.833
10	0.648	0.746	0.794
12	0.591	0.712	0.777
14	0.544	0.645	0.715
16	0.506	0.601	0.665
18	0.475	0.564	0.625
20	0.450	0.534	0.591
22	0.428	0.508	0.562
24	0.409	0.485	0.537
26	0.392	0.465	0.515
28	0.377	0.448	0.496
30	0.364	0.432	0.478

Table 8.8 Critical values for the correlation coefficient *r*, Pearson's product moment. Reject the null hypothesis if your value is equal to or greater than the tabulated value.

Degrees of freedom	Significance	
	5%	1%
1	0.997	1
2	0.95	0.99
3	0.878	0.959
4	0.811	0.917
5	0.754	0.874
6	0.707	0.834
7	0.666	0.798
8	0.632	0.765
9	0.602	0.735
10	0.576	0.708
12	0.532	0.661
14	0.497	0.623
16	0.468	0.59
18	0.444	0.561
20	0.423	0.537
22	0.404	0.515
24	0.388	0.496
26	0.374	0.478
28	0.361	0.463
30	0.349	0.449
35	0.325	0.418
40	0.304	0.393
45	0.288	0.372
50	0.273	0.354
60	0.25	0.325
70	0.232	0.302
80	0.217	0.283
90	0.205	0.267
100	0.195	0.254
125	0.174	0.228
150	0.159	0.208
200	0.138	0.181
300	0.113	0.148
400	0.098	0.128
500	0.088	0.115
1000	0.062	0.081

Table 9.6 Critical values for the chi-squared test (also used with the Kruskal–Wallis test). Reject the null hypothesis if your value is greater than the tabulated value.

Degrees of freedom	Significance level			
	5%	2%	1%	0.01%
1	3.841	5.412	6.635	10.830
2	5.991	7.824	9.210	13.820
3	7.815	9.837	11.341	16.270
4	9.488	11.668	13.277	18.470
5	11.070	13.388	15.086	20.510
6	12.592	15.033	16.812	22.460
7	14.067	16.622	18.475	24.320
8	15.507	18.168	20.090	26.130
9	16.909	19.679	21.666	27.880
10	18.307	21.161	23.209	29.590

Table 10.1 Table of critical values for F at $p = 0.05$. The columns are degrees of freedom for between groups (no. groups = k, df = $k - 1$) and rows are degrees of freedom for within groups (df = $n - k$). Reject the null hypothesis if your calculated value exceeds the tabulated value.

k/n	2	3	4	5	6
2	19.0	19.2	19.2	19.3	19.3
3	9.55	9.28	9.12	9.01	8.94
4	6.94	6.59	6.39	6.26	6.16
5	5.79	5.41	5.19	5.05	4.95
6	5.14	4.76	4.53	4.39	4.28
7	4.74	4.35	4.12	3.97	3.87
8	4.46	4.07	3.84	3.69	3.58
9	4.26	3.86	3.63	3.48	3.37
10	4.10	3.71	3.48	3.33	3.22
15	3.68	3.29	3.06	2.90	2.79
20	3.49	3.10	2.87	2.71	2.60
30	3.32	2.92	2.69	2.53	2.42

Table 10.3 Table of critical values of t used in the Tukey HSD post-hoc test. Reject the null hypothesis if your calculated value is greater than the tabulated value.

5%	2%	1%	0.1%
4.303	6.965	9.925	31.598

Table 10.16 Values of Q, the Studentized range.

Number of groups	Significance	
	5%	1%
2	2.772	3.643
3	3.314	4.120
4	3.633	4.403
5	3.858	4.603
6	4.030	4.757

Index

Page numbers in **bold** indicate glossary terms.